Sobotta
Atlas of Human Anatomy
Volume 1

Sobotta

Atlas of
Human Anatomy

Edited by
JOCHEN STAUBESAND

Volume 1
Head, Neck, Upper Limbs, Skin

Eleventh English Edition
with Nomenclature in English

Translated and edited by Anna N. Taylor

672 Illustrations, most in color

Urban & Schwarzenberg · Baltimore–Munich

This book was founded by Johannes SOBOTTA, former Professor of Anatomy and Director of the Anatomical Institute of the University of Bonn.

Address of the English translator and editor:

Anna N. TAYLOR, Ph.D.
Professor of Anatomy
Department of Anatomy and Cell Biology
UCLA School of Medicine
Los Angeles, CA 90024-1763

Address of the German editor:

Professor Dr. med. Jochen STAUBESAND
Direktor des Anatomischen Instituts (Lehrstuhl I)
der Albert-Ludwigs-Universität, Freiburg
Albertstraße 17, 7800 Freiburg i. Br.

This atlas consists of two separate volumes:

Vol. 1: Head, Neck, Upper Limbs, Skin

Vol. 2: Thorax, Abdomen, Pelvis, Lower Limbs

---

10 9 8 7 6 5 4 3 2 1

Library of Congress Cataloging – in – Publication Data

Atlas der Anatomie des Menschen. English
Sobotta atlas of human anatomy. – 11th English ed. with nomenclature in English/edited by Jochen Staubesand; translated and edited by Anna N. Taylor. p. cm.
Translation of: Atlas der Anatomie des Menschen. 19th ed. 1988.
Includes index.
Contents: Vol. 1. Head, neck, upper limbs, skin – Vol. 2. Thorax, abdomen, pelvis, lower limbs.
  ISBN 0-8067-1711-4 (v.1): ISBN 3-541-71711-4 (v. 1)
  ISBN 0-8067-1721-1 (v. 2): ISBN 3-541-71721-1 (v. 2)
1. Human Anatomy-Atlases. I. Sobotta, Johannes, 1869–1945.
II. Staubesand, Jochen. III. Taylor, Anna N. IV. Title. V. Title: Atlas of human anatomy.
  [DNLM: 1. Anatomy-Atlases. QS 17 A88125]
QM25.A7913 1990
611'.0022'2–dc20
DNLM/DLC
for Library of Congress   90-12583 CIP

Printed in Germany by R. Oldenbourg, Munich
© Urban & Schwarzenberg 1990

ISBN 0-8067-1711-4 Baltimore
ISBN 3-541-71711-4 Munich

German Editions:

1. Edition: 1904–1907 J.F. Lehmanns Verlag, München

2.–11. Edition: 1913–1944 J.F. Lehmanns Verlag, München

12. Edition: 1948 and following editions Urban & Schwarzenberg, München

13. Edition: 1953

14. Edition: 1956

15. Edition: 1957

16. Edition: 1967 (ISBN 3-541-02816-5)

17. Edition: 1972 (ISBN 3-541-02817-3)

18. Edition: 1982 (ISBN 3-541-02818-1)

19. Edition: 1988 (ISBN 3-541-02819-X)

Foreign Editions:

Arabic Edition
Al Ahram, Cairo, Egypt

English Edition (with nomenclature in English)
Urban & Schwarzenberg

English Edition (with nomenclature in Latin)
Urban & Schwarzenberg

French Edition
Urban & Schwarzenberg

Greek Edition
Gregory Parisianos, Athens, Greece

Indonesian Edition
CV. EGC Medical Publisher, Jarkarta, Indonesia

Italian Edition
USES, Florence, Italy

Japanese Edition
Igaku-Shoin Ltd., Tokyo, Japan

Portuguese Edition
Editora Guanabara Kougan, Rio de Janeiro, Brazil

Spanish Edition
Editorial Medica Panamericana, Buenos Aires, Argentina

Turkish Edition
Urban & Schwarzenberg

Preface

The primary objective of the editor in preparing a new edition of the atlas, originally produced by Johannes SOBOTTA in 1903, was to improve both the contents and their didactic value. It was considered essential both to increase the number of illustrations and to update them by including pictures taken using modern imaging techniques, such as sonography, computed tomography (CT) and magnetic resonance imaging (MRI). Sobotta's concept was to produce an atlas that would "serve the needs of both the medical student and of the practicing physician." This has clearly been achieved.

For various reasons it was deemed necessary to change the terminology to conform to that of the 5th edition of the official NOMINA ANATOMICA published in 1983. This entailed an immense amount of unforeseen, additional work resulting in numerous changes (some considered superfluous or even illogical) and led to strong provisos by the editor concerning the use of current terminology.

With clinical terminology continuing to drift away from the Nomina Anatomica, numerous additional editorial annotations and footnotes were required.

Changes in the text and in the illustrations, plus the changes in terminology, meant that few figures could be transferred directly to the new edition. However, reduction in size of some of the older illustrations together with improved utilization of the space devoted to text enabled the addition of 324 new figures and the inclusion of many excellent plates produced by modern imaging methods without incurring an increase either in the size of the two volumes or their cost.

The 19th edition has benefitted greatly from the expert advice and contribution of illustrations by the following colleagues: Prof. Dr. W.S. ALTARAS (Giessen), Dr. L. BAUMEISTER (Freiburg/Br.), Prof. Dr. Ch. BECK (Freiburg/Br.), Dr. J.C. DEMBSKI (Koblenz), Prof. Dr. J. DÜKER (Freiburg/Br.), Dr. A. FRANKENSCHMIDT (Freiburg/Br.), Dr. H. FRIEDBURG (Freiburg/Br.), Dr. G. GREEVEN (Neuwied), Prof. Dr. T. GRIMM (Würzburg), Prof. Dr. H.-G. HILLEMANNS (Freiburg/Br.), PD Dr. B. HÖGEMANN (Münster), Prof. Dr. Hj. JUST (Freiburg/Br.), Prof. Dr. G.W. KAUFMANN (Freiburg/Br.), Dr. E. KRUSE (Marburg), Prof. Dr. D. KUHN (Marbach/Neckar), Dr. M. LUDWIG (Bonn), Prof.

Dr. M. v. LÜDINGHAUSEN (Würzburg), Prof. Dr. h.c. G. MACKENSEN (Freiburg/Br.), Dr. M.T. MCNAMARA (Monte Carlo, Monaco), Prof. Dr. E.E. PETERSEN (Freiburg/Br.), Prof. Dr. T.H. RAKOSI (Freiburg/Br.), PD Dr. W.S. RAU (Freiburg/Br.), Prof. Dr. G.-M. v. REUTERN (Freiburg/Br.), Dr. G. RILLING (Fribourg/Switzerland), Dr. A. SCHEIBE (Breisach), Prof. Dr. W. SCHILLI (Freiburg/Br.), Prof. Dr. H. SCHILLINGER (Freiburg/Br.), Prof. Dr. H.-M. SCHMIDT (Bonn), Prof. Dr. W. SEEGER (Freiburg/Br.), Prof. Dr. R. ÜNSÖLD (Düsseldorf), Prof. Dr. Chr. WALTHER (Ulm), Prof. Dr. W. WENZ (Freiburg/Br.), Dr. S. ZULEGER (Freiburg/Br.).

Heartfelt thanks also to the numerous students who, through their revisions and suggestions, have supplied valuable ideas. I owe thanks to the members of my staff for improvements of the text, many suggestions, assistance in the terminological changes, and in the difficult proof reading. Here I wish to give special mention to the Academic Director, Dr. F. PLATZ (who also revised the chapter "Arterial Supply Areas"), Superior Academic Councillor, Dr. H. FLÖEL, Academic Councillors, Dr. H. ARNOLD-SCHMIEBUSCH, Dr. G. ADELMANN, Dr. F. STEEL, Dr. J. KERL, Mr. F. KULVELIS, and Mr. R. HACKLÄNDER as well as Dr. P. POSEL, Munich.

The publisher, Urban & Schwarzenberg, has spared no effort to further improve the quality of the reproductions and graphic arrangement of the new edition. May I express my gratitude to all persons involved. In addition to the publisher, Mr. Michael URBAN, I wish above all to give thanks to the production manager, Mr. Peter MAZZETTI, and his assistant, Ms. Renate HAUSDORF, for their sympathetic cooperation.

My special thanks to Ms. M. ENGLER, my assistant for many years, who with consideration, conscientiousness, and perseverance has assisted in three editions of this book. Especially, I wish to emphasize her proven commitment, her competence and diligent, untiring help in all phases of development, and in the shaping of the new edition.

My sincere thanks to all those who have contributed to the publication of this book even if they have not been mentioned here.

Freiburg, August 1988   J. STAUBESAND

Preface to the 11th English Edition

As a student and teacher of Anatomy, I have often found the multiplicity of terms for so many structures bothersome. My primary objective, therefore, in undertaking the translation of this classic, anatomical atlas was to produce an English edition which would contribute to a standardization of the nomenclature. As such, Professor Carmine D. CLEMENTE's edition of *Gray's Anatomy* (Lea & Febiger, 1985) and his atlas, *Anatomy* (Urban & Schwarzenberg, 1987), proved to be invaluable resources. I have utilized these two books, adhering to their anglicized forms of the terms in the Paris NOMINA ANATOMICA, as my primary references for terminology and for the organization and contents of the tables and textual material.

My colleagues in the Department of Anatomy and Cell Biology at the University of California at Los Angeles were most helpful whenever I reached an impasse. Those to whom I owe special thanks, in addition to Prof. CLEMENTE, include: Professors Anthony ADINOLFI, Earl ELDRED, Roger GORSKI, Frances GROVER, Elizabeth LOMAX, David MAXWELL, Ynez O'NEILL, and Charles SAWYER. I am also grateful to Mr. Braxton D. MITCHELL, President of Urban & Schwarzenberg, Baltimore, for entrusting me with this project and to his staff, Ms. Kathleen MILLET, Ms. Starr BELSKY and Ms. Mary KIDD, for their superb editorial assistance. I would not have been prepared for this project were it not for my parents, who maintained a bilingual home, and my teachers of French and Latin, Ms. Theresa LaMARCA and Ms. Nellie ROSEBAUGH, in the model foreign language program of the Cleveland Public Schools.

Finally and most importantly, I wish to thank my husband and partner in all aspects of this project, Kenneth C. TAYLOR, for his invaluable contributions throughout.

Los Angeles, California   Anna Newman TAYLOR

Abbreviations

ant.	= anterior	ext.	= external	med.	= medial	surf.	= surface
a. or aa.	= artery or arteries	exten.	= extensor	n. or nn.	= nerve or nerves	sut.	= suture
		flex.	= flexor			transv.	= transverse
art.	= articulation	inf.	= inferior	obl.	= oblique	tuberc.	= tubercle
br. or brr.	= branch or branches	int.	= internal	post.	= posterior	tuberos.	= tuberosity
		inteross.	= interosseous	proc.	= process(es)	v. or vv.	= vein or veins
caud.	= caudal	lat.	= lateral	prot.	= protuberance	vent.	= ventral
cran.	= cranial	lig. or ligg.	= ligament or ligaments	prox.	= proximal	vert.	= vertebra
dist.	= distal			r. or rr.	= ramus or rami		
div.	= division	m. or mm.	= muscle or muscles	sup.	= superior		
dors.	= dorsal			superf.	= superficial		

BNA = Basle Nomina Anatomica (1895)
INA = Jena Nomina Anatomica (1936)
PNA = Paris Nomina Anatomica (1st: 1955, 2nd: 1961/63, 3rd: 1966/68, 4th: 1977, 5th: 1983, 6th: 1989)

In the figure legends, the anatomical terms are generally given in unabbreviated form.

Terms of Direction and Position

The following terms designate the position of organs and parts of the body and their relationship to each other with reference to the anatomical position, i.e., the human body standing erect, feet together, arms at sides with thumbs pointing away from the body. Where the English and Latin terms differ, the Latin term is indicated in parentheses.

General Terms

anterior – posterior = in front – behind (e.g., anterior and posterior tibial arteries)

ventral – dorsal (ventralis – dorsalis) = toward the belly – toward the back (synonymous with anterior – posterior)

superior – inferior = above – below (e.g., superior and inferior nasal conchae)

cranial – caudal (cranialis – caudalis) = toward the head – toward the tail (synonymous with superior – inferior)

right – left (dexter – sinister) = (e.g., right and left common iliac arteries)

internal – external (internus – externus) = located inside – located outside

superficial – deep (superficialis – profundus) = relative depth from the surface (e.g., flexor digitorum superficialis and profundus muscles)

middle (medius, intermedius) = in the middle between two other structures (e.g., the middle nasal concha is located between the superior and inferior nasal conchae)

median (medianus) = located in the midline (e.g., anterior median fissure of the spinal cord); the median plane is a vertical plane which divides the body into right and left halves

medial – lateral (medialis – lateralis) = located near to the midline of the body – located away from the midline of the body (e.g., medial and lateral inguinal fovea)

sagittal (sagittalis) = located in a sagittal plane, which is any vertical plane parallel to the median plane (e.g., sagittal suture of the skull)

frontal (frontalis) = located in a frontal plane, which is any vertical plane perpendicular to the sagittal plane (e.g., frontal process of the maxilla)

transverse (transversalis) = located in a transverse plane, which is a horizontal plane perpendicular to both sagittal and frontal planes (e.g., fascia transversalis)

longitudinal (longitudinalis) = parallel to the long axis (e.g., superior longitudinal muscle of the tongue)

proximal – distal (proximalis – distalis) = located toward or away from the attached end of a limb or the origin of a structure (e.g., proximal and distal radioulnar joints)

Terms for the Limbs

For the arm:

radial – ulnar (radialis – ulnaris) = on the radial side – on the ulnar side (e.g., radial and ulnar arteries)

For the hand:

palmar – dorsal (palmaris – dorsalis) = towards the palm of the hand – towards the back of the hand (e.g., palmar aponeurosis, dorsal interosseus muscle)

For the leg:

tibial – fibular (tibialis – fibularis) = on the tibial side – on the fibular side (e.g., anterior tibial artery)

For the foot:

plantar – dorsal (plantaris – dorsalis) = towards the sole of the foot – towards the upper surface of the foot (e.g., lateral and medial plantar arteries, dorsalis pedis artery)

Table of Contents

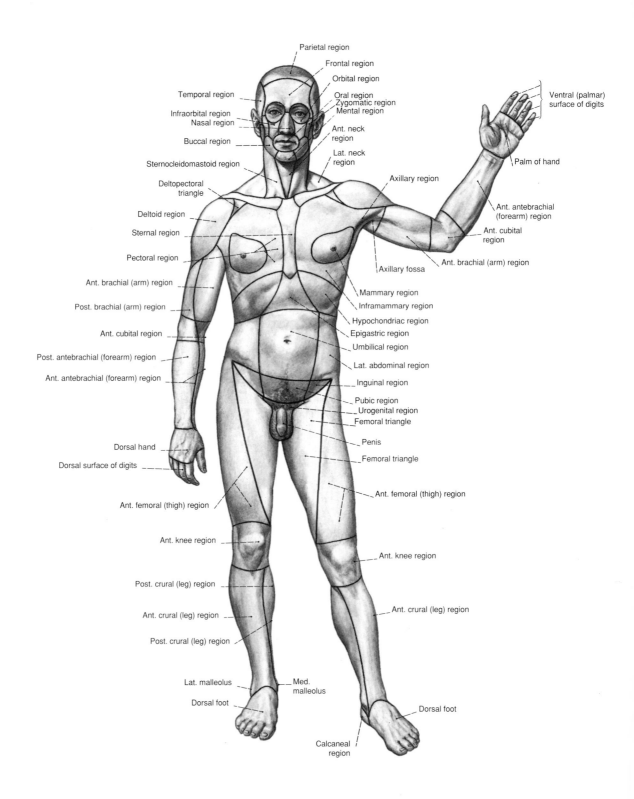

Parietal region
Frontal region
Orbital region
Temporal region
Oral region
Zygomatic region
Mental region
Infraorbital region
Nasal region
Ant. neck region
Buccal region
Lat. neck region
Sternocleidomastoid region
Axillary region
Deltopectoral triangle
Deltoid region
Ant. antebrachial (forearm) region
Sternal region
Ant. cubital region
Pectoral region
Ant. brachial (arm) region
Axillary fossa
Ant. brachial (arm) region
Post. brachial (arm) region
Mammary region
Inframammary region
Ant. cubital region
Hypochondriac region
Post. antebrachial (forearm) region
Epigastric region
Ant. antebrachial (forearm) region
Umbilical region
Lat. abdominal region
Inguinal region
Pubic region
Urogenital region
Femoral triangle
Dorsal hand
Penis
Dorsal surface of digits
Femoral triangle
Ventral (palmar) surface of digits
Palm of hand
Ant. femoral (thigh) region
Ant. femoral (thigh) region
Ant. knee region
Ant. knee region
Post. crural (leg) region
Ant. crural (leg) region
Ant. crural (leg) region
Post. crural (leg) region
Lat. malleolus
Med. malleolus
Dorsal foot
Dorsal foot
Calcaneal region

Fig. 1. Anterior body regions (cf. Fig. 77).

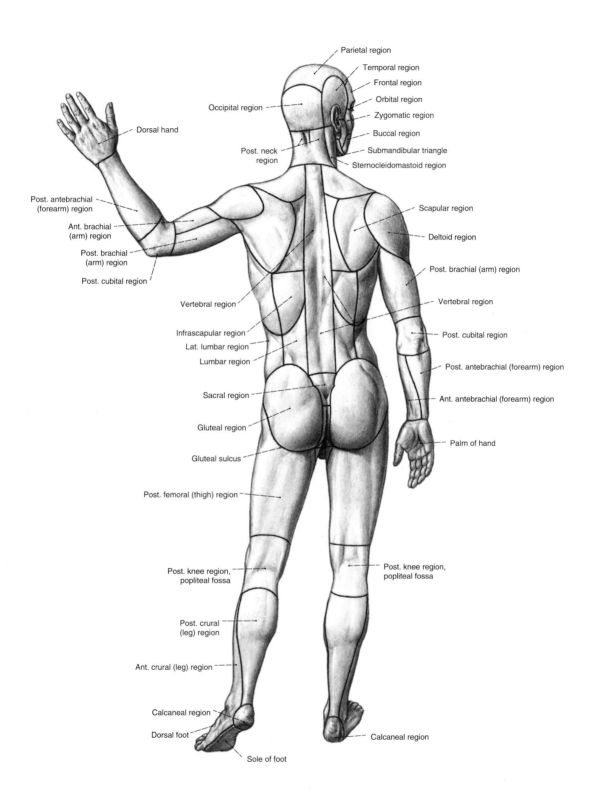

Parietal region
Temporal region
Frontal region
Orbital region
Zygomatic region
Buccal region
Submandibular triangle
Sternocleidomastoid region

Occipital region
Post. neck region

Dorsal hand

Scapular region
Deltoid region
Post. brachial (arm) region
Vertebral region
Post. cubital region
Post. antebrachial (forearm) region
Ant. antebrachial (forearm) region
Palm of hand

Post. antebrachial (forearm) region
Ant. brachial (arm) region
Post. brachial (arm) region
Post. cubital region

Vertebral region
Infrascapular region
Lat. lumbar region
Lumbar region
Sacral region
Gluteal region
Gluteal sulcus
Post. femoral (thigh) region

Post. knee region, popliteal fossa
Post. knee region, popliteal fossa

Post. crural (leg) region
Ant. crural (leg) region

Calcaneal region
Dorsal foot
Sole of foot
Calcaneal region

Fig. 2. Posterior body regions.

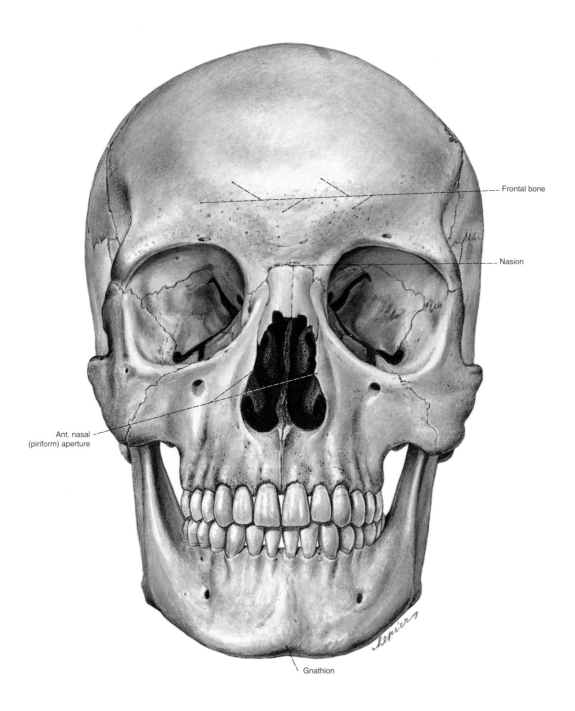

Frontal bone

Nasion

Ant. nasal
(piriform) aperture

Gnathion

Fig. 3. The skull. Anterior aspect. Norma frontalis: Facial aspect with **frontal bone** and midpoint of the root of the nose **(nasion)** in the horizontal plane (cf. Reid's base line) and the most prominent point of the border of the lower jaw **(gnathion)** in the vertical midsagittal plane.

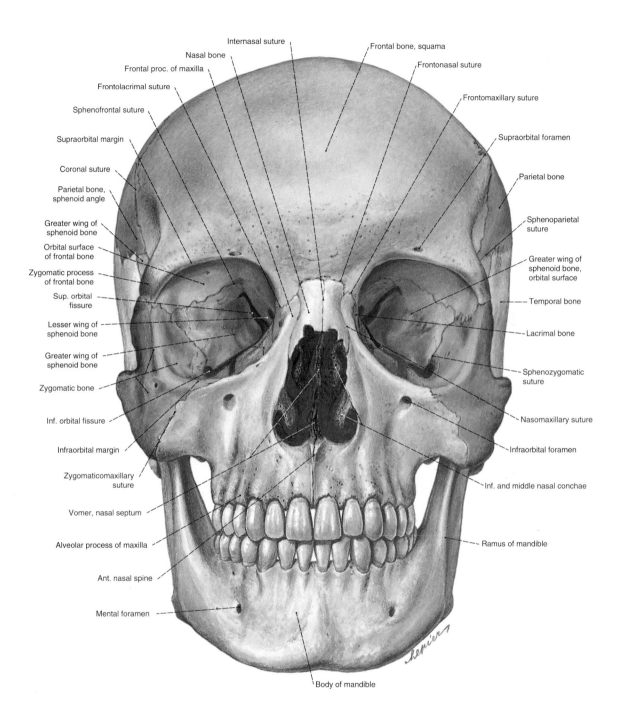

Internasal suture

Nasal bone

Frontal proc. of maxilla

Frontolacrimal suture

Sphenofrontal suture

Supraorbital margin

Coronal suture

Parietal bone, sphenoid angle

Greater wing of sphenoid bone

Orbital surface of frontal bone

Zygomatic process of frontal bone

Sup. orbital fissure

Lesser wing of sphenoid bone

Greater wing of sphenoid bone

Zygomatic bone

Inf. orbital fissure

Infraorbital margin

Zygomaticomaxillary suture

Vomer, nasal septum

Alveolar process of maxilla

Ant. nasal spine

Mental foramen

Frontal bone, squama

Frontonasal suture

Frontomaxillary suture

Supraorbital foramen

Parietal bone

Sphenoparietal suture

Greater wing of sphenoid bone, orbital surface

Temporal bone

Lacrimal bone

Sphenozygomatic suture

Nasomaxillary suture

Infraorbital foramen

Inf. and middle nasal conchae

Ramus of mandible

Body of mandible

Frontal bone – lavender
Nasal bone – light gray
Maxillary bone – yellow
Zygomatic bone – orange
Sphenoid bone – green
Temporal bone – light gray
Inferior nasal concha – gray
Mandible – light blue
Lacrimal bone – red
Vomer – red
Middle nasal concha (ethmoid bone) – dark yellow
Parietal bone – brown

Fig. 4. The skull. Anterior aspect. The bones of the skull are depicted by different colors.

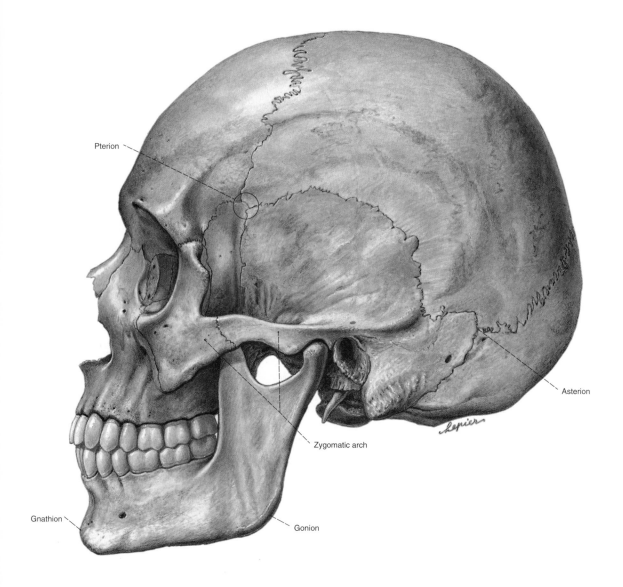

Pterion

Asterion

Zygomatic arch

Gnathion

Gonion

Fig. 5.   The skull. Lateral aspect. Norma lateralis: Viewed from the side in the vertical midsagittal and horizontal (cf. REID's base line) planes with **pterion** (= the point on the lateral wall of the skull where the squamous parts of the frontal and temporal bones meet the parietal bone and the greater wing of the sphenoid), **asterion** (= the point of intersection of the lambdoidal, occipitomastoid and parietomastoid sutures), and the **zygomatic arch,** which forms the lateral wall of the temporal fossa and its continuation into the infratemporal fossa (see Fig. 70).

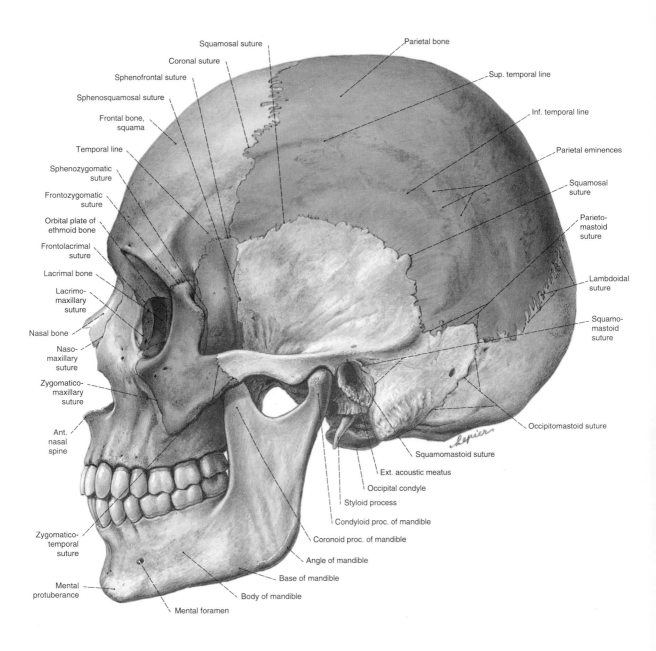

Squamosal suture

Coronal suture

Sphenofrontal suture

Sphenosquamosal suture

Frontal bone, squama

Temporal line

Sphenozygomatic suture

Frontozygomatic suture

Orbital plate of ethmoid bone

Frontolacrimal suture

Lacrimal bone

Lacrimo-maxillary suture

Nasal bone

Naso-maxillary suture

Zygomatico-maxillary suture

Ant. nasal spine

Zygomatico-temporal suture

Mental protuberance

Mental foramen

Mental foramen

Body of mandible

Base of mandible

Angle of mandible

Coronoid proc. of mandible

Condyloid proc. of mandible

Styloid process

Occipital condyle

Ext. acoustic meatus

Squamomastoid suture

Occipitomastoid suture

Squamo-mastoid suture

Lambdoidal suture

Parieto-mastoid suture

Squamosal suture

Parietal eminences

Inf. temporal line

Sup. temporal line

Parietal bone

Lepier

Frontal bone – lavender
Parietal bone – brown
Occipital bone – light blue
Nasal bone – light gray
Lacrimal bone – red
Zygomatic bone – orange
Sphenoid bone – green
Temporal bone – light gray
Maxilla – yellow
Mandible – light blue

Fig. 6. The skull. Lateral aspect. The cranial and facial bones of the head are depicted by different colors.

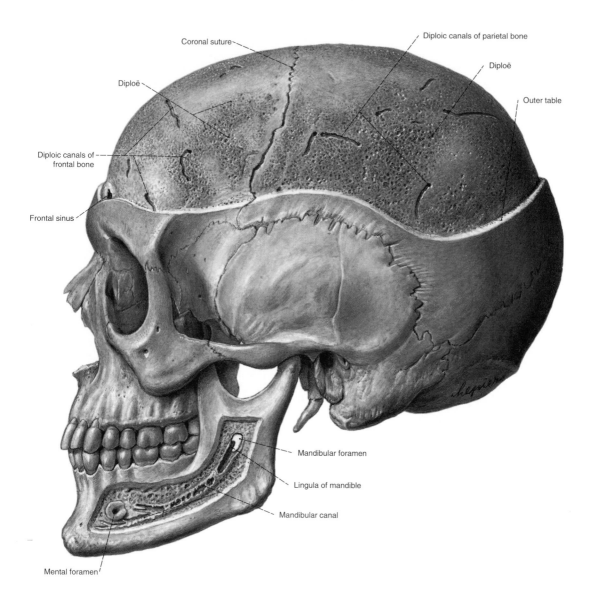

Coronal suture

Diploic canals of parietal bone

Diploë

Diploë

Outer table

Diploic canals of frontal bone

Frontal sinus

Mandibular foramen

Lingula of mandible

Mandibular canal

Mental foramen

Fig. 7. The skull. Lateral aspect. Stratified structure of the calvarium after removal of the outer table of bone. Diploë with diploic canals (cf. Fig. 79). Mandibular canal exposed.

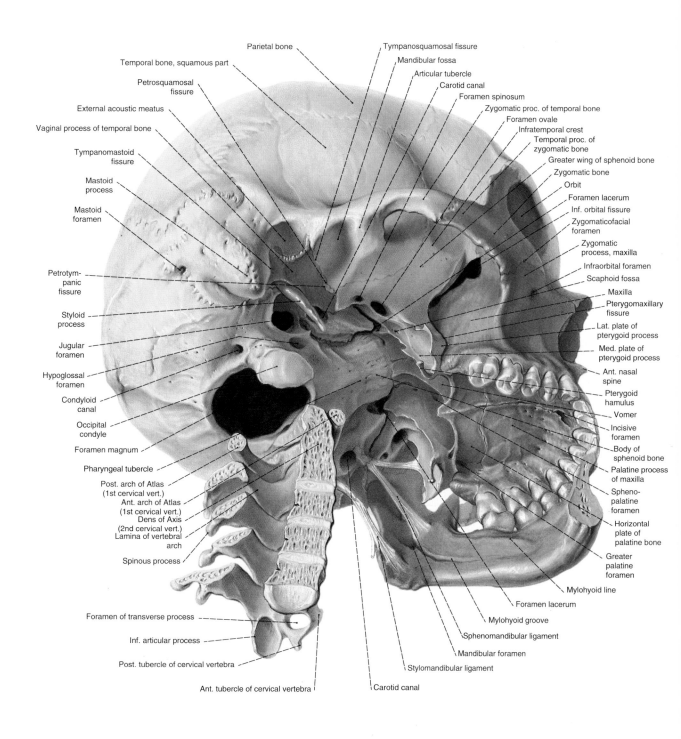

Parietal bone

Temporal bone, squamous part

Petrosquamosal
fissure

External acoustic meatus

Vaginal process of temporal bone

Tympanomastoid
fissure

Mastoid
process

Mastoid
foramen

Petrotym-
panic
fissure

Styloid
process

Jugular
foramen

Hypoglossal
foramen

Condyloid
canal

Occipital
condyle

Foramen magnum

Pharyngeal tubercle

Post. arch of Atlas
(1st cervical vert.)

Ant. arch of Atlas
(1st cervical vert.)

Dens of Axis
(2nd cervical vert.)

Lamina of vertebral
arch

Spinous process

Foramen of transverse process

Inf. articular process

Post. tubercle of cervical vertebra

Ant. tubercle of cervical vertebra

Tympanosquamosal fissure

Mandibular fossa

Articular tubercle

Carotid canal

Foramen spinosum

Zygomatic proc. of temporal bone

Foramen ovale

Infratemporal crest

Temporal proc. of
zygomatic bone

Greater wing of sphenoid bone

Zygomatic bone

Orbit

Foramen lacerum

Inf. orbital fissure

Zygomaticofacial
foramen

Zygomatic
process, maxilla

Infraorbital foramen

Scaphoid fossa

Maxilla

Pterygomaxillary
fissure

Lat. plate of
pterygoid process

Med. plate of
pterygoid process

Ant. nasal
spine

Pterygoid
hamulus

Vomer

Incisive
foramen

Body of
sphenoid bone

Palatine process
of maxilla

Spheno-
palatine
foramen

Horizontal
plate of
palatine bone

Greater
palatine
foramen

Mylohyoid line

Foramen lacerum

Mylohyoid groove

Sphenomandibular ligament

Mandibular foramen

Stylomandibular ligament

Carotid canal

Fig. 8. The skull, with bisected mandible and upper cervical
vertebral column viewed from below and from the side.

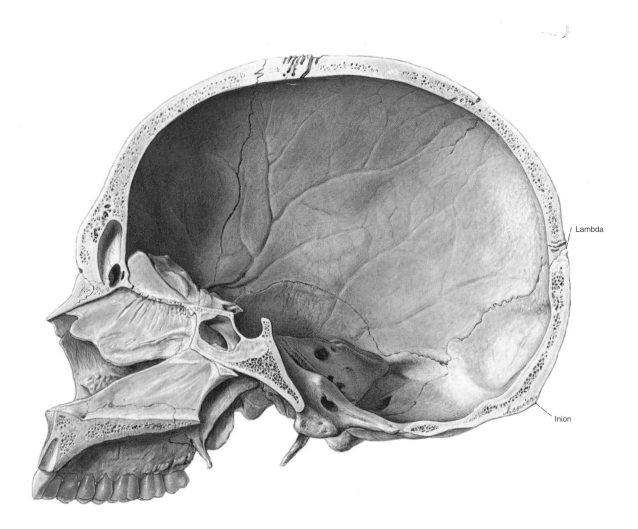

Lambda

Inion

Fig. 9. The skull. Paramedian sagittal section, medial aspect.
Inion = a craniometric term designating the external occipital
protuberance. **Lambda** = the point of intersection of the lambdoidal and sagittal sutures (cf. Fig. 17). On the internal surface
of the cranial bones, observe the arterial grooves [sulci] in
which the branches of the middle meningeal artery pass
between the dura mater of the brain and bone (cf. Fig. 18).

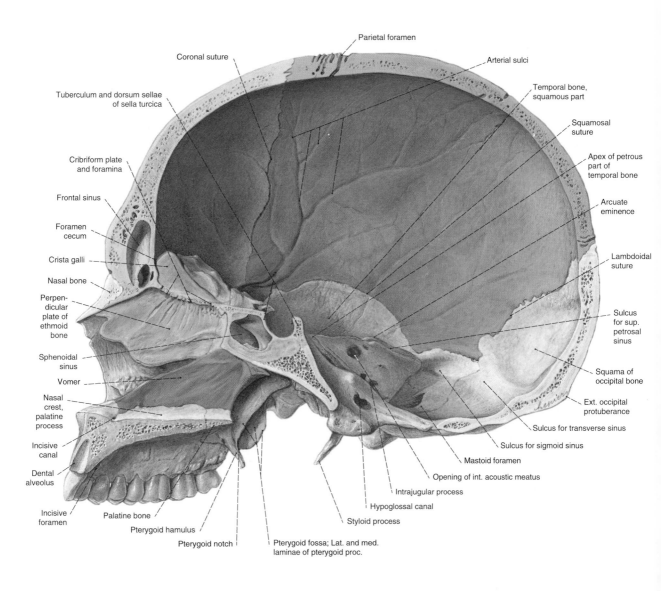

Parietal foramen

Coronal suture

Arterial sulci

Temporal bone, squamous part

Tuberculum and dorsum sellae of sella turcica

Squamosal suture

Apex of petrous part of temporal bone

Cribriform plate and foramina

Arcuate eminence

Frontal sinus

Foramen cecum

Crista galli

Lambdoidal suture

Nasal bone

Perpendicular plate of ethmoid bone

Sulcus for sup. petrosal sinus

Sphenoidal sinus

Vomer

Squama of occipital bone

Nasal crest, palatine process

Ext. occipital protuberance

Incisive canal

Sulcus for transverse sinus

Sulcus for sigmoid sinus

Dental alveolus

Mastoid foramen

Opening of int. acoustic meatus

Incisive foramen

Palatine bone

Intrajugular process

Hypoglossal canal

Pterygoid hamulus

Styloid process

Pterygoid notch

Pterygoid fossa; Lat. and med. laminae of pterygoid proc.

Frontal bone – lavender
Parietal bone – brown
Occipital bone – light blue
Nasal bone – light gray
Ethmoid bone – yellow
Sphenoid bone – green
Temporal bone – gray
Maxilla – light yellow
Vomer – red
Palatine bone – light blue

Fig. 10. The skull. Paramedian sagittal section, medial aspect. The bones of the skull are depicted by different colors.

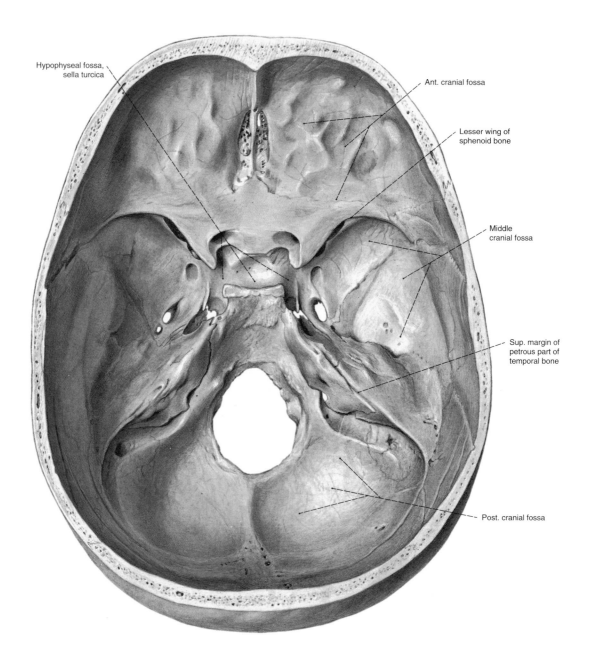

Hypophyseal fossa,
sella turcica

Ant. cranial fossa

Lesser wing of
sphenoid bone

Middle
cranial fossa

Sup. margin of
petrous part of
temporal bone

Post. cranial fossa

Fig. 11.   Interior of the base of the skull, showing anterior,
middle and posterior cranial fossae.

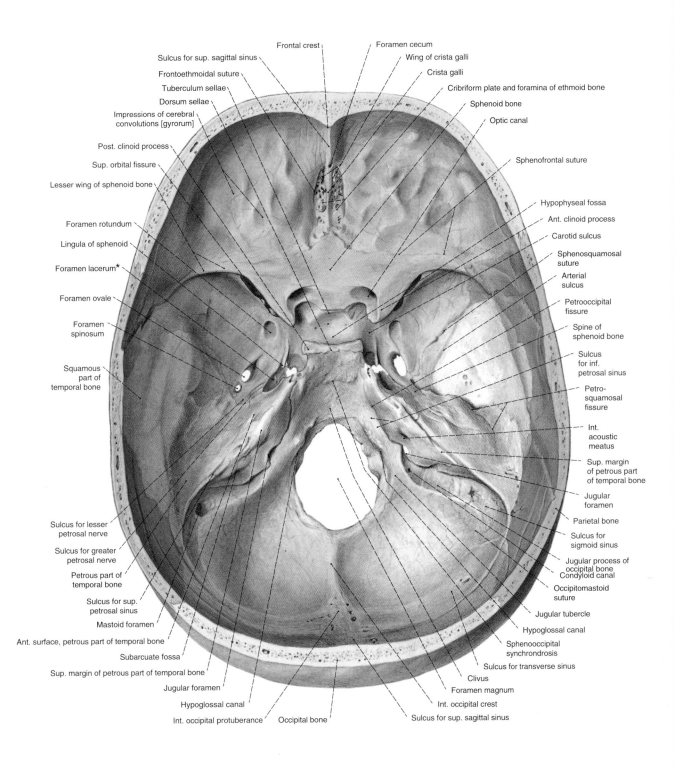

Frontal crest | Foramen cecum
Sulcus for sup. sagittal sinus | Wing of crista galli
Frontoethmoidal suture | Crista galli
Tuberculum sellae | Cribriform plate and foramina of ethmoid bone
Dorsum sellae | Sphenoid bone
Impressions of cerebral convolutions [gyrorum] | Optic canal
Post. clinoid process | Sphenofrontal suture
Sup. orbital fissure | Hypophyseal fossa
Lesser wing of sphenoid bone | Ant. clinoid process
Foramen rotundum | Carotid sulcus
Lingula of sphenoid | Sphenosquamosal suture
Foramen lacerum* | Arterial sulcus
Foramen ovale | Petrooccipital fissure
Foramen spinosum | Spine of sphenoid bone
Squamous part of temporal bone | Sulcus for inf. petrosal sinus
Sulcus for lesser petrosal nerve | Petro-squamosal fissure
Sulcus for greater petrosal nerve | Int. acoustic meatus
Petrous part of temporal bone | Sup. margin of petrous part of temporal bone
Sulcus for sup. petrosal sinus | Jugular foramen
Mastoid foramen | Parietal bone
Ant. surface, petrous part of temporal bone | Sulcus for sigmoid sinus
Subarcuate fossa | Jugular process of occipital bone
Sup. margin of petrous part of temporal bone | Condyloid canal
Jugular foramen | Occipitomastoid suture
Hypoglossal canal | Jugular tubercle
Int. occipital protuberance | Hypoglossal canal
Occipital bone | Sphenooccipital synchrondrosis
| Sulcus for transverse sinus
| Clivus
| Foramen magnum
| Int. occipital crest
| Sulcus for sup. sagittal sinus

Frontal bone – lavender
Parietal bone – brown
Ethmoid bone – yellow
Sphenoid bone – green
Temporal bone – gray
Occipital bone – light blue

Fig. 12. Interior of the base of the skull (from a young person as indicated by the residual spheno-occipital synchondrosis). The bones of the skull are depicted by different colors.

* The foramen lacerum is visible only in the dried skull, while *in vivo* this aperture is covered by a layer of fibrocartilage.

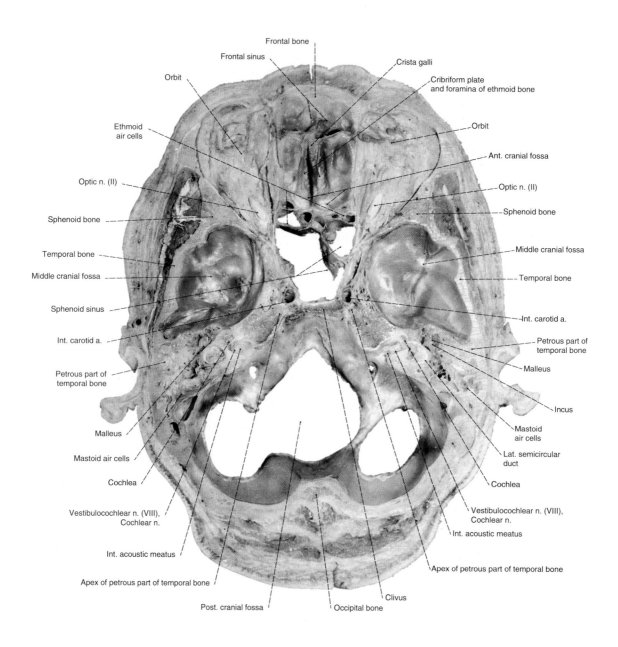

Fig. 13. Horizontal section from the head of a 54-year-old woman taken parallel to and about 1 cm cranial to a plane through the outer corner of the eye (canthus) and the midpoint of the external acoustic meatus. (From: Dr. G. RILLING, Institute of Anatomy and Embryology, University of Fribourg, Switzerland), (cf. Fig. 231).

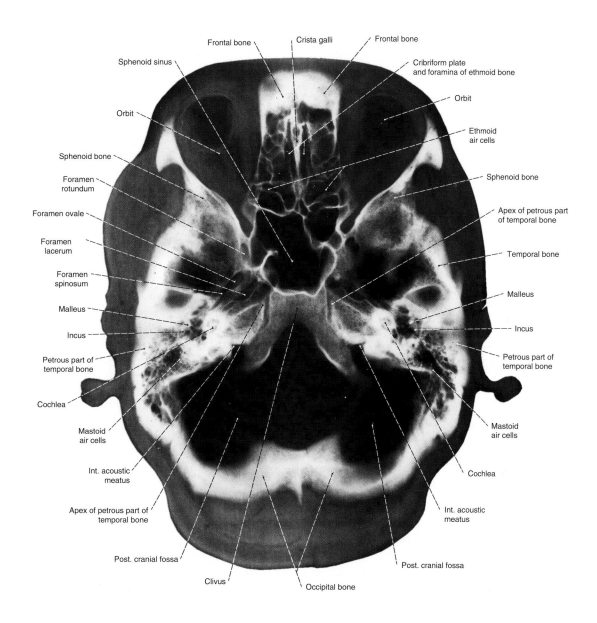

Fig. 14. Radiograph of the cranial section illustrated in Fig. 13, taken with the underneath surface of the section placed on the platen. (From: Dr. G. RILLING, Institute of Anatomy and Embryology, University of Fribourg, Switzerland.)

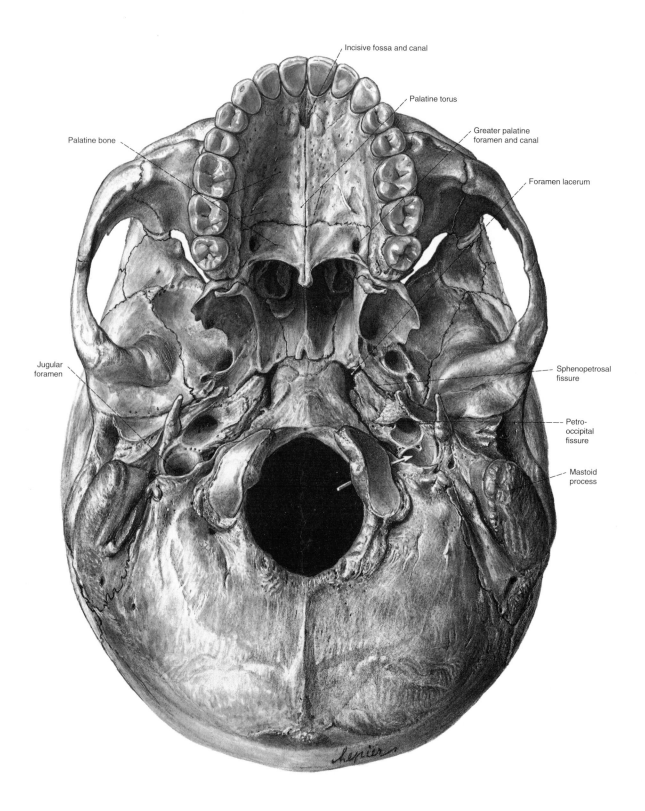

Incisive fossa and canal

Palatine torus

Greater palatine
foramen and canal

Foramen lacerum

Palatine bone

Sphenopetrosal
fissure

Petro-
occipital
fissure

Jugular
foramen

Mastoid
process

Fig. 15. Exterior of the base of the skull [Norma basalis]. The arrow is in the left hypoglossal canal.

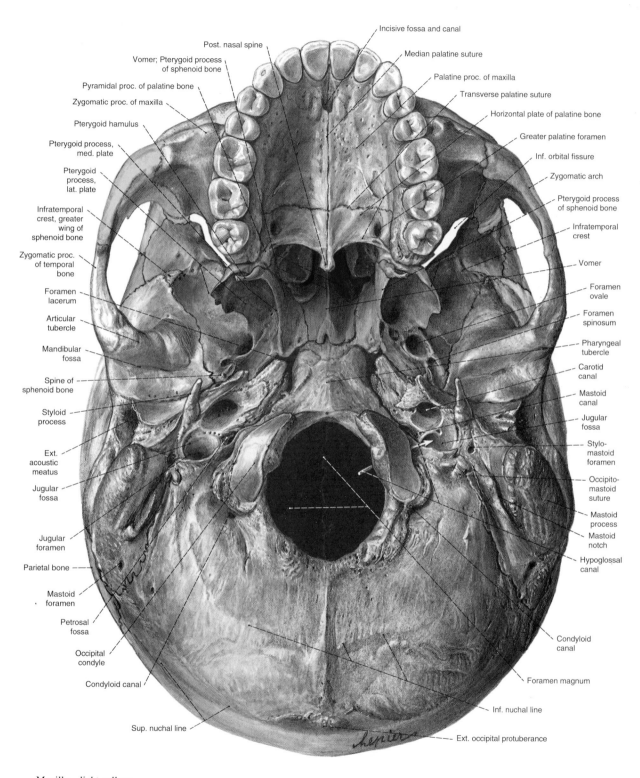

Incisive fossa and canal
Post. nasal spine
Median palatine suture
Vomer; Pterygoid process of sphenoid bone
Palatine proc. of maxilla
Pyramidal proc. of palatine bone
Transverse palatine suture
Zygomatic proc. of maxilla
Horizontal plate of palatine bone
Pterygoid hamulus
Greater palatine foramen
Pterygoid process, med. plate
Inf. orbital fissure
Pterygoid process, lat. plate
Zygomatic arch
Infratemporal crest, greater wing of sphenoid bone
Pterygoid process of sphenoid bone
Zygomatic proc. of temporal bone
Infratemporal crest
Foramen lacerum
Vomer
Articular tubercle
Foramen ovale
Mandibular fossa
Foramen spinosum
Spine of sphenoid bone
Pharyngeal tubercle
Styloid process
Carotid canal
Ext. acoustic meatus
Mastoid canal
Jugular fossa
Jugular fossa
Stylo-mastoid foramen
Jugular foramen
Occipito-mastoid suture
Parietal bone
Mastoid process
Mastoid foramen
Mastoid notch
Petrosal fossa
Hypoglossal canal
Occipital condyle
Condyloid canal
Condyloid canal
Foramen magnum
Sup. nuchal line
Inf. nuchal line
Ext. occipital protuberance

Maxilla – light yellow
Frontal bone – lavender
Parietal bone – brown
Palatine bone – light blue
Vomer – red
Zygomatic bone – orange
Sphenoid bone – green
Temporal bone – gray
Occipital bone – light blue

Fig. 16. Exterior of the base of the skull [Norma basalis]. The arrow is in the hypoglossal canal. The bones of the skull are depicted by different colors.

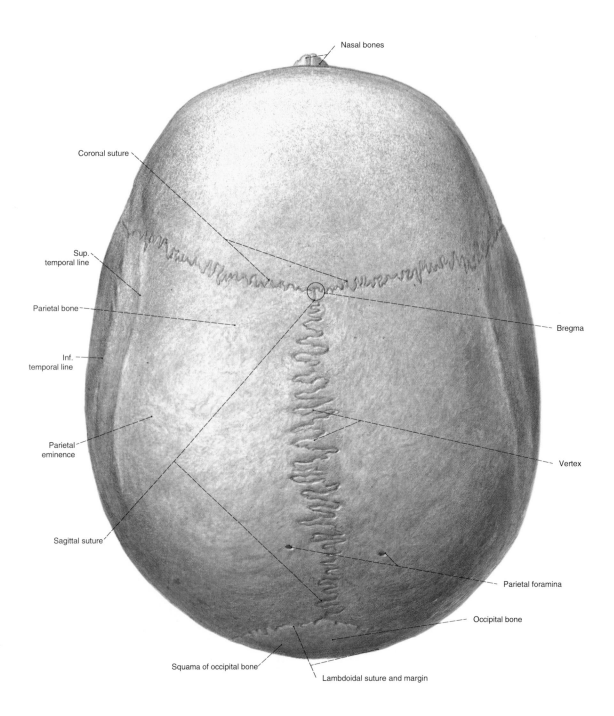

Nasal bones

Coronal suture

Sup.
temporal line

Parietal bone

Inf.
temporal line

Parietal
eminence

Sagittal suture

Bregma

Vertex

Parietal foramina

Occipital bone

Squama of occipital bone

Lambdoidal suture and margin

Fig. 17. The calvaria, viewed from above. The French term, calotte, which was formerly used (BNA, INA) for skull cap, has not been retained in recent editions of the NOMINA ANATOMICA.

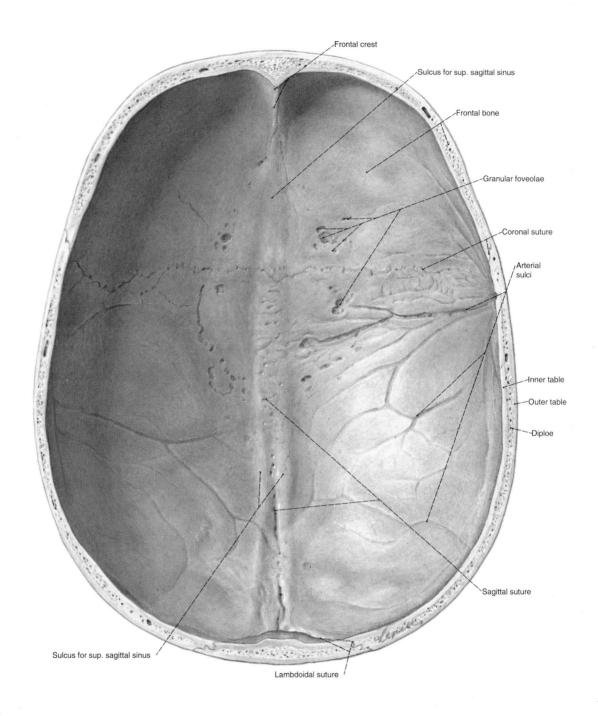

Frontal crest

Sulcus for sup. sagittal sinus

Frontal bone

Granular foveolae

Coronal suture

Arterial
sulci

Inner table

Outer table

Diploe

Sagittal suture

Sulcus for sup. sagittal sinus

Lambdoidal suture

Fig. 18.   The calvaria, internal aspect. Note the relatively deep, frequently ramifying bony grooves (sulci) for the branches of the middle meningeal artery; the flat or deep pits (foveolae) for the arachnoid granulations (PACCHIONIAN bodies) which may penetrate to the outer table (lamina) of the skull; the laminated structure of the cranial bones with outer table, diploë, and inner table; and the ossification of the cranial sutures.

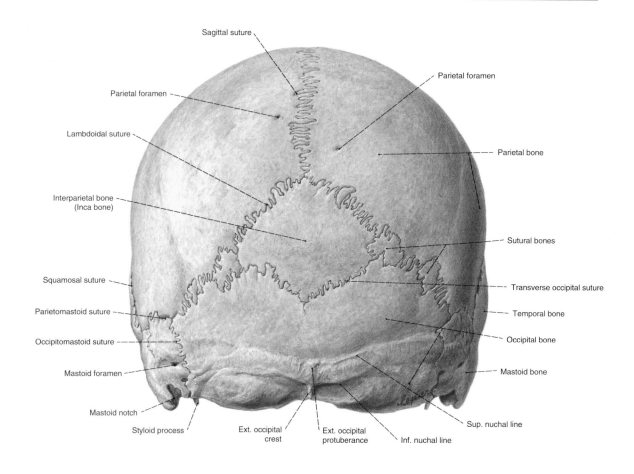

Fig. 19. The skull, dorsal aspect. Cranial sutures not fully ossified. The sutural bones and the interparietal bone ("Inca bone") are present. Openings (foramina) for passage of the emissary veins are named according to the cranial bones in which they are located. Those most consistently present are the mastoid foramen, the condyloid canal (Fig. 16) and the parietal foramen (which may be present only as an external opening).

The Skull at Birth

The skull of the newborn differs from the adult skull in numerous ways. At birth, the calvaria contains soft membranous sites (fontanelles) between the developing bones which later become sutures. There are two unpaired and two paired sets of fontanelles. The largest is the rhomboid-shaped anterior fontanelle. It is situated at the junction of the frontal, coronal and sagittal sutures, which form later, and connects the frontal and parietal bones. The smaller, posterior fontanelle is triangular in shape. It is located at the junction of the sagittal suture (which also forms later) with the occipital bone. The anterolateral or sphenoid fontanelles are situated bilaterally between the sphenoid angle of the parietal bone and the parietal border of the sphenoid bone, at the site of the later-forming sphenoparietal suture. They are rather rectangularly shaped. The posterolateral or mastoid fontanelles are situated bilaterally between the mastoid angle of the parietal bone and the petrous portion of the temporal bone at the site of the later developing parietomastoid suture. Like the anterior fontanelles, they are irregularly shaped and relatively large.

In the newborn, the occipital bone is not a single bone. It is divided into four ossified pieces – the squama, two lateral parts, and the basilar part. The posterior intraoccipital synchondrosis connects the squama with the two lateral parts, and the anterior intraoccipital synchondroses unite each lateral part with the basilar part. The tympanic part of the temporal bone is an incomplete ring, the tympanic annulus, which is open at the top. The mastoid process does not protrude. The petrosquamous fissure separates the squamous and the petrous parts. The eminences of the parietal and the frontal bones are very prominent. The frontal bone is separated into two halves by the frontal suture. The spheno-occipital synchondrosis is found between the sphenoid and occipital bones, and the intersphenoid synchondrosis lies within the body of the sphenoid. The maxilla and mandible are short because the alveolar processes have not yet formed. The incisive suture of the hard palate forms the boundary between the maxilla and the incisive bone. Between the two halves of the mandible, which as yet are incompletely joined, remnants of a suture are visible. The ramus of the mandible is virtually an elongation of its body, forming an obtuse mandibular angle.

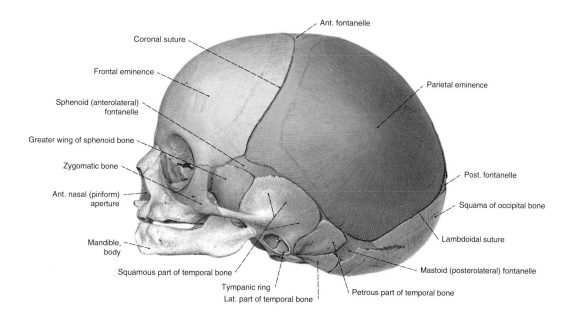

Ant. fontanelle

Coronal suture

Frontal eminence

Sphenoid (anterolateral) fontanelle

Greater wing of sphenoid bone

Zygomatic bone

Ant. nasal (piriform) aperture

Mandible, body

Squamous part of temporal bone

Tympanic ring

Lat. part of temporal bone

Parietal eminence

Post. fontanelle

Squama of occipital bone

Lambdoidal suture

Mastoid (posterolateral) fontanelle

Petrous part of temporal bone

Lacrimal bone, vomer – red
Frontal bone – lavender
Parietal bone – brown
Occipital bone – light blue
Nasal bone, temporal bone, mandible – gray
Maxilla, incisive bone – light yellow
Zygomatic bone – orange
Sphenoid bone – green

Fig. 20. The skull at birth, with fontanelles. Lateral aspect. The bones of the skull are depicted by different colors.

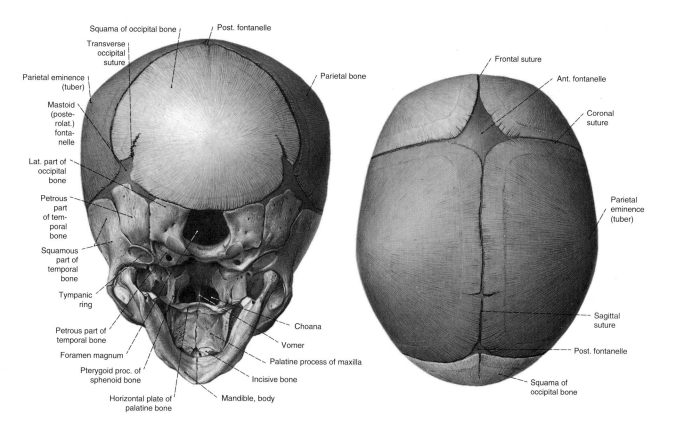

Squama of occipital bone

Post. fontanelle

Transverse occipital suture

Parietal bone

Parietal eminence (tuber)

Mastoid (posterolat.) fontanelle

Lat. part of occipital bone

Petrous part of temporal bone

Squamous part of temporal bone

Tympanic ring

Petrous part of temporal bone

Foramen magnum

Pterygoid proc. of sphenoid bone

Horizontal plate of palatine bone

Choana

Vomer

Palatine process of maxilla

Incisive bone

Mandible, body

Frontal suture

Ant. fontanelle

Coronal suture

Parietal eminence (tuber)

Sagittal suture

Post. fontanelle

Squama of occipital bone

Fig. 21. The skull at birth, with fontanelles. Posterior-inferior aspect. The bones of the skull are depicted by different colors.

Fig. 22. The skull at birth, with fontanelles. Viewed from above. The bones of the skull are depicted by different colors.

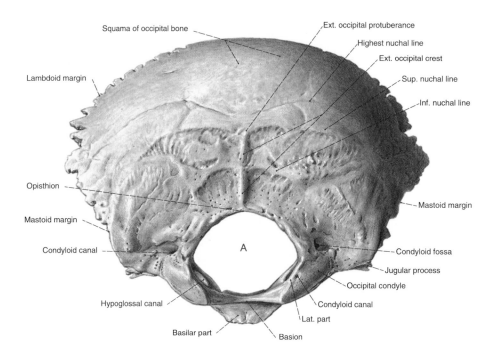

Fig. 23. Occipital bone. Outer, posterior surface.

A = Foramen magnum

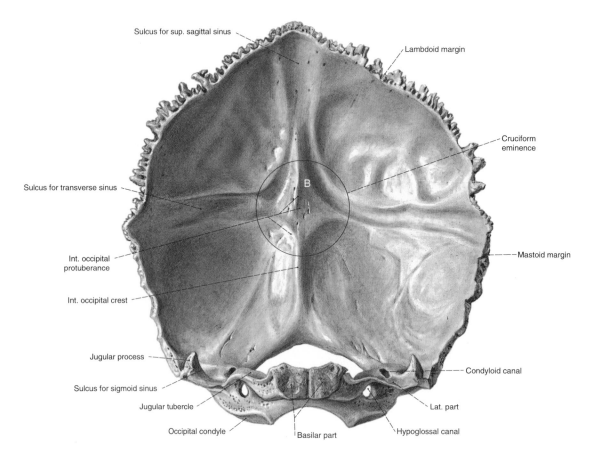

Fig. 24. Occipital bone. Inner, anterior surface.

B = Squama of occipital bone

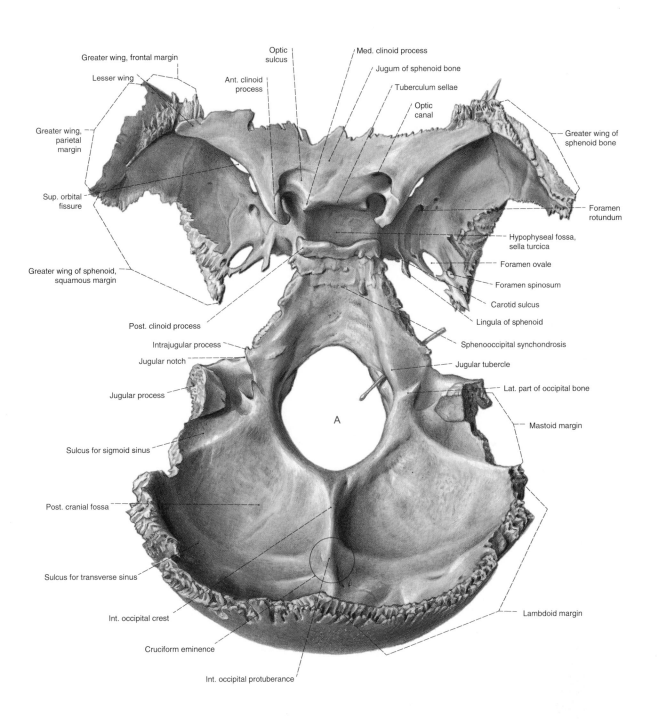

Greater wing, frontal margin

Lesser wing

Greater wing, parietal margin

Sup. orbital fissure

Greater wing of sphenoid, squamous margin

Post. clinoid process

Intrajugular process

Jugular notch

Jugular process

Sulcus for sigmoid sinus

Post. cranial fossa

Sulcus for transverse sinus

Int. occipital crest

Cruciform eminence

Int. occipital protuberance

Optic sulcus

Ant. clinoid process

Med. clinoid process

Jugum of sphenoid bone

Tuberculum sellae

Optic canal

Greater wing of sphenoid bone

Foramen rotundum

Hypophyseal fossa, sella turcica

Foramen ovale

Foramen spinosum

Carotid sulcus

Lingula of sphenoid

Sphenooccipital synchondrosis

Jugular tubercle

Lat. part of occipital bone

Mastoid margin

Lambdoid margin

A

Fig. 25. Occipital and sphenoid bones. Internal aspect. Remnants of the spheno-occipital synchondrosis are still visible as the sphenoid and occipital bones are not completely joined. A probe is in the right hypoglossal canal.

A = foramen magnum

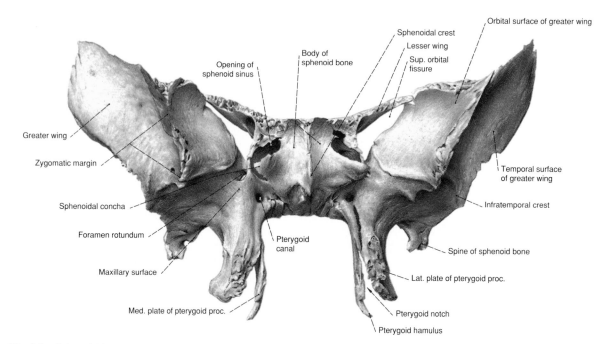

Fig. 26. Sphenoid bone. Anterior aspect.

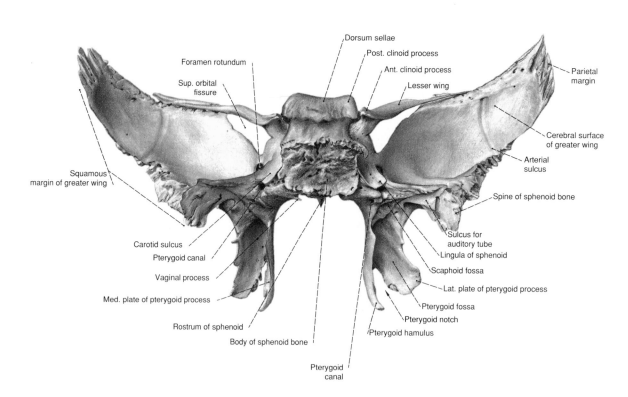

Fig. 27. Sphenoid bone. Posterior aspect. Figs. 26 and 27 are from a young person with an unossified spheno-occipital synchondrosis.

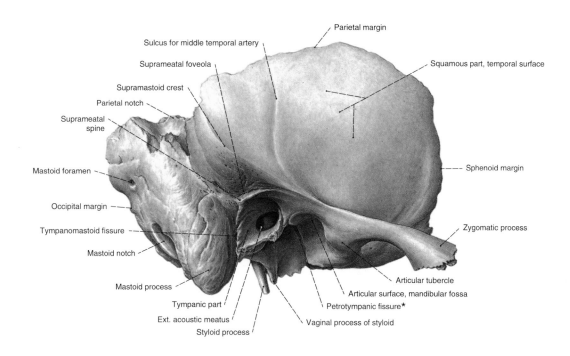

Fig. 28. Right temporal bone. External aspect.

* Glaserian fissure

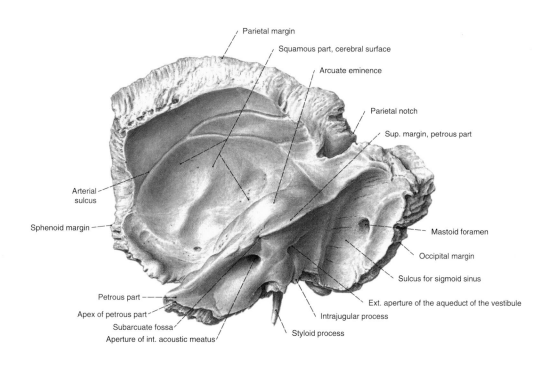

Fig. 29. Right temporal bone. Internal aspect.

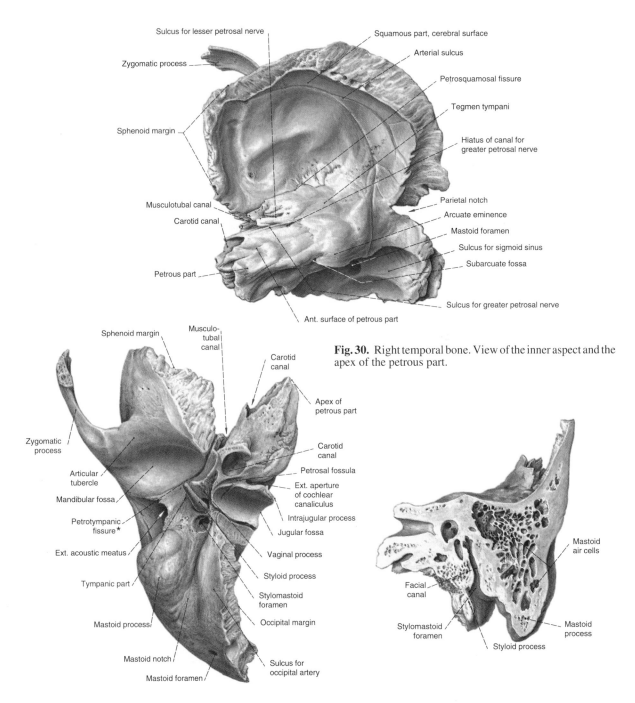

Sulcus for lesser petrosal nerve

Zygomatic process

Sphenoid margin

Musculotubal canal

Carotid canal

Petrous part

Squamous part, cerebral surface

Arterial sulcus

Petrosquamosal fissure

Tegmen tympani

Hiatus of canal for greater petrosal nerve

Parietal notch

Arcuate eminence

Mastoid foramen

Sulcus for sigmoid sinus

Subarcuate fossa

Sulcus for greater petrosal nerve

Ant. surface of petrous part

Fig. 30. Right temporal bone. View of the inner aspect and the apex of the petrous part.

Sphenoid margin

Musculo-tubal canal

Zygomatic process

Articular tubercle

Mandibular fossa

Petrotympanic fissure*

Ext. acoustic meatus

Tympanic part

Mastoid process

Mastoid notch

Mastoid foramen

Carotid canal

Apex of petrous part

Carotid canal

Petrosal fossula

Ext. aperture of cochlear canaliculus

Intrajugular process

Jugular fossa

Vaginal process

Styloid process

Stylomastoid foramen

Occipital margin

Sulcus for occipital artery

Mastoid air cells

Facial canal

Stylomastoid foramen

Mastoid process

Styloid process

Fig. 31. Right temporal bone, viewed from below.

* GLASERIAN fissure

Fig. 32. Oblique coronal section through the right mastoid process of the petrous part of the temporal bone. Note the extent of the mastoid air cells which hollow out the bone.

Mastoid Process and Mastoid Air Cells

The external surface of the conical mastoid process is irregular due to its muscular attachments. Internally, a network of air cells (mastoid cells) communicates with the tympanic cavity (Fig. 32). The size and shape of the mastoid process depend largely upon the extent of the air cells. Small or almost completely absent mastoid cells (as, for example, when sclerosed) result in a small mastoid process.

The proximity of the mastoid air cells and the mastoid antrum with the sigmoid sinus where it enters the mastoid portion of the temporal bone (Figs. 12, 30, 574) can, as a result of inflammation of the middle ear, lead to compression of the wall of the sinus and thrombosis.

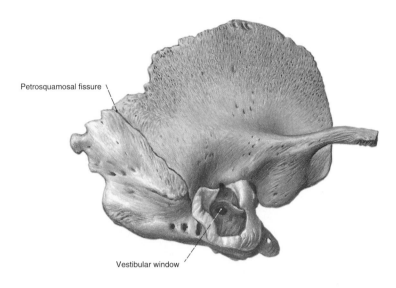

Petrosquamosal fissure

Vestibular window

Fig. 33. Right temporal bone of a newborn. Lateral, external aspect.

Squamous part – light green
Petrous part – light yellow
Tympanic part – gray

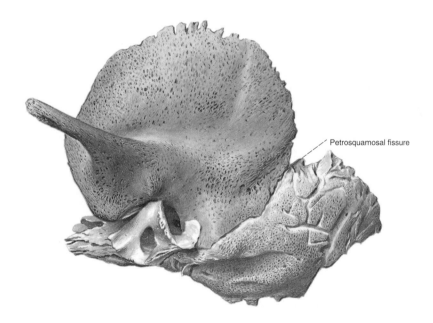

Petrosquamosal fissure

Fig. 34. Left temporal bone of a newborn. Lateral, external aspect. The parts of the bone are depicted by different colors, as in Fig. 33. The specimen is somewhat more enlarged than that shown in Fig. 33.

The temporal bone in the newborn (Figs. 33 and 34) differs considerably from the adult. The tympanic part appears as the tympanic ring, which is not joined at the top. The mastoid process is virtually absent because the air cells have not yet developed.

The petrosquamous fissure forms a distinct boundary between these two principal parts of the bone. In the first year of life, the furrowed plate-like tympanic part begins to sprout from the still incomplete tympanic ring.

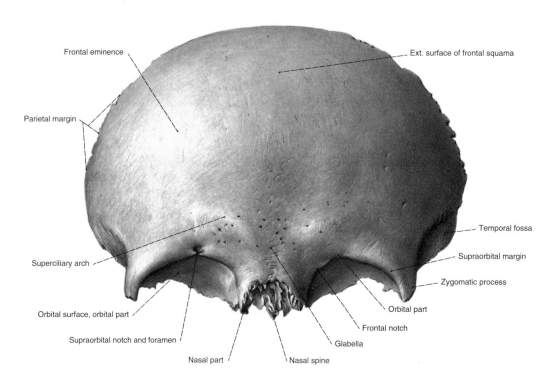

Frontal eminence

Ext. surface of frontal squama

Parietal margin

Temporal fossa

Superciliary arch

Supraorbital margin

Zygomatic process

Orbital surface, orbital part

Orbital part

Supraorbital notch and foramen

Frontal notch

Nasal part

Glabella

Nasal spine

Fig. 35. Frontal bone. External surface.

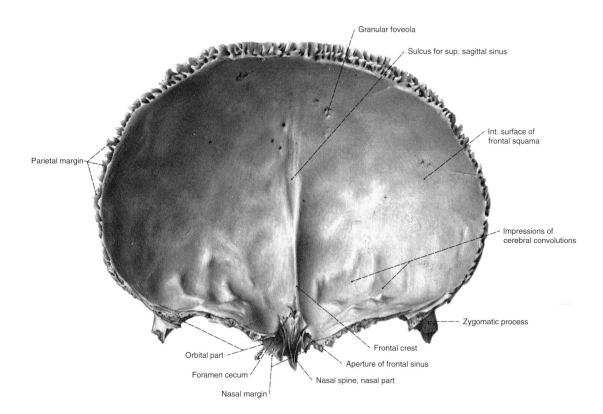

Granular foveola

Sulcus for sup. sagittal sinus

Int. surface of frontal squama

Parietal margin

Impressions of cerebral convolutions

Zygomatic process

Frontal crest

Orbital part

Aperture of frontal sinus

Foramen cecum

Nasal spine, nasal part

Nasal margin

Fig. 36. Frontal bone. Internal surface.

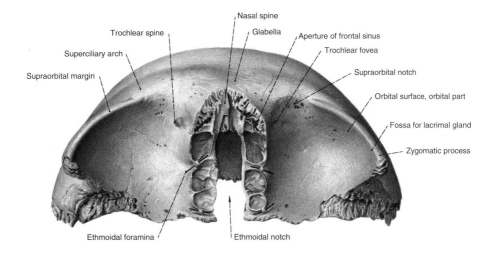

Fig. 37. Frontal bone, viewed from below. Probes are in the ethmoid foramina.

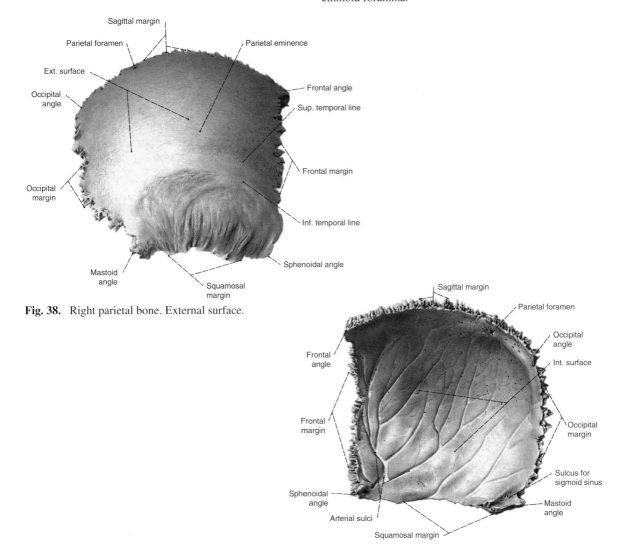

Fig. 38. Right parietal bone. External surface.

Fig. 39. Right parietal bone. Internal surface.

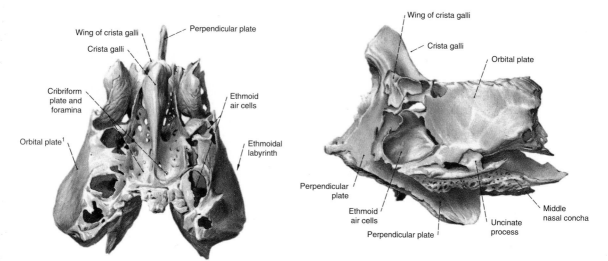

Fig. 40. Ethmoid bone, viewed from above.

¹ Lamina papyracea, so named because of the paper-thin nature of this bony plate

Fig. 41. Ethmoid bone, viewed from the left side.

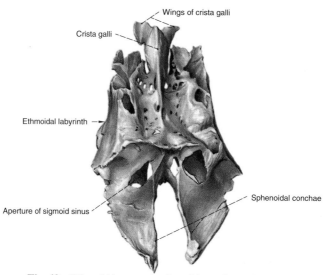

Fig. 42. Ethmoid bone and sphenoid conchae. Viewed from above and posterior.

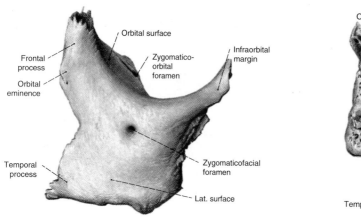

Fig. 43. Right zygomatic bone. Lateral, external surface.

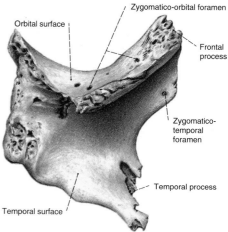

Fig. 44. Right zygomatic bone. Medial, internal surface.

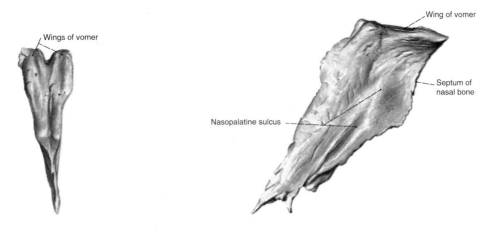

Fig. 45. Vomer. Posterior aspect.

Fig. 46. Vomer, viewed from the left side.

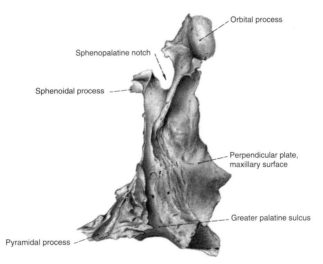

Fig. 47. Right palatine bone. Lateral aspect.

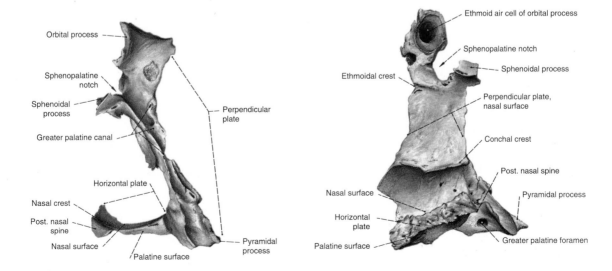

Fig. 48. Right palatine bone. Posterior aspect.

Fig. 49. Right palatine bone. Medial aspect.

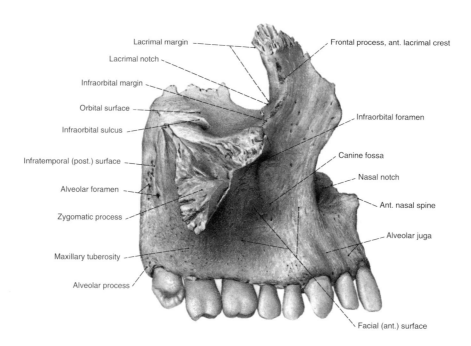

Lacrimal margin

Lacrimal notch

Infraorbital margin

Orbital surface

Infraorbital sulcus

Infratemporal (post.) surface

Alveolar foramen

Zygomatic process

Maxillary tuberosity

Alveolar process

Frontal process, ant. lacrimal crest

Infraorbital foramen

Canine fossa

Nasal notch

Ant. nasal spine

Alveolar juga

Facial (ant.) surface

Fig. 50. Right maxilla. Lateral aspect.

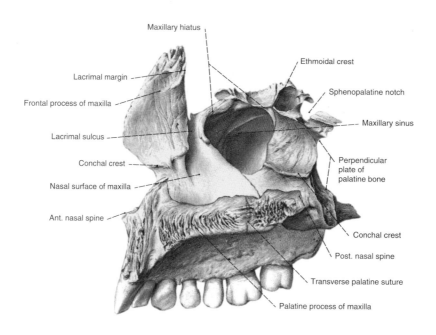

Maxillary hiatus

Lacrimal margin

Frontal process of maxilla

Lacrimal sulcus

Conchal crest

Nasal surface of maxilla

Ant. nasal spine

Ethmoidal crest

Sphenopalatine notch

Maxillary sinus

Perpendicular plate of palatine bone

Conchal crest

Post. nasal spine

Transverse palatine suture

Palatine process of maxilla

Fig. 51. Right maxilla and right palatine bone. Medial aspect.

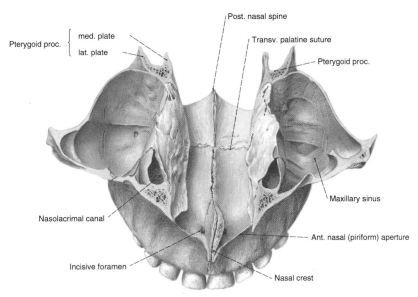

Pterygoid proc. { med. plate lat. plate

Post. nasal spine

Transv. palatine suture

Pterygoid proc.

Maxillary sinus

Nasolacrimal canal

Ant. nasal (piriform) aperture

Incisive foramen

Nasal crest

Maxilla – light yellow
Palatine bone – light blue
Sphenoid bone – light green
Inferior nasal concha – gray

Fig. 52. Hard palate, maxillary sinus, inferior nasal concha. Viewed from above (from the nasal cavity).

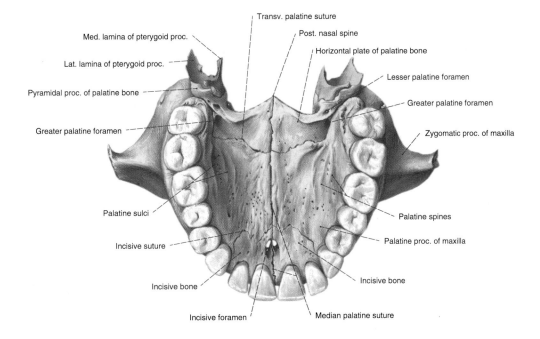

Transv. palatine suture

Med. lamina of pterygoid proc.

Post. nasal spine

Horizontal plate of palatine bone

Lat. lamina of pterygoid proc.

Lesser palatine foramen

Pyramidal proc. of palatine bone

Greater palatine foramen

Greater palatine foramen

Zygomatic proc. of maxilla

Palatine sulci

Palatine spines

Incisive suture

Palatine proc. of maxilla

Incisive bone

Incisive bone

Incisive foramen

Median palatine suture

Fig. 53. Hard palate. Viewed from below (from the oral cavity).

The incisive bone or premaxilla: Medial and anterior, between the palatine processes of both maxillae, is a separate bone in most vertebrates, including man. It contains the sockets of the incisor teeth and develops embryonically from its own center of ossification. In humans during fetal life, the incisive bone fuses almost completely with the adjacent maxilla. The incisive suture separating it from the palate is generally seen only in infancy. Though the premaxilla had been described by VESALIUS in the early 16th century, its existence in man was the subject of controversy for centuries until GOETHE re-discovered its presence in 1784. In GOETHE's time the presence or absence of the premaxilla was viewed as a major difference between man and apes. GOETHE erroneously extended the theory that parts of the skull are analogous to the structure of the vertebrae ("vertebral theory") by including the palatine bone, the maxilla and the premaxilla as vertebral components.
JOHANN WOLFGANG V. GOETHE, 1749-1832
ANDREAS VESALIUS, Professor of Anatomy at the University of Padua, 1514-1564

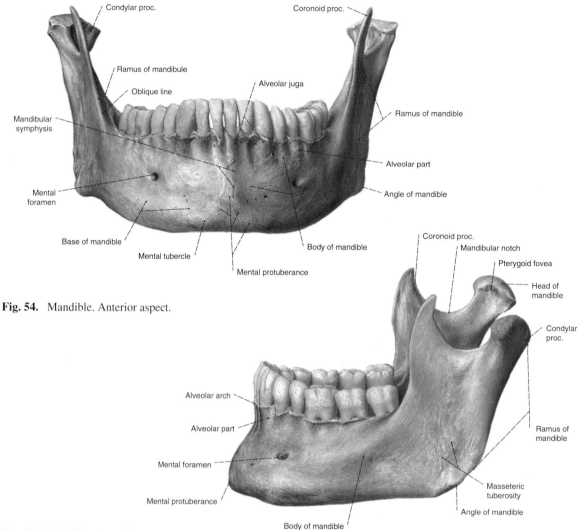

Condylar proc.

Coronoid proc.

Ramus of mandibule

Oblique line

Alveolar juga

Ramus of mandible

Mandibular symphysis

Alveolar part

Mental foramen

Angle of mandible

Base of mandible

Body of mandible

Mental tubercle

Mental protuberance

Fig. 54. Mandible. Anterior aspect.

Coronoid proc.

Mandibular notch

Pterygoid fovea

Head of mandible

Condylar proc.

Alveolar arch

Alveolar part

Mental foramen

Ramus of mandible

Mental protuberance

Masseteric tuberosity

Angle of mandible

Body of mandible

Fig. 55. Mandible. Lateral aspect.

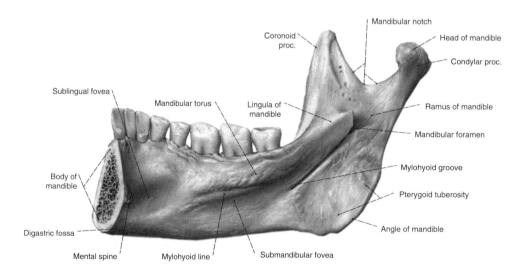

Mandibular notch

Coronoid proc.

Head of mandible

Condylar proc.

Sublingual fovea

Mandibular torus

Lingula of mandible

Ramus of mandible

Mandibular foramen

Mylohyoid groove

Body of mandible

Pterygoid tuberosity

Digastric fossa

Angle of mandible

Mental spine

Mylohyoid line

Submandibular fovea

Fig. 56. Right half of the mandible. Medial aspect.

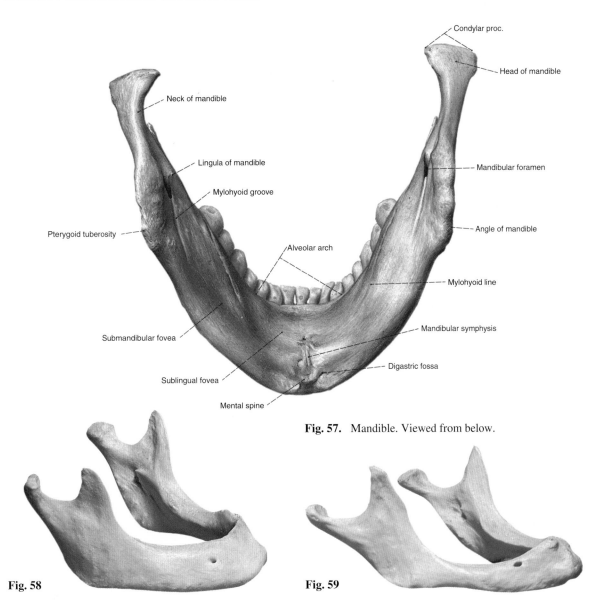

Condylar proc.

Head of mandible

Neck of mandible

Mandibular foramen

Lingula of mandible

Mylohyoid groove

Angle of mandible

Pterygoid tuberosity

Alveolar arch

Mylohyoid line

Submandibular fovea

Mandibular symphysis

Sublingual fovea

Digastric fossa

Mental spine

Fig. 57. Mandible. Viewed from below.

Fig. 58

Fig. 59

Figs. 58, 59. The mandibular body of the elderly can shrink to one-third of its original height. This process of bone resorption depends upon the onset of dental loss and the extent of prostheses. The bone can shrink relatively uniformly (Fig.

58) or preferentially in the regions of the molars (Fig. 59) or incisors. [From L. HUPFAUF (Ed.): Total prostheses. Dental Care Practice, Vol. 7. Urban & Schwarzenberg. Munich-Vienna-Baltimore, 1987].

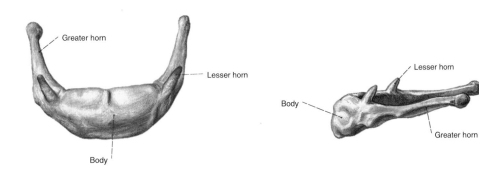

Greater horn

Lesser horn

Lesser horn

Body

Body

Greater horn

Fig. 60. Hyoid bone. Viewed from above and anterior.

Fig. 61. Hyoid bone. Lateral aspect.

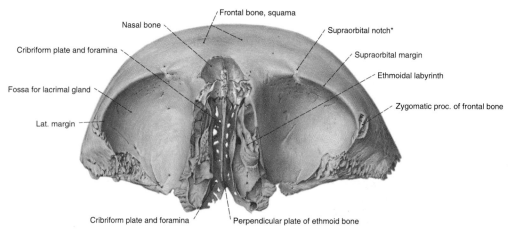

Fig. 62. Frontal bone, ethmoid bone, and nasal bones. Viewed from below (cf. Figs. 40–42).

* May exist as a supraorbital foramen

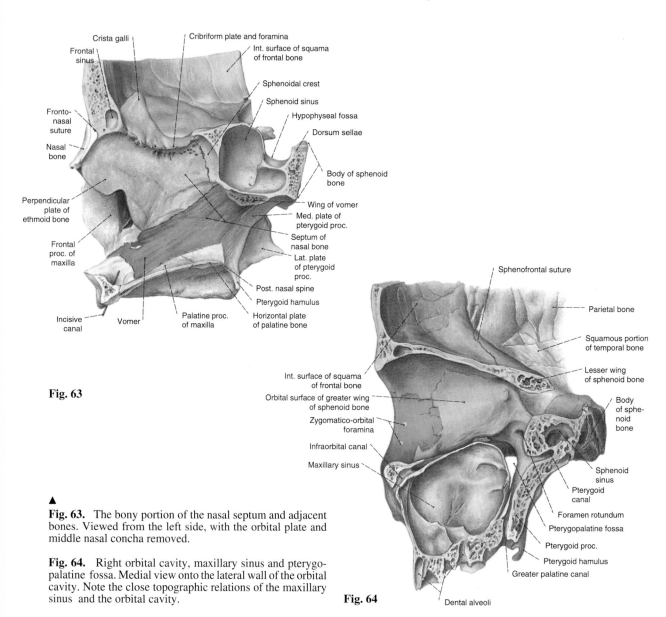

Fig. 63

Fig. 63. The bony portion of the nasal septum and adjacent bones. Viewed from the left side, with the orbital plate and middle nasal concha removed.

Fig. 64. Right orbital cavity, maxillary sinus and pterygo-palatine fossa. Medial view onto the lateral wall of the orbital cavity. Note the close topographic relations of the maxillary sinus and the orbital cavity.

Fig. 64

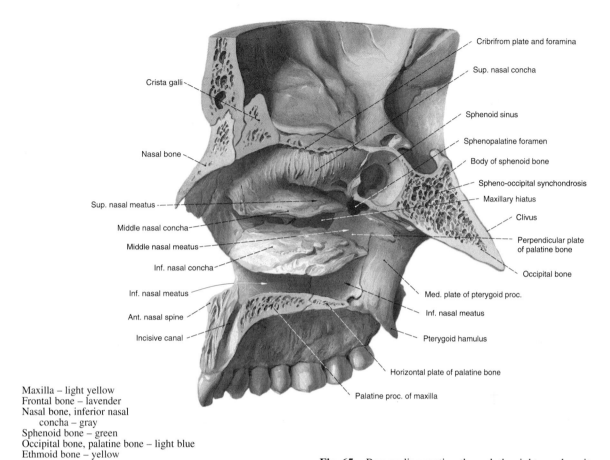

Crista galli

Nasal bone

Sup. nasal meatus

Middle nasal concha

Middle nasal meatus

Inf. nasal concha

Inf. nasal meatus

Ant. nasal spine

Incisive canal

Cribrifrom plate and foramina

Sup. nasal concha

Sphenoid sinus

Sphenopalatine foramen

Body of sphenoid bone

Spheno-occipital synchondrosis

Maxillary hiatus

Clivus

Perpendicular plate
of palatine bone

Occipital bone

Med. plate of pterygoid proc.

Inf. nasal meatus

Pterygoid hamulus

Horizontal plate of palatine bone

Palatine proc. of maxilla

Maxilla – light yellow
Frontal bone – lavender
Nasal bone, inferior nasal
    concha – gray
Sphenoid bone – green
Occipital bone, palatine bone – light blue
Ethmoid bone – yellow
Vomer – red
Lacrimal bone – red
Parietal bone – orange

Fig. 65. Paramedian section through the right nasal cavity, with adjacent bones. Medial aspect.

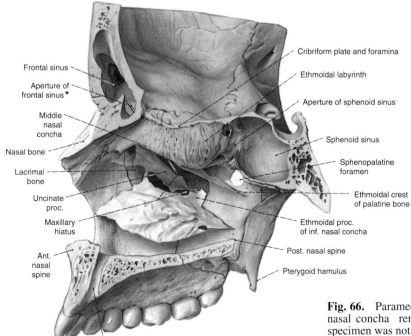

Frontal sinus

Aperture of
frontal sinus *

Middle
nasal
concha

Nasal bone

Lacrimal
bone

Uncinate
proc.

Maxillary
hiatus

Ant.
nasal
spine

Incisive canal

Cribriform plate and foramina

Ethmoidal labyrinth

Aperture of sphenoid sinus

Sphenoid sinus

Sphenopalatine
foramen

Ethmoidal crest
of palatine bone

Ethmoidal proc.
of inf. nasal concha

Post. nasal spine

Pterygoid hamulus

Fig. 66. Paramedian section as in Fig. 65 with the middle nasal concha removed (the superior nasal concha of this specimen was not distinct from the middle concha).

* The probe marks the connection of the frontal sinus with the middle nasal meatus.

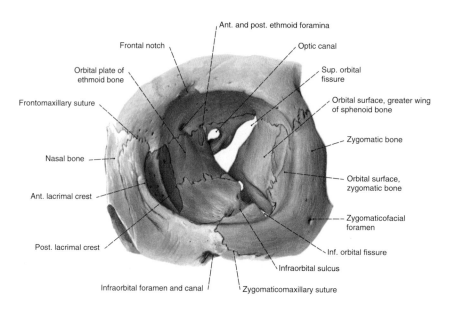

Ant. and post. ethmoid foramina

Frontal notch

Optic canal

Orbital plate of
ethmoid bone

Sup. orbital
fissure

Frontomaxillary suture

Orbital surface, greater wing
of sphenoid bone

Zygomatic bone

Nasal bone

Orbital surface,
zygomatic bone

Ant. lacrimal crest

Zygomaticofacial
foramen

Post. lacrimal crest

Inf. orbital fissure

Infraorbital sulcus

Infraorbital foramen and canal

Zygomaticomaxillary suture

Fig. 67. The left orbital cavity. Anterior view. The probe is
in the infraorbital canal. The component bones are depicted by
different colors.

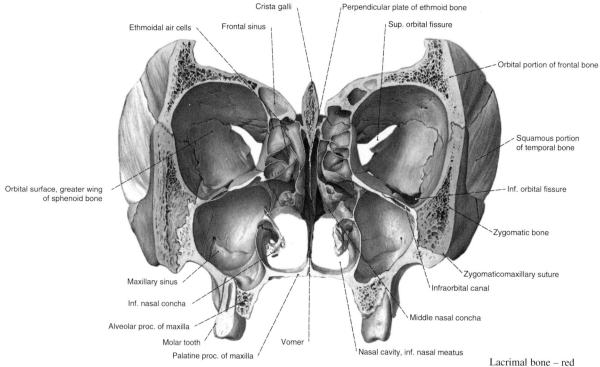

Crista galli

Perpendicular plate of ethmoid bone

Ethmoidal air cells

Frontal sinus

Sup. orbital fissure

Orbital portion of frontal bone

Squamous portion
of temporal bone

Orbital surface, greater wing
of sphenoid bone

Inf. orbital fissure

Zygomatic bone

Zygomaticomaxillary suture

Infraorbital canal

Maxillary sinus

Inf. nasal concha

Middle nasal concha

Alveolar proc. of maxilla

Molar tooth

Vomer

Palatine proc. of maxilla

Nasal cavity, inf. nasal meatus

Lacrimal bone – red
Nasal bone – gray
Frontal bone – lavender
Palatine bone – light blue
Ethmoid bone – yellow
Vomer – red
Zygomatic bone – orange
Maxilla – light yellow
Inferior nasal concha,
    temporal bone – gray
Sphenoid bone – green

Fig. 68. Coronal section through the anterior part of the facial
skeleton: Orbital cavities, nasal cavities, maxillary sinus,
ethmoid air cells. The component bones are depicted by different colors.

Frontal bone – lavender
Lacrimal bone – red
Sphenoid bone – green
Nasal bone – gray
Palatine bone – light blue
Ethmoid bone – yellow
Maxilla – light yellow
Zygomatic bone – orange
Temporal bone – gray

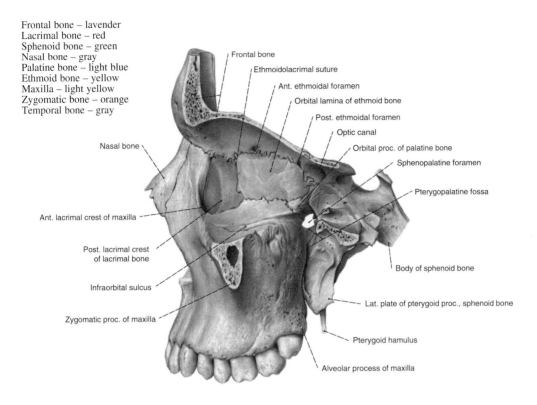

Frontal bone
Ethmoidolacrimal suture
Ant. ethmoidal foramen
Orbital lamina of ethmoid bone
Post. ethmoidal foramen
Optic canal
Orbital proc. of palatine bone
Sphenopalatine foramen
Pterygopalatine fossa
Body of sphenoid bone
Lat. plate of pterygoid proc., sphenoid bone
Pterygoid hamulus
Alveolar process of maxilla

Nasal bone
Ant. lacrimal crest of maxilla
Post. lacrimal crest of lacrimal bone
Infraorbital sulcus
Zygomatic proc. of maxilla

Fig. 69. The medial wall of the left orbital cavity and the lateral aspect of the pterygopalatine fossa. The component bones are depicted by different colors.

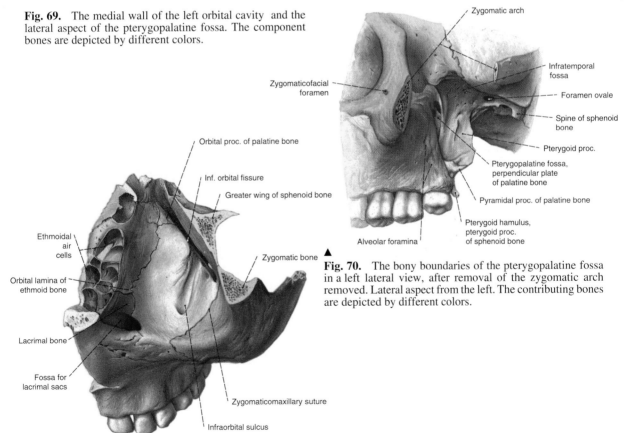

Zygomaticofacial foramen
Orbital proc. of palatine bone
Inf. orbital fissure
Greater wing of sphenoid bone
Zygomatic bone
Ethmoidal air cells
Orbital lamina of ethmoid bone
Lacrimal bone
Fossa for lacrimal sacs
Zygomaticomaxillary suture
Infraorbital sulcus

Zygomatic arch
Infratemporal fossa
Foramen ovale
Spine of sphenoid bone
Pterygoid proc.
Pterygopalatine fossa, perpendicular plate of palatine bone
Pyramidal proc. of palatine bone
Pterygoid hamulus, pterygoid proc. of sphenoid bone
Alveolar foramina

▲
Fig. 70. The bony boundaries of the pterygopalatine fossa in a left lateral view, after removal of the zygomatic arch removed. Lateral aspect from the left. The contributing bones are depicted by different colors.

Fig. 71. The floor of the left orbital cavity, with the roof of the cavity removed.

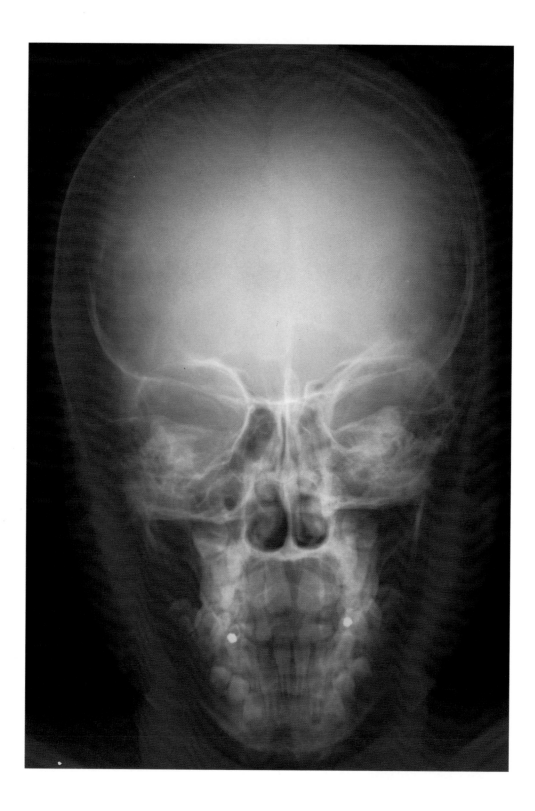

Fig. 72. Radiograph of the skull. Sagittal projection. (From:
Dr. G. GREEVEN, St. Elizabeth Hospital, Neuwied.)

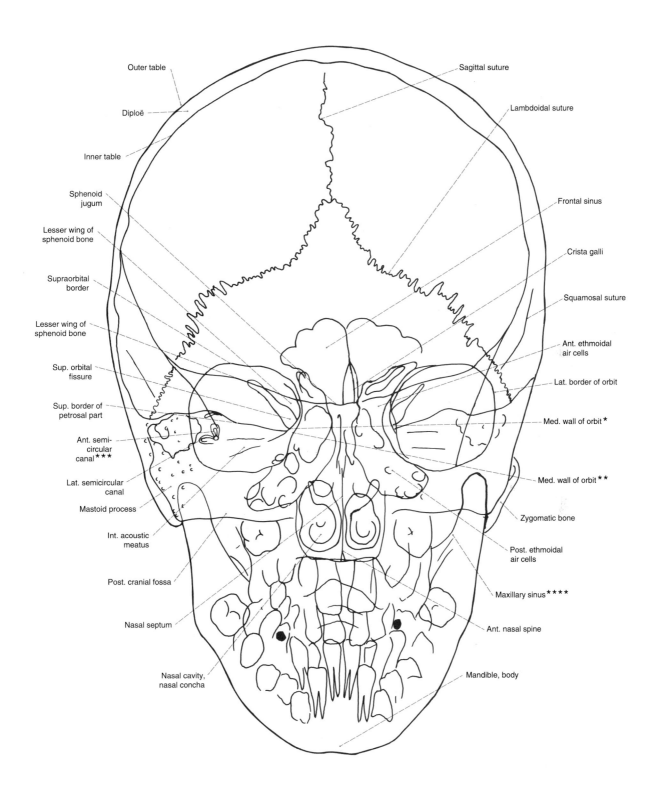

Fig. 73.   Line drawing of Fig. 72.

* Posterior part of the medial wall of the orbital cavity
** Anterior part of the medial wall of the orbital cavity
*** Also known as the superior semicircular canal
**** The contour designates the anterolateral wall of the maxillary sinus.

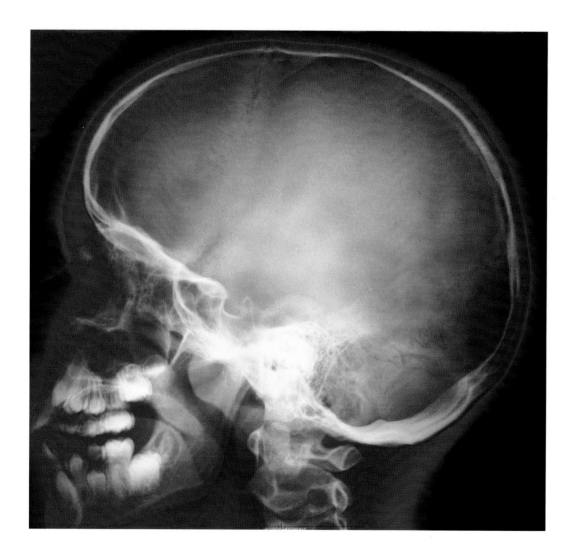

Fig. 74. Radiograph of the skull of an 11-year-old child; bitemporal projection. (From: Dr. G. GREEVEN, St. Elizabeth Hospital, Neuwied.)

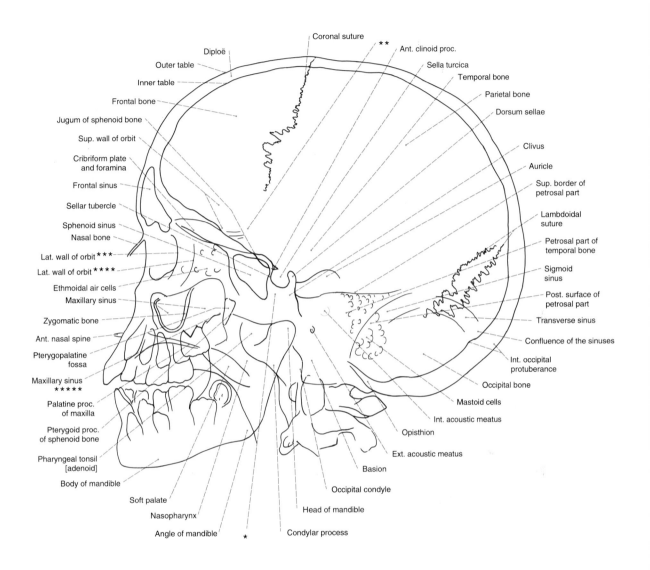

Fig. 75. Line drawing of Fig. 74.

    * Posterior nonaerated portion of the sphenoid sinus
   ** Typical thickening due to branching of the wing of the sphenoid bone
 *** Adjacent side
**** Distal side
***** Posterior wall of the maxillary sinus

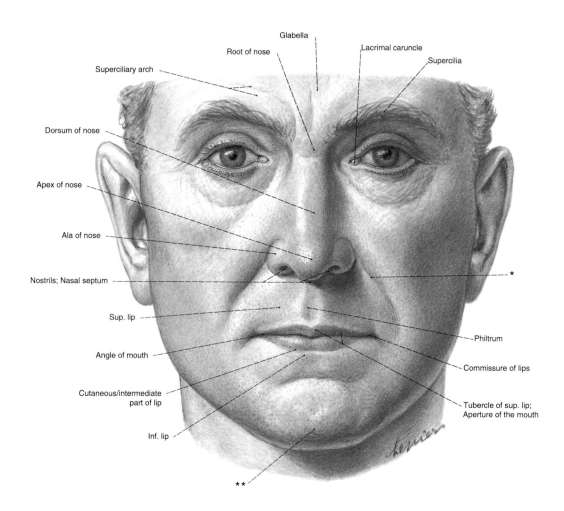

Glabella

Root of nose

Lacrimal caruncle

Supercilia

Superciliary arch

Dorsum of nose

Apex of nose

Ala of nose

Nostrils; Nasal septum

Sup. lip

Angle of mouth

Cutaneous/intermediate part of lip

Inf. lip

Philtrum

Commissure of lips

Tubercle of sup. lip; Aperture of the mouth

*

**

Fig. 76. The face. Anterior view.

 * Nasolabial sulcus
** Chin or mentum

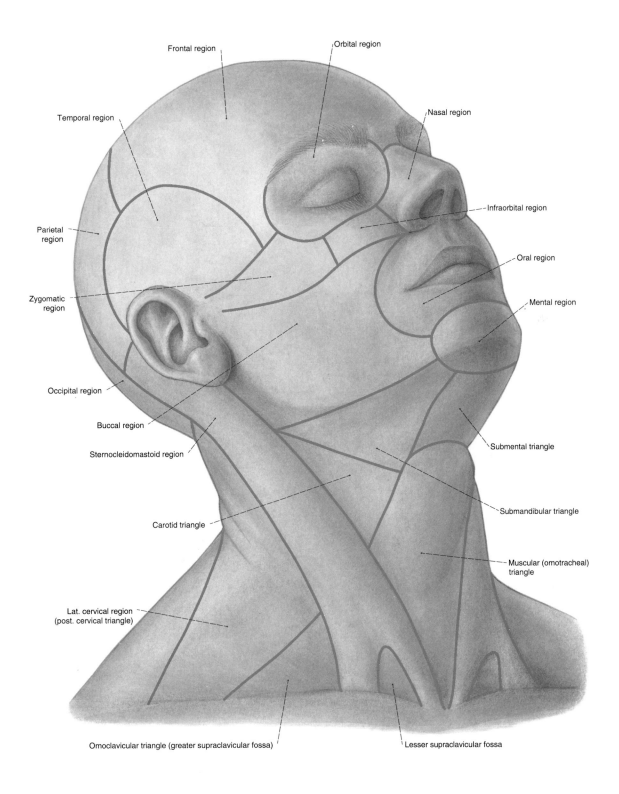

Frontal region

Orbital region

Nasal region

Temporal region

Parietal region

Infraorbital region

Oral region

Zygomatic region

Mental region

Occipital region

Buccal region

Sternocleidomastoid region

Submental triangle

Submandibular triangle

Carotid triangle

Muscular (omotracheal) triangle

Lat. cervical region (post. cervical triangle)

Omoclavicular triangle (greater supraclavicular fossa)

Lesser supraclavicular fossa

Fig. 77. The regions of the face, head and neck.

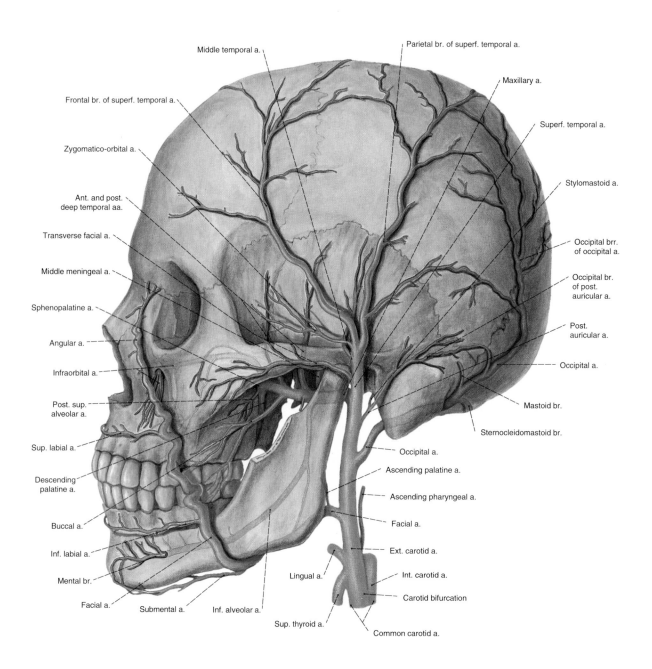

Fig. 78. Schematic diagram of the external carotid artery and its branches in the head.

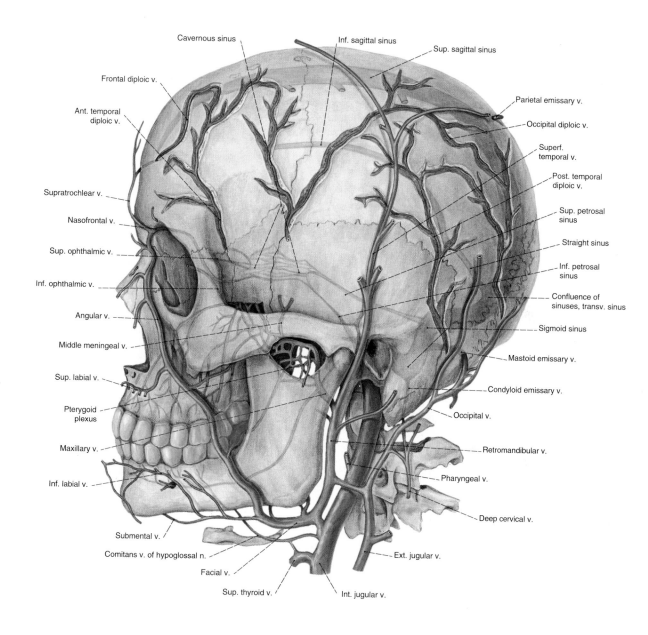

Fig. 79. Schematic diagram of the superficial veins of the head and face, showing connections to deeper veins and dural sinuses (in light blue).

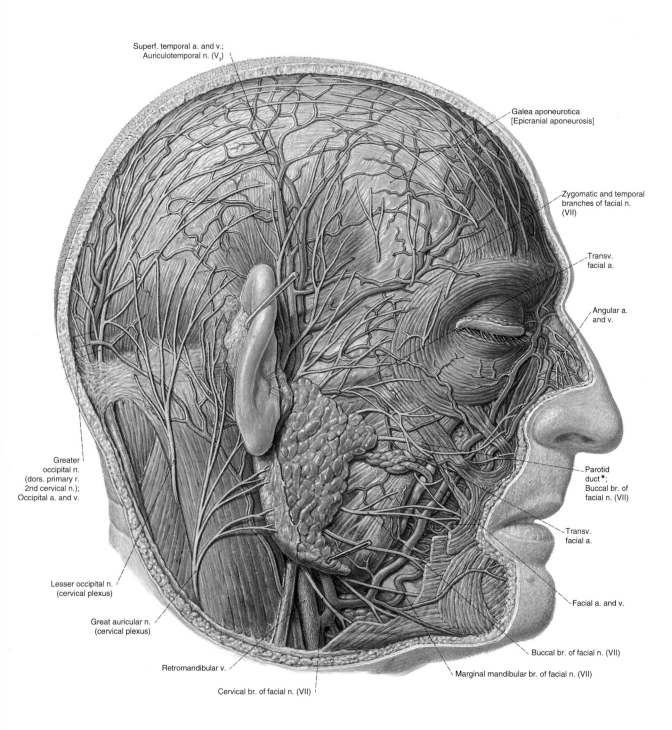

Superf. temporal a. and v.;
Auriculotemporal n. (V₃)

Galea aponeurotica
[Epicranial aponeurosis]

Zygomatic and temporal
branches of facial n.
(VII)

Transv.
facial a.

Angular a.
and v.

Greater
occipital n.
(dors. primary r.
2nd cervical n.);
Occipital a. and v.

Parotid
duct*;
Buccal br. of
facial n. (VII)

Transv.
facial a.

Lesser occipital n.
(cervical plexus)

Great auricular n.
(cervical plexus)

Facial a. and v.

Retromandibular v.

Buccal br. of facial n. (VII)

Marginal mandibular br. of facial n. (VII)

Cervical br. of facial n. (VII)

Fig. 80. Superficial nerves and blood vessels of the head and
face and the parotid gland. Lateral aspect.

* Clinical eponym: STENSEN's duct

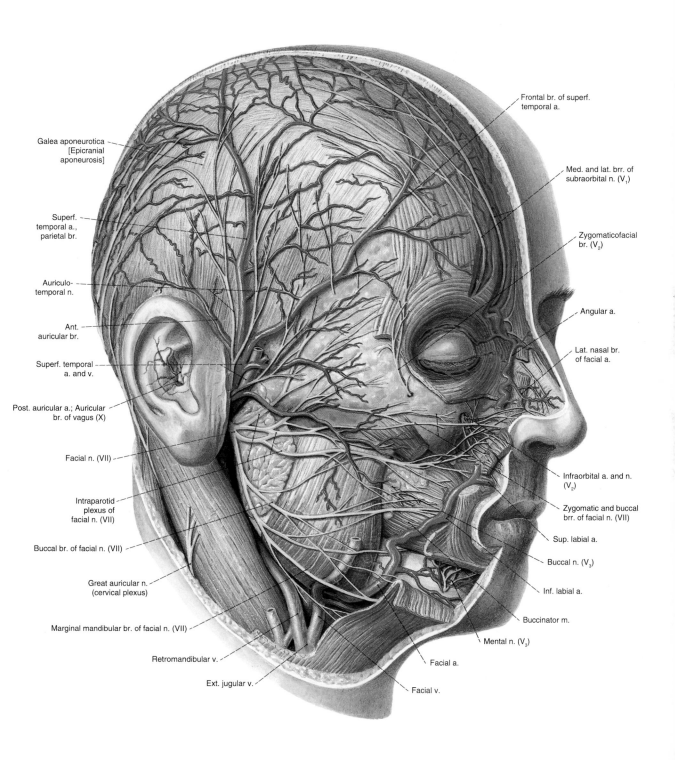

Galea aponeurotica [Epicranial aponeurosis]

Superf. temporal a., parietal br.

Auriculo- temporal n.

Ant. auricular br.

Superf. temporal a. and v.

Post. auricular a.; Auricular br. of vagus (X)

Facial n. (VII)

Intraparotid plexus of facial n. (VII)

Buccal br. of facial n. (VII)

Great auricular n. (cervical plexus)

Marginal mandibular br. of facial n. (VII)

Retromandibular v.

Ext. jugular v.

Frontal br. of superf. temporal a.

Med. and lat. brr. of subraorbital n. (V₁)

Zygomaticofacial br. (V₂)

Angular a.

Lat. nasal br. of facial a.

Infraorbital a. and n. (V₂)

Zygomatic and buccal brr. of facial n. (VII)

Sup. labial a.

Buccal n. (V₃)

Inf. labial a.

Buccinator m.

Mental n. (V₃)

Facial a.

Facial v.

Fig. 81. Superficial nerves and arteries of the head and deeper nerves and arteries of the face after removal of the superficial part of the parotid gland to expose the parotid plexus of the facial nerve.

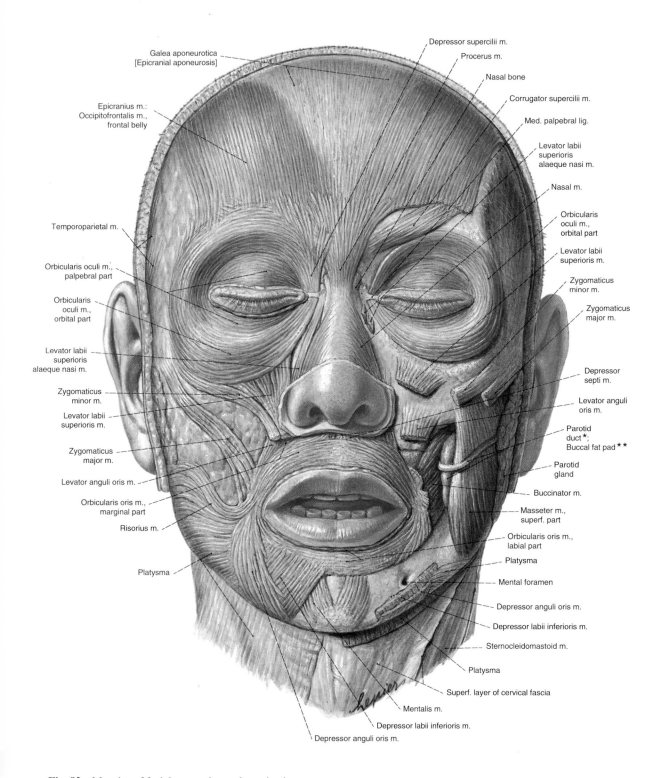

Galea aponeurotica
[Epicranial aponeurosis]

Depressor supercilii m.

Procerus m.

Nasal bone

Corrugator supercilii m.

Med. palpebral lig.

Epicranius m.:
Occipitofrontalis m.,
frontal belly

Levator labii
superioris
alaeque nasi m.

Nasal m.

Temporoparietal m.

Orbicularis
oculi m.,
orbital part

Levator labii
superioris m.

Zygomaticus
minor m.

Orbicularis oculi m.,
palpebral part

Orbicularis
oculi m.,
orbital part

Zygomaticus
major m.

Levator labii
superioris
alaeque nasi m.

Depressor
septi m.

Levator anguli
oris m.

Zygomaticus
minor m.

Parotid
duct*;
Buccal fat pad**

Levator labii
superioris m.

Parotid
gland

Zygomaticus
major m.

Buccinator m.

Levator anguli oris m.

Masseter m.,
superf. part

Orbicularis oris m.,
marginal part

Orbicularis oris m.,
labial part

Risorius m.

Platysma

Mental foramen

Platysma

Depressor anguli oris m.

Depressor labii inferioris m.

Sternocleidomastoid m.

Platysma

Superf. layer of cervical fascia

Mentalis m.

Depressor labii inferioris m.

Depressor anguli oris m.

Fig. 82. Muscles of facial expression and mastication, anterior view. The more superficial layer is shown on the right side of the face while the deeper muscles are on the left.

* Stensen's duct
** Bichat's fat pad

Facial Muscles: Muscles of the Mouth

Name	Origin	Insertion	Innervation	Function
1. **Levator labii superioris alaeque nasi muscle**	Frontal process of maxilla	Greater alar cartilage and skin of nose, upper lips	Facial nerve (VII)	Facial expression: Move lips, nasal alae, cheeks, skin of chin
2. **Levator labii superioris muscle**	Infraorbital margin, maxilla, zygomatic bone	Upper lip		
3. **Zygomaticus minor muscle**	Molar surface of zygomatic bone	Angle of mouth		
4. **Zygomaticus major muscle**	Lateral surface of zygomatic bone	Angle of mouth		
5. **Risorius muscle** (part of the Platysma or No. 6)	Masseteric fascia	Angle of mouth		
6. **Depressor anguli oris muscle**	Oblique line of mandible	Angle of mouth and lower lip		
7. **Levator anguli oris muscle**	Canine fossa of maxilla	Upper lip musculature and angle of mouth		
8. **Depressor labii inferioris muscle**	Oblique line of mandible	Lower lip		
9. **Orbicularis oris muscle**	Consists of marginal part (from other facial muscles) and labial part (from skin of lip)	Upper and lower lips, philtrum of upper lip		
10. **Buccinator muscle**	Outer surface of alveolar processes of maxilla and mandible, pterygomandibular raphe	Angle of mouth and lips		
11. **Mentalis muscle**	Incisive fossa of mandible	Skin of chin		
12. **Transverse menti muscle**	Mandible	Angle of mouth		

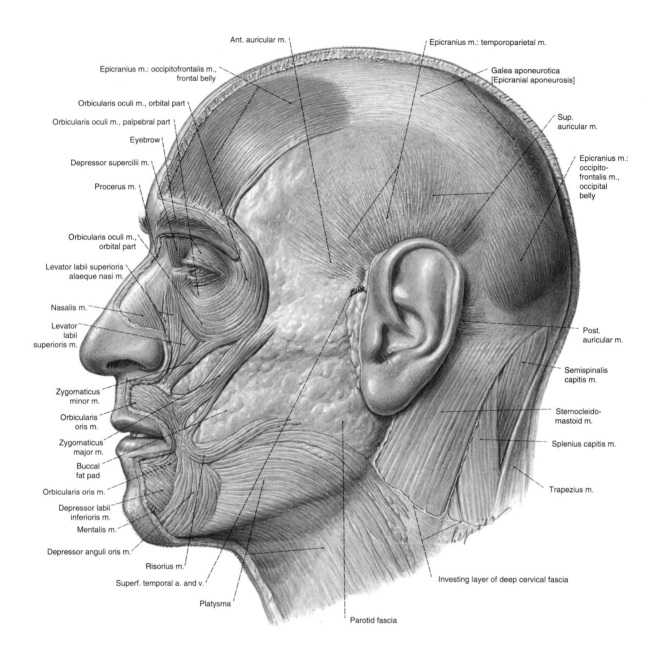

Ant. auricular m.

Epicranius m.: temporoparietal m.

Epicranius m.: occipitofrontalis m., frontal belly

Galea aponeurotica [Epicranial aponeurosis]

Orbicularis oculi m., orbital part

Sup. auricular m.

Orbicularis oculi m., palpebral part

Epicranius m.: occipitofrontalis m., occipital belly

Eyebrow

Depressor supercilii m.

Procerus m.

Orbicularis oculi m., orbital part

Levator labii superioris alaeque nasi m.

Nasalis m.

Post. auricular m.

Levator labii superioris m.

Semispinalis capitis m.

Zygomaticus minor m.

Sternocleido-mastoid m.

Orbicularis oris m.

Splenius capitis m.

Zygomaticus major m.

Buccal fat pad

Trapezius m.

Orbicularis oris m.

Depressor labii inferioris m.

Mentalis m.

Depressor anguli oris m.

Risorius m.

Superf. temporal a. and v.

Investing layer of deep cervical fascia

Platysma

Parotid fascia

Fig. 83. Facial muscles, superficial layer. Lateral view.

Facial Muscles.
Muscle of the Scalp: the Epicranius

Name	Origin	Insertion	Innervation	Function
1. **Occipitofrontalis muscle, frontal belly**	Supraorbital margin	Galea aponeurotica	Facial nerve (VII)	Move scalp
2. **Occipitofrontalis muscle, occipital belly**	Supreme nuchal line	Galea aponeurotica		
3. **Temporoparietalis muscle**	Temporal fascia near ear, superior lamina	Skin or temporal fascia above and in front of ear		

Muscles of the Nose

Name	Origin	Insertion	Innervation	Function
1. **Nasalis muscle**				
Transverse part	Maxilla above canine teeth	Aponeurosis over bridge of nose	Facial nerve (VII)	Dilate and contract nostrils
Alar part	Maxilla above lateral incisors	Alar cartilage of nose		
2. **Depressor septi muscle**	Maxilla above medial incisors	Nasal septum, cartilaginous part		

Muscles of the Eyelids

Name	Origin	Insertion	Innervation	Function
1. **Orbicularis oculi muscle**				
Orbital part	Frontal process of maxilla, medial angle of eye, medial palpebral ligament	Surrounds orbital opening as a sphincter, some fibers go to eyebrow	Facial nerve (VII), temporal and zygomatic branches	Close eyelids, compress lacrimal sac, move eyebrows
Palpebral part	Medial palpebral ligament	Lateral palpebral raphe		
Lacrimal part	Posterior lacrimal crest	Superior and inferior tarsi medial to puncta lacrimalia		
2. **Depressor supercilii muscle**	Nasal part of frontal bone	Skin of eyebrow	Facial nerve (VII), temporal and zygomatic branches	Act upon skin of forehead and eyebrows
3. **Corrugator supercilii muscle**	Nasal part of frontal bone	Skin of eyebrow		
4. **Procerus muscle**	Bridge of nose and lateral nasal cartilage	Skin of forehead between eyebrows		

Extrinsic Muscles of the Ear

Name	Origin	Insertion	Innervation	Function
1. **Auricularis anterior muscle**	Temporal fascia, superficial lamina	Spine of the helix	Facial nerve (VII)	Move the auricula
2. **Auricularis superior muscle**	Galea aponeurotica	Root of auricula		
3. **Auricularis posterior muscle**	Mastoid process of temporal bone, tendon of sterno-cleidomastoid muscle	Auricula at convexity of concha		

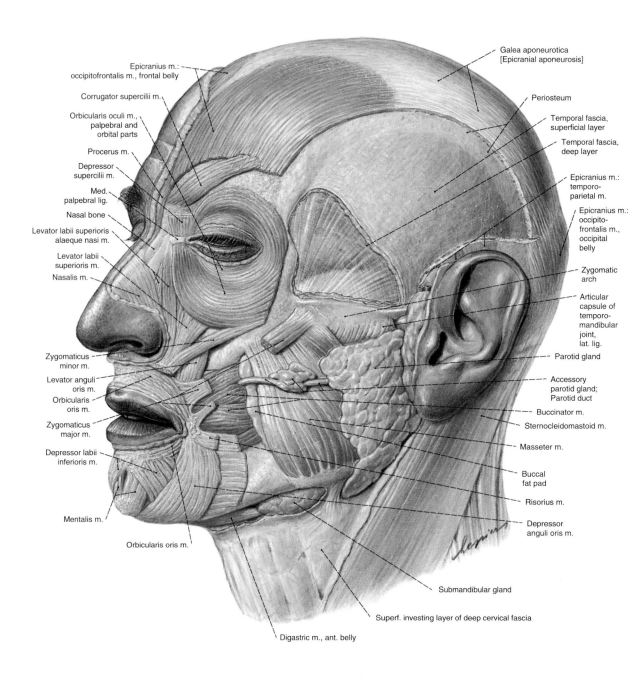

Galea aponeurotica
[Epicranial aponeurosis]

Epicranius m.:
occipitofrontalis m., frontal belly

Periosteum

Corrugator supercilii m.

Temporal fascia,
superficial layer

Orbicularis oculi m.,
palpebral and
orbital parts

Temporal fascia,
deep layer

Procerus m.

Epicranius m.:
temporo-
parietal m.

Depressor
supercilii m.

Med.
palpebral lig.

Epicranius m.:
occipito-
frontalis m.,
occipital
belly

Nasal bone

Levator labii superioris
alaeque nasi m.

Zygomatic
arch

Levator labii
superioris m.

Articular
capsule of
temporo-
mandibular
joint,
lat. lig.

Nasalis m.

Zygomaticus
minor m.

Parotid gland

Levator anguli
oris m.

Accessory
parotid gland;
Parotid duct

Orbicularis
oris m.

Buccinator m.

Sternocleidomastoid m.

Zygomaticus
major m.

Masseter m.

Depressor labii
inferioris m.

Buccal
fat pad

Risorius m.

Mentalis m.

Depressor
anguli oris m.

Orbicularis oris m.

Submandibular gland

Superf. investing layer of deep cervical fascia

Digastric m., ant. belly

Fig. 84. Muscles of facial expression and mastication, parotid
gland and duct (Stensen's duct), and submandibular gland.
Lateral view. The superficial layers of the temporal fascia
and the masseter and parotid fasciae have been partially or
completely removed.

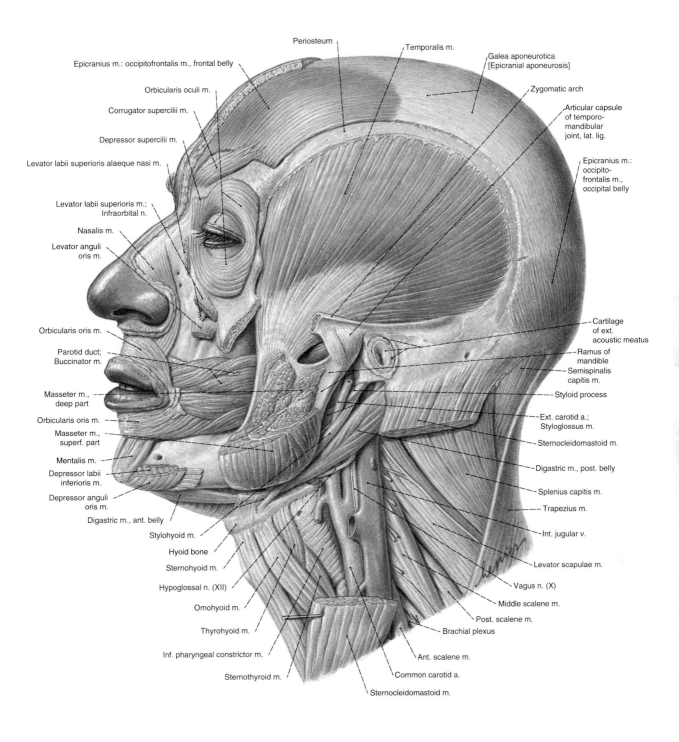

Periosteum
Temporalis m.
Galea aponeurotica
[Epicranial aponeurosis]

Epicranius m.: occipitofrontalis m., frontal belly

Zygomatic arch

Orbicularis oculi m.

Articular capsule
of temporo-
mandibular
joint, lat. lig.

Corrugator supercilii m.

Depressor supercilii m.

Epicranius m.:
occipito-
frontalis m.,
occipital belly

Levator labii superioris alaeque nasi m.

Levator labii superioris m.;
Infraorbital n.

Nasalis m.

Levator anguli
oris m.

Orbicularis oris m.

Cartilage
of ext.
acoustic meatus

Parotid duct;
Buccinator m.

Ramus of
mandible

Semispinalis
capitis m.

Masseter m.,
deep part

Styloid process

Orbicularis oris m.

Ext. carotid a.;
Styloglossus m.

Masseter m.,
superf. part

Sternocleidomastoid m.

Mentalis m.

Digastric m., post. belly

Depressor labii
inferioris m.

Splenius capitis m.

Depressor anguli
oris m.

Trapezius m.

Digastric m., ant. belly

Int. jugular v.

Stylohyoid m.

Hyoid bone

Levator scapulae m.

Sternohyoid m.

Hypoglossal n. (XII)

Vagus n. (X)

Omohyoid m.

Middle scalene m.

Thyrohyoid m.

Post. scalene m.

Brachial plexus

Inf. pharyngeal constrictor m.

Ant. scalene m.

Sternothyroid m.

Common carotid a.

Sternocleidomastoid m.

Fig. 85. Lateral view of the head and the upper neck and some muscles of the face and of mastication. The external ear and part of the zygomatic arch have been removed. A portion of the sternocleidomastoid muscle has been removed to expose the large cervical vessels. The masseter muscle has been partly removed to display the tendinous infiltrations of its muscle bundles. Several facial muscles have been sectioned near their origins.

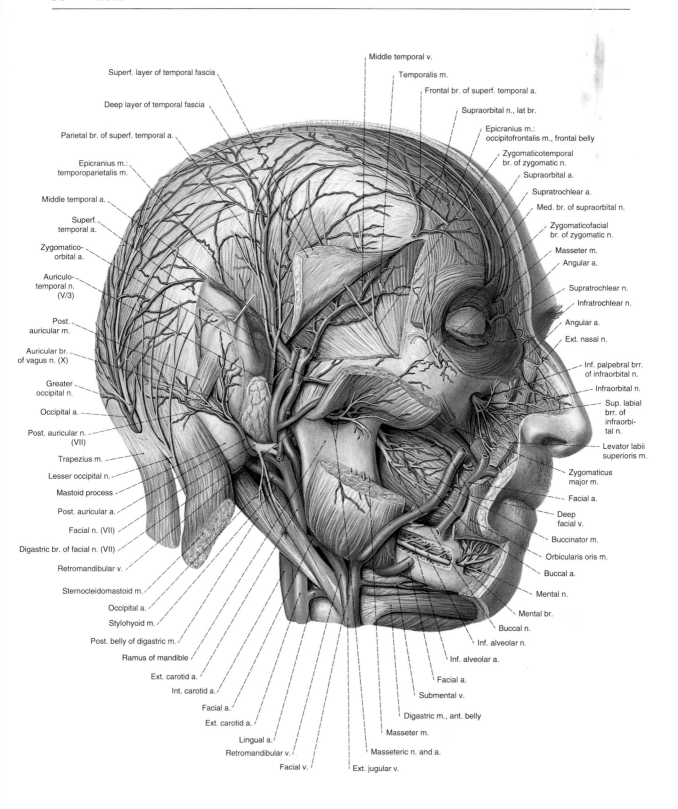

Middle temporal v.

Superf. layer of temporal fascia

Temporalis m.

Deep layer of temporal fascia

Frontal br. of superf. temporal a.

Parietal br. of superf. temporal a.

Supraorbital n., lat. br.

Epicranius m.:
occipitofrontalis m., frontal belly

Epicranius m.:
temporoparietalis m.

Zygomaticotemporal
br. of zygomatic n.

Middle temporal a.

Supraorbital a.

Superf.
temporal a.

Supratrochlear a.

Zygomatico-
orbital a.

Med. br. of supraorbital n.

Zygomaticofacial
br. of zygomatic n.

Auriculo-
temporal n.
(V/3)

Masseter m.

Angular a.

Post.
auricular m.

Supratrochlear n.

Infratrochlear n.

Auricular br.
of vagus n. (X)

Angular a.

Ext. nasal n.

Greater
occipital n.

Inf. palpebral brr.
of infraorbital n.

Occipital a.

Infraorbital n.

Post. auricular n.
(VII)

Sup. labial
brr. of
infraorbi-
tal n.

Trapezius m.

Levator labii
superioris m.

Lesser occipital n.

Zygomaticus
major m.

Mastoid process

Facial a.

Post. auricular a.

Deep
facial v.

Facial n. (VII)

Buccinator m.

Digastric br. of facial n. (VII)

Orbicularis oris m.

Retromandibular v.

Buccal a.

Sternocleidomastoid m.

Mental n.

Occipital a.

Mental br.

Stylohyoid m.

Buccal n.

Post. belly of digastric m.

Inf. alveolar n.

Ramus of mandible

Inf. alveolar a.

Ext. carotid a.

Facial a.

Int. carotid a.

Submental v.

Facial a.

Digastric m., ant. belly

Ext. carotid a.

Masseter m.

Lingual a.

Masseteric n. and a.

Retromandibular v.

Facial v.

Ext. jugular v.

Fig. 86. Nerves and blood vessels of the face. Lateral view. The masseter muscle has been severed and reflected. The superficial and deep layers of the temporal fascia have been reflected from the upper border of the zygomatic arch. The facial muscles, parotid gland and the facial nerve (VII) have been extensively removed. The mandibular canal is partially exposed.

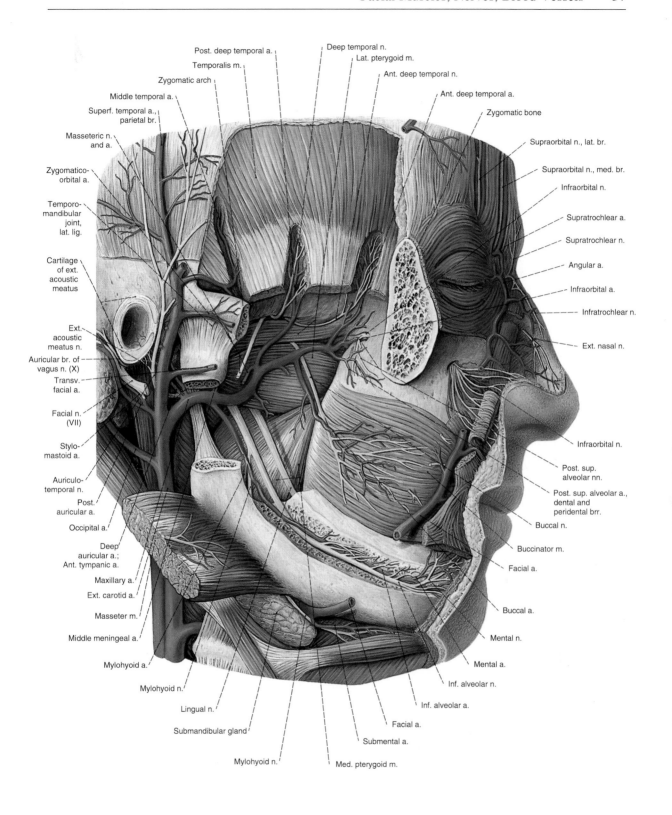

Post. deep temporal a.
Temporalis m.
Zygomatic arch
Middle temporal a.
Superf. temporal a.,
parietal br.
Masseteric n.
and a.
Zygomatico-
orbital a.
Temporo-
mandibular
joint,
lat. lig.
Cartilage
of ext.
acoustic
meatus
Ext.
acoustic
meatus n.
Auricular br. of
vagus n. (X)
Transv.
facial a.
Facial n.
(VII)
Stylo-
mastoid a.
Auriculo-
temporal n.
Post.
auricular a.
Occipital a.
Deep
auricular a.;
Ant. tympanic a.
Maxillary a.
Ext. carotid a.
Masseter m.
Middle meningeal a.
Mylohyoid a.
Mylohyoid n.
Lingual n.
Submandibular gland
Mylohyoid n.

Deep temporal n.
Lat. pterygoid m.
Ant. deep temporal n.
Ant. deep temporal a.
Zygomatic bone
Supraorbital n., lat. br.
Supraorbital n., med. br.
Infraorbital n.
Supratrochlear a.
Supratrochlear n.
Angular a.
Infraorbital a.
Infratrochlear n.
Ext. nasal n.
Infraorbital n.
Post. sup.
alveolar nn.
Post. sup. alveolar a.,
dental and
peridental brr.
Buccal n.
Buccinator m.
Facial a.
Buccal a.
Mental n.
Mental a.
Inf. alveolar n.
Inf. alveolar a.
Facial a.
Submental a.
Med. pterygoid m.

Fig. 87. Nerves and blood vessels of the face, deep layer:
maxillary artery and its branches. Specimen similar to the one
in Fig. 86, except that the insertion of the temporalis muscle
together with the coronoid process of the ramus of the mandible have been removed. The course of the inferior alveolar
nerve and artery in the mandibular canal is completely visible.

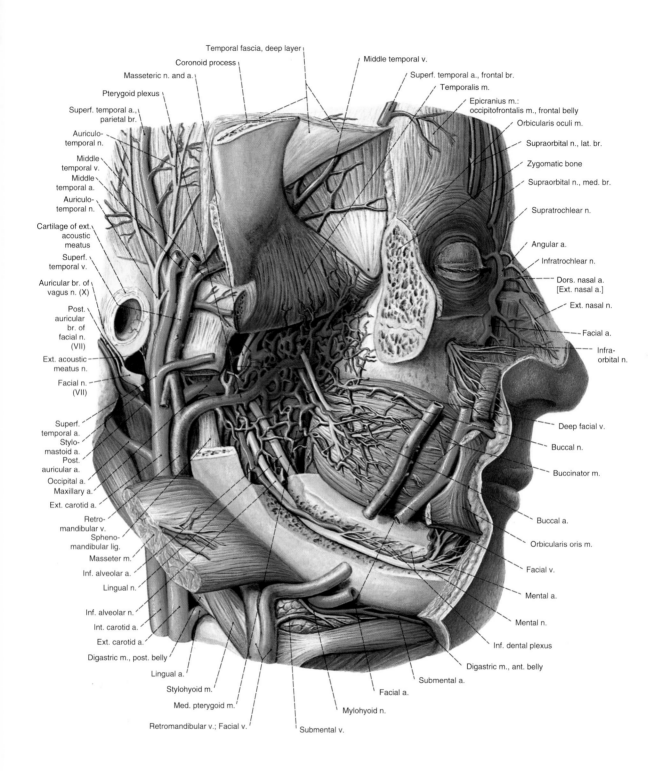

Fig. 88. Nerves and blood vessels of the face. Lateral view. Zygomatic arch and ramus of the mandible have been removed. The mandibular canal is completely exposed. The insertion of the temporalis muscle together with the coronoid process of the ramus of the mandible have been reflected upwards. Note the extensive pterygoid plexus of veins between the temporalis and the medial and lateral pterygoid muscles.

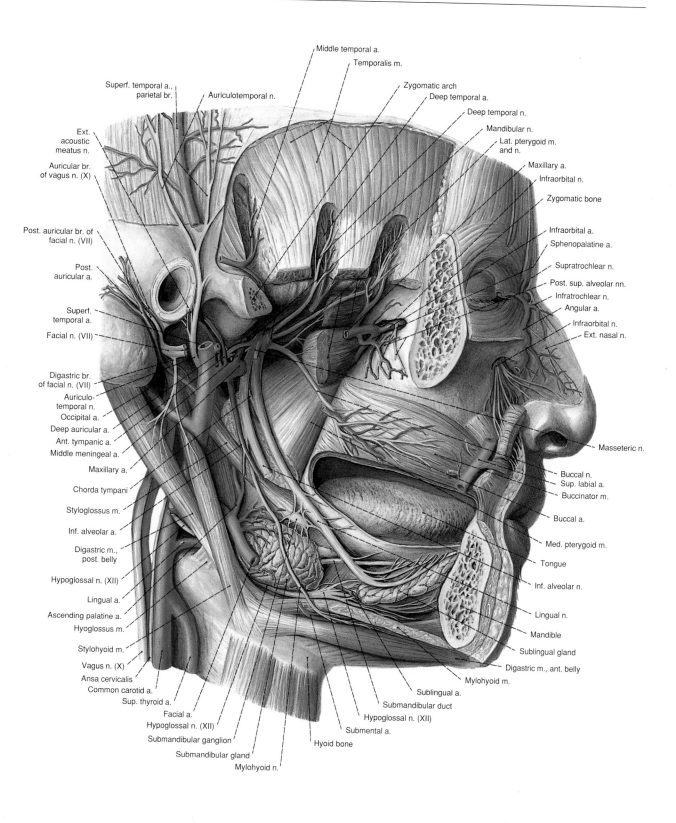

Fig. 89. Nerves and blood vessels of the face, deep layer: Mandibular nerve and its branches. Specimen similar to the one in Fig. 88, except that the condyloid process of the mandible has been disjointed and removed together with the rest of the right mandible. The lower half of the buccinator muscle and the underlying oral mucosa have been removed.

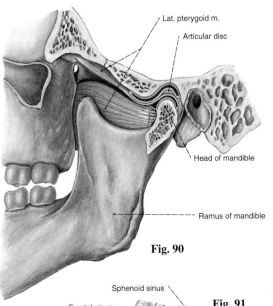

Lat. pterygoid m.

Articular disc

Head of mandible

Ramus of mandible

Fig. 90

◀ **Fig. 90.** The temporomandibular joint, in sagittal section, with articular disc and S-shaped articular surface of the mandibular fossa of the temporal bone. Lateral view.

Fig. 91. The right side of the oral cavity, with muscles of the floor of the mouth, viewed internally from the left. Paramedian sagittal section through the facial portion of the skull, the hyoid bone, and the upper cervical vertebrae (I–V). The oral mucosa has been partially stripped. Pharynx and larynx have been removed.

VI-XII = exit points of cranial nerves VI–XII through the cranial dura mater, and C I and C II = exit points of both upper cervical nerves through the spinal dura mater.

Fig. 92. The temporomandibular joint and some of the muscles of mastication (cf. Figs. 85 and 91). The joint has been opened in the sagittal plane to expose the articular disc. The condyloid process and the head and neck of the mandible are represented as being transparent to show the origin, course and insertion of the lateral and medial pterygoid muscles. ▶

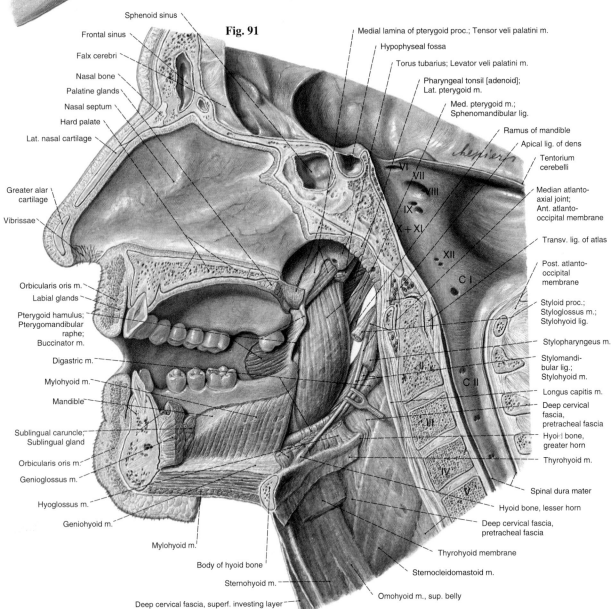

Fig. 91

Sphenoid sinus
Frontal sinus
Falx cerebri
Nasal bone
Palatine glands
Nasal septum
Hard palate
Lat. nasal cartilage
Greater alar cartilage
Vibrissae
Orbicularis oris m.
Labial glands
Pterygoid hamulus; Pterygomandibular raphe; Buccinator m.
Digastric m.
Mylohyoid m.
Mandible
Sublingual caruncle; Sublingual gland
Orbicularis oris m.
Genioglossus m.
Hyoglossus m.
Geniohyoid m.
Mylohyoid m.
Body of hyoid bone
Sternohyoid m.
Deep cervical fascia, superf. investing layer

Medial lamina of pterygoid proc.; Tensor veli palatini m.
Hypophyseal fossa
Torus tubarius; Levator veli palatini m.
Pharyngeal tonsil [adenoid]; Lat. pterygoid m.
Med. pterygoid m.; Sphenomandibular lig.
Ramus of mandible
Apical lig. of dens
Tentorium cerebelli
Median atlanto-axial joint; Ant. atlanto-occipital membrane
Transv. lig. of atlas
Post. atlanto-occipital membrane
Styloid proc.; Styloglossus m.; Stylohyoid lig.
Stylopharyngeus m.
Stylomandibular lig.; Stylohyoid m.
Longus capitis m.
Deep cervical fascia, pretracheal fascia
Hyoid bone, greater horn
Thyrohyoid m.
Spinal dura mater
Hyoid bone, lesser horn
Deep cervical fascia, pretracheal fascia
Thyrohyoid membrane
Sternocleidomastoid m.
Omohyoid m., sup. belly

VI VII VIII IX X + XI XII
C I
C II
II III IV V

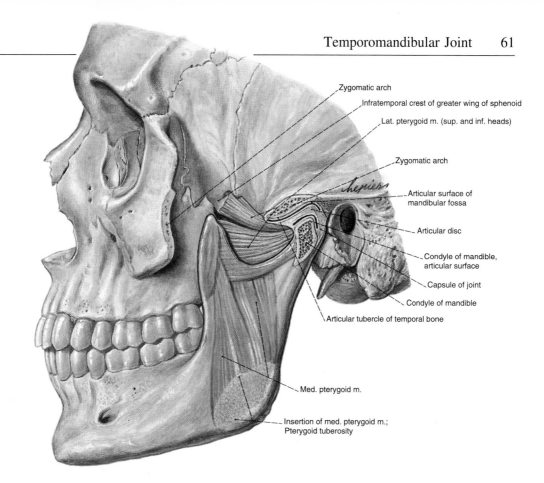

Fig. 92

Labels on figure:
- Zygomatic arch
- Infratemporal crest of greater wing of sphenoid
- Lat. pterygoid m. (sup. and inf. heads)
- Zygomatic arch
- Articular surface of mandibular fossa
- Articular disc
- Condyle of mandible, articular surface
- Capsule of joint
- Condyle of mandible
- Articular tubercle of temporal bone
- Med. pterygoid m.
- Insertion of med. pterygoid m.; Pterygoid tuberosity

Muscles of Mastication

The muscles of mastication are the only cranial muscles that resemble true skeletal muscles. They are strong, distinct, separate muscles with well-defined fasciae and osseous origins and insertions, and comprise the four muscles (masseter, temporalis, and both pterygoid muscles) that move the mandible in the temporomandibular joint.

Name	Origin	Insertion	Innervation	Function
1. Masseter muscle *Superficial part*	Zygomatic process of maxilla, inferior border of zygomatic arch	Angle and lower half of lateral surface of ramus of mandible	Masseteric branch of mandibular division of trigeminal nerve (V_3)	Closes the jaws by elevating the mandible
Intermediate part	Inner surface of zygomatic arch	Ramus of mandible		
Deep part	Posterior part of inferior border and inner surface of zygomatic arch	Superior half of ramus of mandible and lateral surface of coronoid process of mandible		
2. Temporalis muscle	Temporal fossa and temporal fascia	Coronoid process (medial surface, apex, anterior border) and ramus of mandible (anterior border)	Deep temporal nerves from mandibular division of trigeminal nerve (V_3)	Elevates mandible (closes jaws), posterior fibers retract mandible
3. Lateral pterygoid muscle	Superior head: infratemporal crest and lateral surface of great wing of sphenoid; inferior head: lateral surface of lateral pterygoid plate	Condyloid process of mandible (neck); temporomandibular joint (articular disc)	Lateral pterygoid nerve from mandibular division of trigeminal nerve (V_3)	Opens the jaws, protrudes mandible, moves mandible side-to-side
4. Medial pterygoid muscle	Pterygoid fossa; pyramidal process of palatine bone; lateral pterygoid plate	Medial surface of ramus and angle of mandible opposite the masseter muscle	Medial pterygoid nerve from mandibular division of trigeminal nerve (V_3)	Closes the jaws; with lateral pterygoid muscle protrudes and moves mandible side-to-side

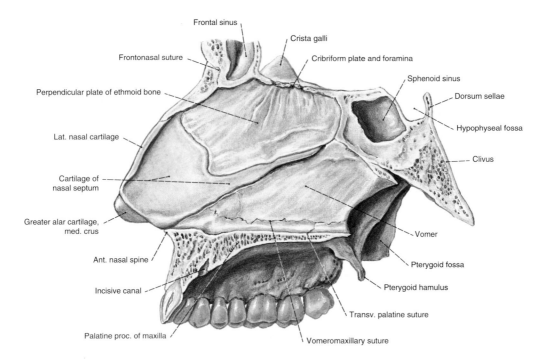

Frontal sinus

Crista galli

Frontonasal suture

Cribriform plate and foramina

Sphenoid sinus

Perpendicular plate of ethmoid bone

Dorsum sellae

Hypophyseal fossa

Lat. nasal cartilage

Clivus

Cartilage of nasal septum

Greater alar cartilage, med. crus

Vomer

Ant. nasal spine

Pterygoid fossa

Incisive canal

Pterygoid hamulus

Transv. palatine suture

Palatine proc. of maxilla

Vomeromaxillary suture

Fig. 93. The bony and cartilaginous nasal septum, and adjacent structures. Medial view.

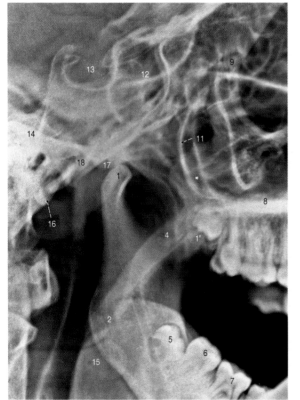

1	=	Condylar process
1'	=	Coronoid proc.
2	=	Angle of mandible
4	=	Soft palate [Velum palatinum]
5	=	Third molar, wisdom tooth [Dens serotinus]
6	=	Second molar tooth
7	=	First molar tooth
8	=	Hard palate
9	=	Lateral margin of orbit
11	=	Maxillary sinus
12	=	Spenoid sinus
13	=	Sella turerca
14	=	Petrosal part of temporal bone
15	=	Root of the tongue
16	=	Styloid process
17	=	Articular tubercle
18	=	Mandibular fossa

Fig. 94. Lateral radiograph of the temporomandibular joint (contact radiograph according to PARMA). (From: R. BIRKNER: Normal Radiologic Patterns and Variances of the Human Skeleton, Urban & Schwarzenberg, Baltimore-Munich, 1978).

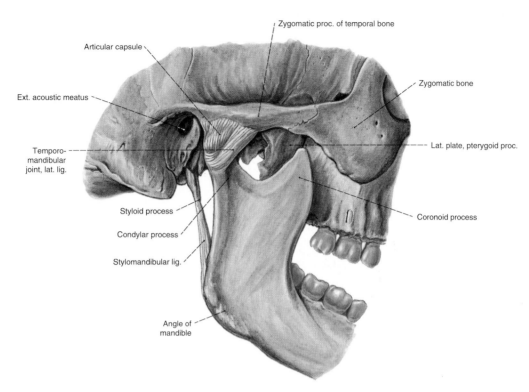

Fig. 95. The right temporomandibular joint, with articular capsule and lateral (temporomandibular) ligament. Lateral view.

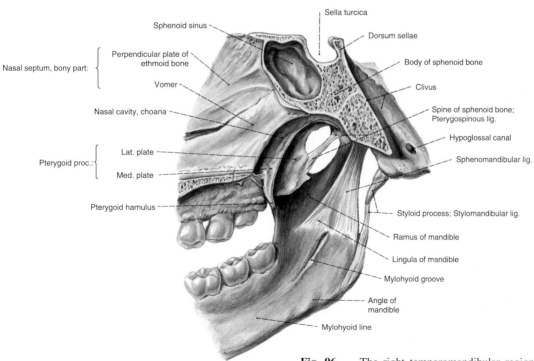

Fig. 96. The right temporomandibular region with the pterygospinous and sphenomandibular ligaments viewed from the medial aspect. A paramedian sagittal section through the sphenoid bone and the hard palate.

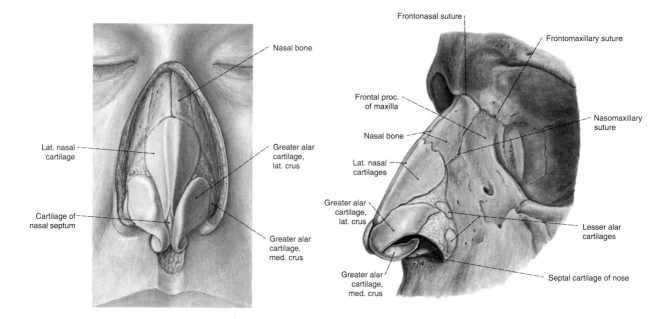

Fig. 97. The bones, cartilages and muscles of the dorsum of the external nose. Anterior view.

Fig. 98. The cartilages and bones of the external nose. Lateral view. Nasal cartilages are blue.

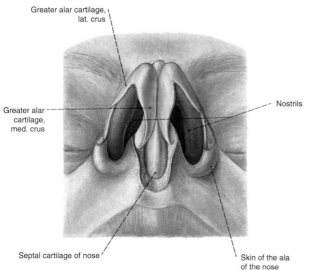

Fig. 99. Framework of the external nose, inferior aspect.

Nasal cavity, an overview: The nasal cavities are paired narrow spaces in the middle of the facial skeleton that extend in the sagittal direction from the anterior nasal (piriform) aperture to the choanae. A midline nasal septum separates the right and left cavities. Three nasal conchae project from the lateral walls of each cavity. Each concha overhangs a nasal passage, the superior, middle or inferior meatus. The floor of each cavity is formed by the palatine process of the maxilla and the roof, by the cribriform plate of the ethmoid bone with foramina for passage of the olfactory nerves. The cavities communicate with four paranasal sinuses that are lined with a ciliated mucous membrane and are filled with air. These are the maxillary (also known by its clinical eponym as the antrum of HIGHMORE), frontal, sphenoidal and ethmoidal sinuses.

The **paranasal sinuses** are extensions of the nasal cavity that are lined with mucous membrane and filled with air: **Maxillary sinus, frontal sinus, sphenoidal sinus** and **ethmoidal sinus air cells** [anterior, middle (form **bulla ethmoidalis**) and posterior groups].

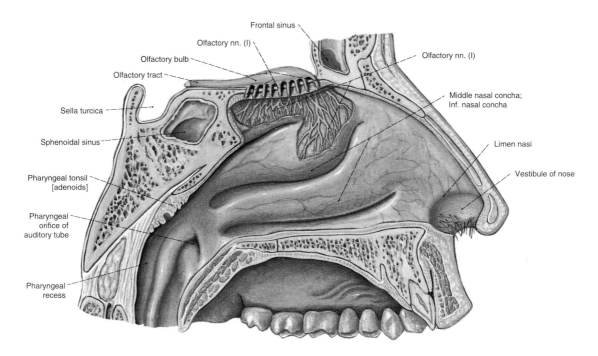

Frontal sinus

Olfactory nn. (I)

Olfactory bulb

Olfactory tract

Sella turcica

Sphenoidal sinus

Pharyngeal tonsil
[adenoids]

Pharyngeal
orifice of
auditory tube

Pharyngeal
recess

Olfactory nn. (I)

Middle nasal concha;
Inf. nasal concha

Limen nasi

Vestibule of nose

Fig. 100. The lateral wall of the left nasal cavity. The medial olfactory nerves which pass to the olfactory bulb have been sectioned. The mucous membrane in the region of the superior and middle nasal conchae overlying the lateral olfactory nerves has been partially removed.

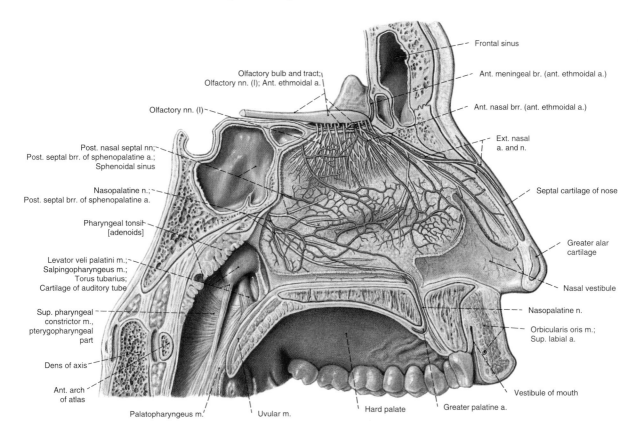

Frontal sinus

Olfactory bulb and tract;
Olfactory nn. (I); Ant. ethmoidal a.

Olfactory nn. (I)

Post. nasal septal nn;
Post. septal brr. of sphenopalatine a.;
Sphenoidal sinus

Nasopalatine n.;
Post. septal brr. of sphenopalatine a.

Pharyngeal tonsil
[adenoids]

Levator veli palatini m.;
Salpingopharyngeus m.;
Torus tubarius;
Cartilage of auditory tube

Sup. pharyngeal
constrictor m.,
pterygopharyngeal
part

Dens of axis

Ant. arch
of atlas

Palatopharyngeus m.   Uvular m.

Ant. meningeal br. (ant. ethmoidal a.)

Ant. nasal brr. (ant. ethmoidal a.)

Ext. nasal
a. and n.

Septal cartilage of nose

Greater alar
cartilage

Nasal vestibule

Nasopalatine n.

Orbicularis oris m.;
Sup. labial a.

Vestibule of mouth

Hard palate   Greater palatine a.

Fig. 101. Arteries and nerves of the nasal septum. Paramedian sagittal section through the facial skeleton with the mucous membrane removed from the nasal septum and nasopharynx.

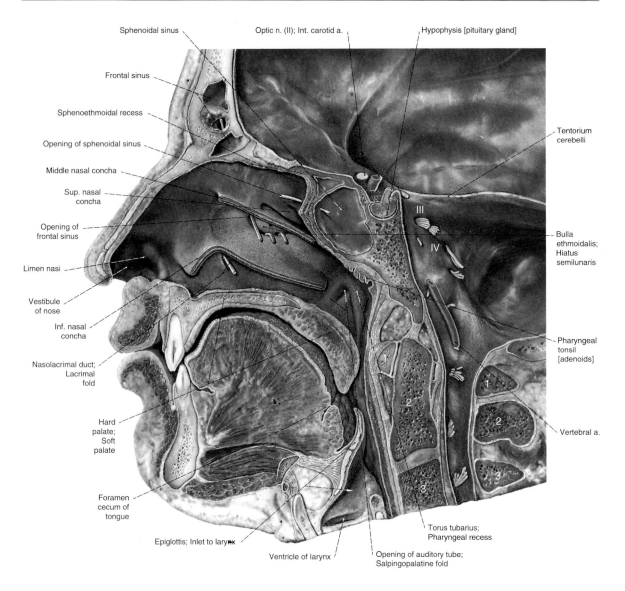

Sphenoidal sinus

Frontal sinus

Sphenoethmoidal recess

Opening of sphenoidal sinus

Middle nasal concha

Sup. nasal concha

Opening of frontal sinus

Limen nasi

Vestibule of nose

Inf. nasal concha

Nasolacrimal duct; Lacrimal fold

Hard palate; Soft palate

Foramen cecum of tongue

Epiglottis; Inlet to larynx

Ventricle of larynx

Optic n. (II); Int. carotid a.

Hypophysis [pituitary gland]

Tentorium cerebelli

Bulla ethmoidalis; Hiatus semilunaris

Pharyngeal tonsil [adenoids]

Vertebral a.

Torus tubarius; Pharyngeal recess

Opening of auditory tube; Salpingopalatine fold

III

IV

1
2
3

Fig. 102. Lateral wall of the right nasal cavity. Paramedian sagittal section through the head. The middle and the inferior nasal conchae have been cut off near their base. Probes are in the openings of the paranasal sinuses and in the nasolacrimal duct.
1 = atlas [C1], 2 = axis [C2], 3 = 3rd cervical vertebra; III = oculomotor nerve, IV = trochlear nerve.

Nasal nerves: for **motor** functions, fibers from the facial nerve; for **sensory** functions, fibers from the trigeminal nerve: anterior ethmoidal nerve with branches for the nasal septum, lateral nasal wall, tip and wings of the nose (external nasal branch); posterior ethmoidal nerve for sphenoidal sinus and posterior ethmoidal air cells; supraorbital nerve for frontal sinus; posterior superior lateral nasal branches for the two upper conchae and posterior ethmoidal air cells; posterior inferior lateral nasal branches for the inferior concha and the middle and inferior nasal meatus; posterior superior medial branches for the upper part of the septum; superior posterior alveolar branches for the maxillary sinus; external nasal branches for the outside of the nasal wings; olfactory nerves for **olfaction** (Figs. 100, 101).

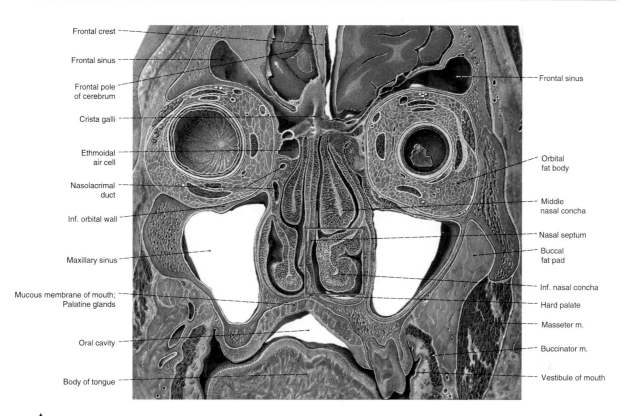

Frontal crest

Frontal sinus

Frontal pole
of cerebrum

Crista galli

Ethmoidal
air cell

Nasolacrimal
duct

Inf. orbital wall

Maxillary sinus

Mucous membrane of mouth;
Palatine glands

Oral cavity

Body of tongue

Frontal sinus

Orbital
fat body

Middle
nasal concha

Nasal septum

Buccal
fat pad

Inf. nasal concha

Hard palate

Masseter m.

Buccinator m.

Vestibule of mouth

▲
Fig. 103. Frontal section through the facial skeleton showing frontal sinus, orbits, nasal cavity, maxillary sinus, and the upper part of the oral cavity with the dorsum of the tongue. The area outlined in red is enlarged in Fig. 104.

Fig. 104. Part of the inferior nasal concha, with mucous membrane and venous plexus.
▼

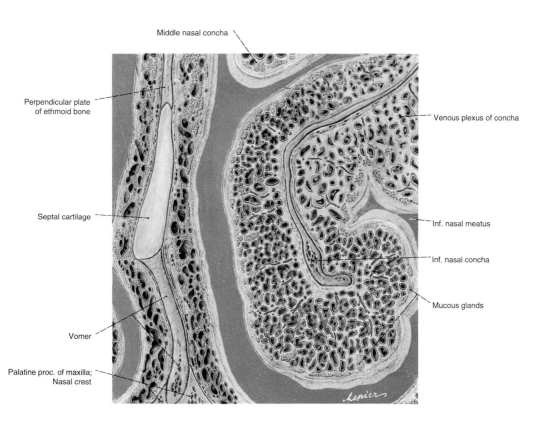

Middle nasal concha

Perpendicular plate
of ethmoid bone

Septal cartilage

Vomer

Palatine proc. of maxilla;
Nasal crest

Venous plexus of concha

Inf. nasal meatus

Inf. nasal concha

Mucous glands

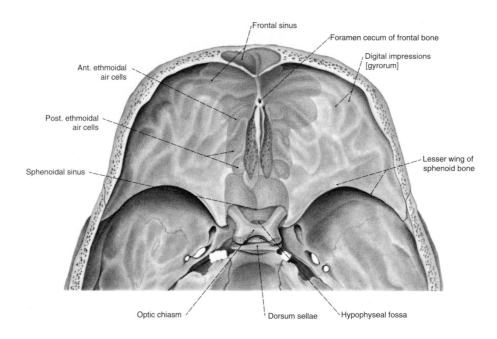

Frontal sinus

Foramen cecum of frontal bone

Digital impressions
[gyrorum]

Ant. ethmoidal
air cells

Post. ethmoidal
air cells

Sphenoidal sinus

Lesser wing of
sphenoid bone

Optic chiasm

Dorsum sellae

Hypophyseal fossa

Fig. 105.   A portion of the paranasal air sinuses and their projection on the anterior cranial fossa.

Fig. 107a.   Radiograph of the paranasal sinuses. Radiation in ▶ occipito-oral direction (with mouth open). (From: Dr. G. GREEVEN, St. Elizabeth Hospital, Neuwied).

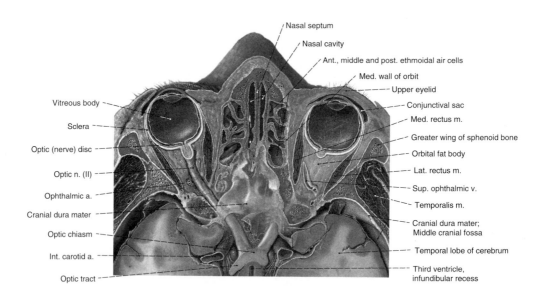

Nasal septum

Nasal cavity

Ant., middle and post. ethmoidal air cells

Med. wall of orbit

Upper eyelid

Vitreous body

Conjunctival sac

Sclera

Med. rectus m.

Optic (nerve) disc

Greater wing of sphenoid bone

Optic n. (II)

Orbital fat body

Ophthalmic a.

Lat. rectus m.

Cranial dura mater

Sup. ophthalmic v.

Optic chiasm

Temporalis m.

Int. carotid a.

Cranial dura mater;
Middle cranial fossa

Optic tract

Temporal lobe of cerebrum

Third ventricle,
infundibular recess

Fig. 106.   Horizontal section through the facial skeleton at the level of the ocular bulbs, the ethmoidal air cells, the upper portion of the nasal cavity, with nasal septum. The optic chiasm is in the lower part of the figure.

Fig 107b.   Line drawing of Fig. 107a.   ▶

* Zygomatic recess of the maxillary sinus
** Innominate line, which ascends from medial and caudal to cranial and lateral, as it passes through the outer half of the orbit. Its lower part is formed by the external surface of the greater wing of the sphenoid and its upper part by the temporal fascia of the frontal bone lying directly behind the zygomatic process.

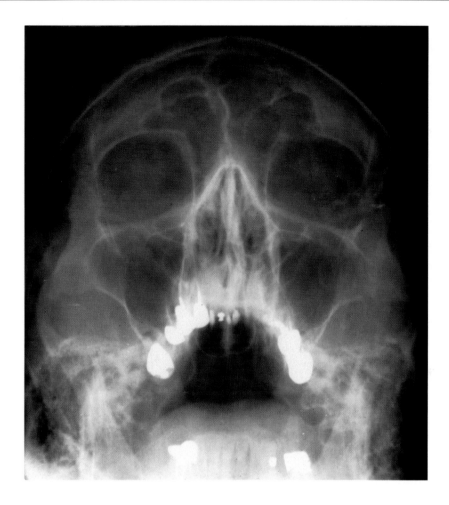

Fig. 107a

Fig. 107b

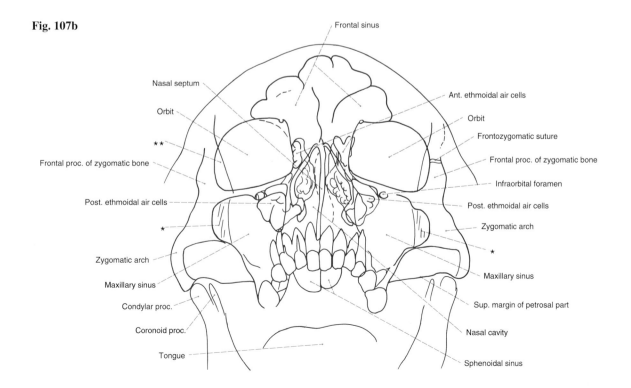

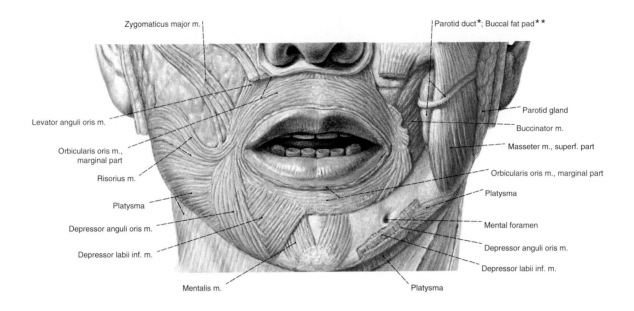

Fig. 108. Muscles of facial expression surrounding the lips.

  * STENSEN's duct
 ** BICHAT's fat pad

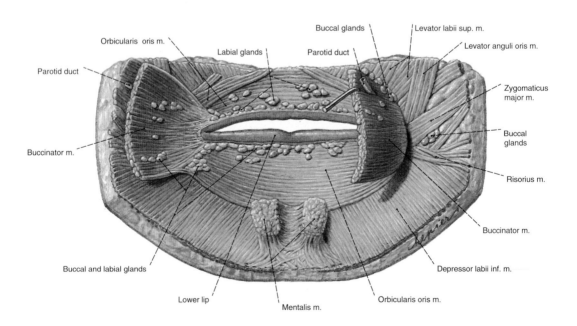

Fig. 109. Facial muscles in the region of the mouth. View from within the oral cavity after the oral mucous membrane has been removed up to the red portion of the lip. The lips and cheeks contain small salivary glands.

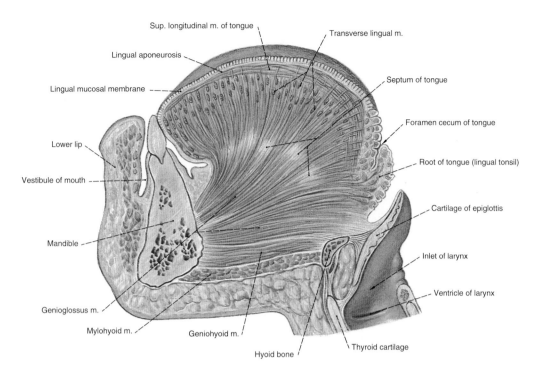

Sup. longitudinal m. of tongue

Transverse lingual m.

Lingual aponeurosis

Septum of tongue

Lingual mucosal membrane

Foramen cecum of tongue

Lower lip

Root of tongue (lingual tonsil)

Vestibule of mouth

Cartilage of epiglottis

Inlet of larynx

Mandible

Ventricle of larynx

Genioglossus m.

Mylohyoid m.   Geniohyoid m.

Thyroid cartilage

Hyoid bone

Fig. 110. Midsagittal section through the tongue, mandible, lower lip and part of the larynx.

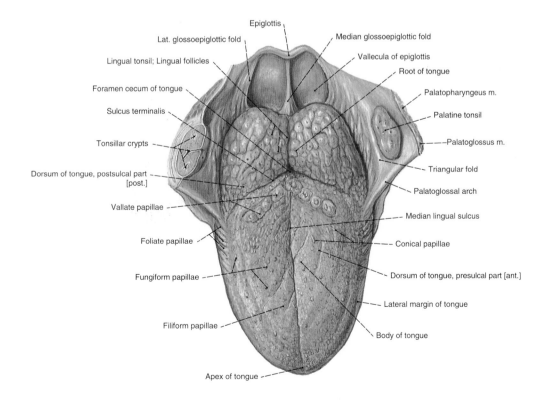

Epiglottis

Lat. glossoepiglottic fold

Median glossoepiglottic fold

Lingual tonsil; Lingual follicles

Vallecula of epiglottis

Foramen cecum of tongue

Root of tongue

Sulcus terminalis

Palatopharyngeus m.

Palatine tonsil

Tonsillar crypts

Palatoglossus m.

Dorsum of tongue, postsulcal part [post.]

Triangular fold

Vallate papillae

Palatoglossal arch

Median lingual sulcus

Foliate papillae

Conical papillae

Fungiform papillae

Dorsum of tongue, presulcal part [ant.]

Lateral margin of tongue

Filiform papillae

Body of tongue

Apex of tongue

Fig. 111. The dorsum of the tongue and portions of the root of the tongue, lingual tonsil and palatine tonsils. Note the papillae on the dorsum of the tongue.

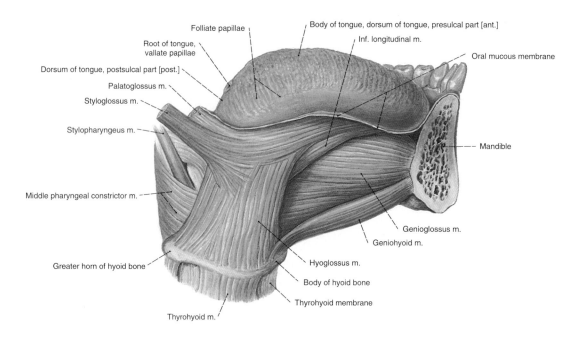

Folliate papillae

Root of tongue,
vallate papillae

Dorsum of tongue, postsulcal part [post.]

Palatoglossus m.

Styloglossus m.

Stylopharyngeus m.

Middle pharyngeal constrictor m.

Greater horn of hyoid bone

Thyrohyoid m.

Body of tongue, dorsum of tongue, presulcal part [ant.]

Inf. longitudinal m.

Oral mucous membrane

Mandible

Genioglossus m.

Geniohyoid m.

Hyoglossus m.

Body of hyoid bone

Thyrohyoid membrane

Fig. 112. The extrinsic muscles of the tongue, superficial layer. Lateral view. The mandible has been sectioned immediately to the right of the midline.

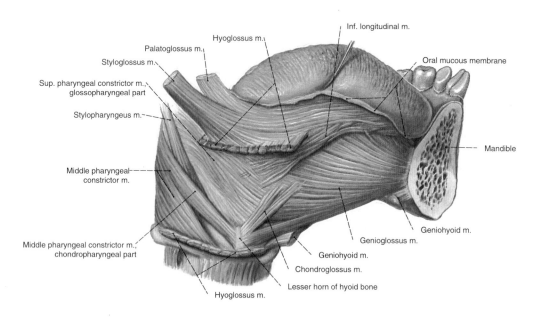

Hyoglossus m.

Palatoglossus m.

Styloglossus m.

Sup. pharyngeal constrictor m.,
glossopharyngeal part

Stylopharyngeus m.

Middle pharyngeal
constrictor m.

Middle pharyngeal constrictor m.;
chondropharyngeal part

Hyoglossus m.

Inf. longitudinal m.

Oral mucous membrane

Mandible

Geniohyoid m.

Genioglossus m.

Geniohyoid m.

Chondroglossus m.

Lesser horn of hyoid bone

Fig. 113. The extrinsic muscles of the tongue, deeper layer. Lateral view. The hyoglossus muscle has been sectioned and the geniohyoid muscle removed.

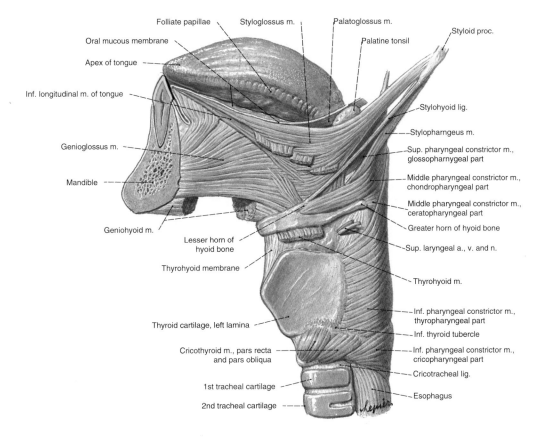

Folliate papillae — Styloglossus m. — Palatoglossus m. — Styloid proc.

Oral mucous membrane — Palatine tonsil

Apex of tongue —

Inf. longitudinal m. of tongue — — Stylohyoid lig.

— Stylopharngeus m.

Genioglossus m. — — Sup. pharyngeal constrictor m., glossopharnygeal part

Mandible — — Middle pharyngeal constrictor m., chondropharyngeal part

— Middle pharyngeal constrictor m., ceratopharyngeal part

Geniohyoid m. — Greater horn of hyoid bone

Lesser horn of hyoid bone — Sup. laryngeal a., v. and n.

Thyrohyoid membrane —

— Thyrohyoid m.

— Inf. pharyngeal constrictor m., thyropharyngeal part

Thyroid cartilage, left lamina — — Inf. thyroid tubercle

Cricothyroid m., pars recta and pars obliqua — — Inf. pharyngeal constrictor m., cricopharyngeal part

— Cricotracheal lig.

1st tracheal cartilage — — Esophagus

2nd tracheal cartilage —

Fig. 114. The extrinsic muscles of the tongue and portions of the laryngeal and pharyngeal musculature. Lateral view.

Extrinsic Muscles of the Tongue

Name	Origin	Insertion	Innervation	Function
1. Genioglossus muscle	Mental spine of mandible	Fans out to body of tongue and lingual aponeurosis	Hypoglossal nerve (XII)	Protrudes and depresses the tongue
2. Hyoglossus muscle	Body and greater horn of hyoid bone	Lateral parts of tongue, lingual aponeurosis		Depresses and retracts the tongue
3. Chondroglossus muscle	Lesser horn of hyoid bone	Lateral parts of tongue, lingual aponeurosis		Same as 2
4. Styloglossus muscle	Styloid process of temporal bone	Enters lateral parts of tongue from above and behind		Retracts and elevates the tongue (in sucking and swallowing)
5. Palatoglossus muscle	Palatal aponeurosis	Upper, posterior parts of tongue	Cranial part of accessory nerve (XI) via vagus nerve (X) to pharyngeal plexus	Constricts isthmus of the fauces

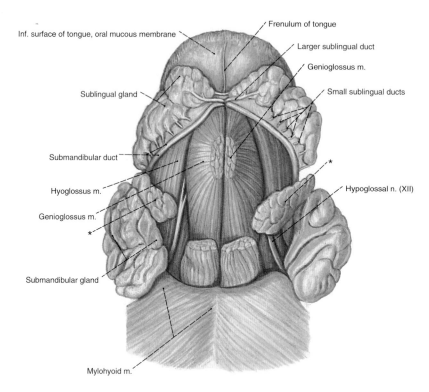

Inf. surface of tongue, oral mucous membrane

Frenulum of tongue

Larger sublingual duct

Genioglossus m.

Sublingual gland

Small sublingual ducts

Submandibular duct

*

Hyoglossus m.

Hypoglossal n. (XII)

Genioglossus m.

*

Submandibular gland

Mylohyoid m.

Fig. 115. View of the tongue from below, with submandibular and sublingual glands.

* The deep process of the submandibular gland which extends anteriorly between the mylohyoid and the hyoglossus muscles.

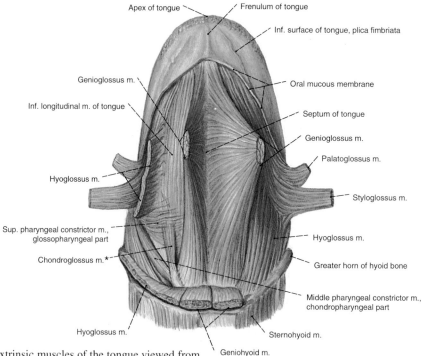

Apex of tongue

Frenulum of tongue

Inf. surface of tongue, plica fimbriata

Genioglossus m.

Oral mucous membrane

Inf. longitudinal m. of tongue

Septum of tongue

Genioglossus m.

Palatoglossus m.

Hyoglossus m.

Styloglossus m.

Sup. pharyngeal constrictor m., glossopharyngeal part

Hyoglossus m.

Chondroglossus m.*

Greater horn of hyoid bone

Middle pharyngeal constrictor m., chondropharyngeal part

Hyoglossus m.

Sternohyoid m.

Geniohyoid m.

Fig. 116. The extrinsic muscles of the tongue viewed from below after removal of the genioglossus muscles from the mandible. On the right side the hyoglossus muscle has been sectioned.

* Part of the hyoglossus muscle arising from the lesser horn of the hyoid bone.

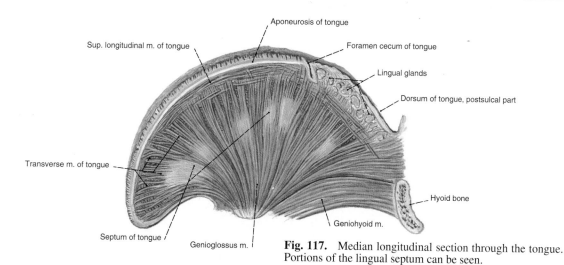

Fig. 117. Median longitudinal section through the tongue. Portions of the lingual septum can be seen.

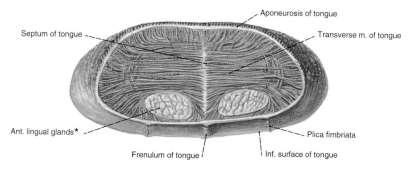

Fig. 118. Coronal section through the tongue at the level of its middle segment.

Fig. 119. Coronal section through the apex (tip) of the tongue with anterior lingual glands.

* NUHN's glands

Intrinsic Muscles of the Tongue
1. Inferior longitudinal muscle, paired, flat cylindrical muscle on the inferior surface of the tongue between the genioglossus and hyoglossus muscles, extending from the root (base) to the apex (tip) of the tongue.
2. Superior longitudinal muscle, unpaired sagittal fiber strands immediately underlying the upper surface of the tongue, which do not form a well-defined muscle but arise as extensions of the hyoglossus and styloglossus muscles.
3. Transverse muscle, transverse fiber strands extending from the lingual septum to the lateral margins of the tongue;

unpaired in front of the anterior end of the septum; posteriorly they merge with the palatoglossus muscle and the glossopharyngeal portion of the superior constrictor muscle of the pharynx.
4. Vertical muscle, collective name for the fibers running vertically from the upper to the inferior surface of the tongue.
These muscles are supplied by the hypoglossal nerve (XII). The function of these muscles corresponds to their course; they alter the shape of the tongue and assist in the act of chewing.

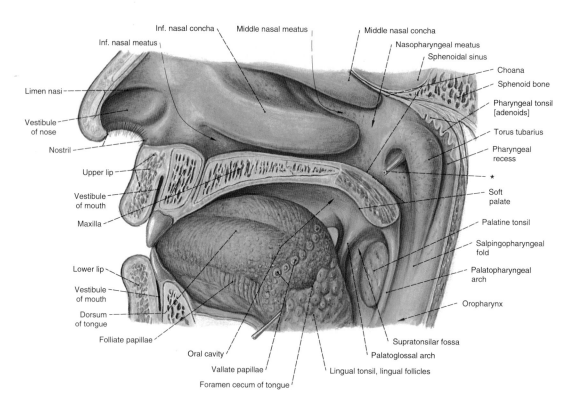

Inf. nasal concha   Middle nasal meatus   Middle nasal concha

Inf. nasal meatus   Nasopharyngeal meatus
                                                                    Sphenoidal sinus
                                                                                Choana
Limen nasi   Sphenoid bone
                                                                                Pharyngeal tonsil
Vestibule   [adenoids]
of nose   Torus tubarius
Nostril   Pharyngeal
                                                                                recess
Upper lip   *
Vestibule   Soft
of mouth   palate
                                                                                Palatine tonsil
Maxilla
                                                                                Salpingopharyngeal
                                                                                fold
Lower lip   Palatopharyngeal
Vestibule   arch
of mouth
                                                                                Oropharynx
Dorsum
of tongue
Folliate papillae   Supratonsilar fossa
                                        Palatoglossal arch
Oral cavity
                                    Lingual tonsil, lingual follicles
Vallate papillae
Foramen cecum of tongue

Fig. 120. Nasopharynx and oropharynx, with palatine arches and palatine tonsil of the right side. Medial view. The head has been sectioned midsagittally, and only the tongue, which has been deflected to the left, is fully visible. The palatine uvula has been cut at its base.

* The probe marks the pharyngeal opening of the auditory tube.

Innervation of the Tongue

The **sensory** nerves distributed to the body of the tongue are carried by the lingual branch of the mandibular nerve (V/3), to the region of the vallate papillae and the root of the tongue by the glossopharyngeal nerve (IX) – possibly along with a thin component of the sensory fibers of the facial nerve [nervus intermedius] (VII) – and to the region of the epiglottic valleculae by the vagus nerve (X).

The **motor** innervation of the intrinsic and extrinsic tongue muscles is supplied by the hypoglossal nerve (XII).

Taste is transmitted from receptors in the **gustatory or-** gans or taste buds to the nucleus of the tractus solitarius in the brain stem: from the body of the tongue via the lingual nerve as the chorda tympani of the facial nerve [nervus intermedius] (VII); from the vallate papillae via the glossopharyngeal nerve (IX). Taste fibers in cranial nerves V, VII, IX and X are carried in the tractus solitarius to its nucleus. Taste receptors are present on the dorsum of the tongue and occasionally on the oral surface of the soft palate and on the posterior surface of the epiglottis. Sweet tastes are localized to the tip of the tongue, bitter to the margins, salt to the tip and margins, and sour to the margins.

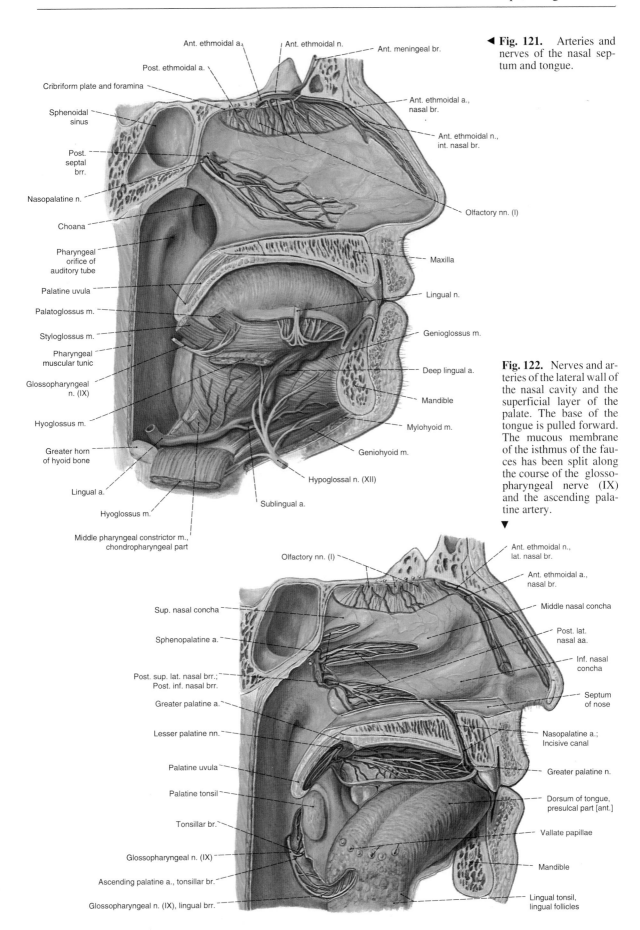

◀ **Fig. 121.** Arteries and nerves of the nasal septum and tongue.

Fig. 122. Nerves and arteries of the lateral wall of the nasal cavity and the superficial layer of the palate. The base of the tongue is pulled forward. The mucous membrane of the isthmus of the fauces has been split along the course of the glossopharyngeal nerve (IX) and the ascending palatine artery.

▼

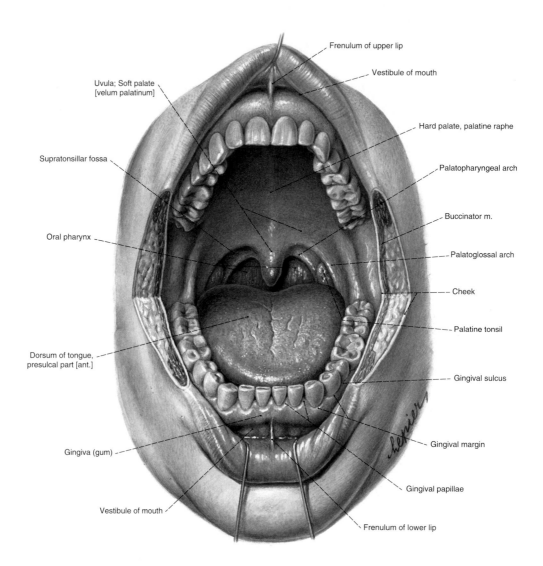

Frenulum of upper lip

Vestibule of mouth

Uvula; Soft palate
[velum palatinum]

Hard palate, palatine raphe

Supratonsillar fossa

Palatopharyngeal arch

Buccinator m.

Oral pharynx

Palatoglossal arch

Cheek

Palatine tonsil

Dorsum of tongue,
presulcal part [ant.]

Gingival sulcus

Gingival margin

Gingiva (gum)

Gingival papillae

Vestibule of mouth

Frenulum of lower lip

Fig. 123. View into the oral cavity, with vestibule of the
mouth, palate, palatine tonsils, dorsum of the tongue and teeth.
The cheeks have been cut, upper and lower lips reflected to
expose the vestibule of the mouth and gums (gingiva). The
gingival sulcus is the shallow groove between the gingival
margin and the teeth where, for example, periodontitis can
produce pockets.

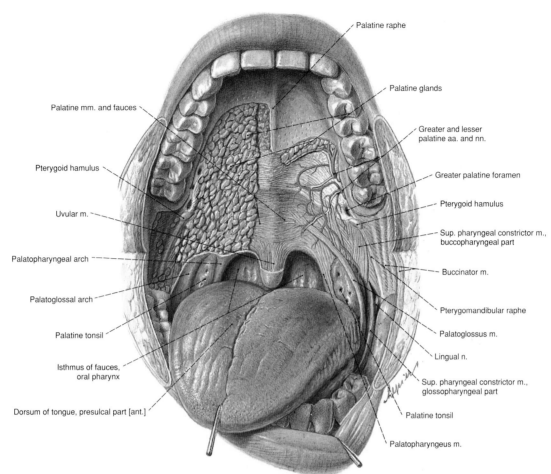

Palatine raphe

Palatine glands

Palatine mm. and fauces

Greater and lesser
palatine aa. and nn.

Pterygoid hamulus

Greater palatine foramen

Uvular m.

Pterygoid hamulus

Sup. pharyngeal constrictor m.,
buccopharyngeal part

Palatopharyngeal arch

Buccinator m.

Palatoglossal arch

Pterygomandibular raphe

Palatine tonsil

Palatoglossus m.

Lingual n.

Isthmus of fauces,
oral pharynx

Sup. pharyngeal constrictor m.,
glossopharyngeal part

Dorsum of tongue, presulcal part [ant.]

Palatine tonsil

Palatopharyngeus m.

Fig. 124. View into the oral cavity, with tongue protracted and palate dissected. Salivary glands form an almost continuous layer in the lateral and dorsal regions of the mucosa of the hard palate with scatterings medially and ventrally. The muscles of the soft palate form several bundles which spread into the dense connective tissue sheet of the palatine aponeurosis. The blood vessels and nerves that supply the palatine mucosa pass through the palatine foramina.

Muscles of the Palate

Name	Origin	Insertion	Innervation	Function
1. Musculus uvulae	Palatine aponeurosis, post. nasal spine of palatine bone	Stroma of the uvula	Cranial part of accessory nerve (XI), via vagus nerve (X) to pharyngeal plexus	Raises uvula
2. Levator veli palatini muscle (forms beneath mucosa of levator ridge)	Temporal bone, inf. surface of petrous part; auditory tube cartilage	Muscles of both sides interdigitate in palatine velum without a tendon	XI via X	Elevates the soft palate against dorsal pharyngeal wall (in deglutition)
3. Tensor veli palatini muscle (overlies fascia of med. pterygoid muscle)	Scaphoid fossa of med. pterygoid plate, spine of sphenoid, auditory tube cartilage	Runs as flat tendon through sulcus of pterygoid hamulus, forming palatine aponeurosis with tendon of opposite side	Mandibular branch of trigeminal nerve (V_3)	Tenses the soft palate, dilates auditory tube (in deglutition)
4. Palatoglossus muscle	Ant. surface of soft palate	Side of tongue, dorsum and transverse muscle of tongue	XI via X	Constricts the fauces (in deglutition), draws root of tongue upwards, lowers soft palate
5. Palatopharyngeus muscle	Soft palate midline	Post. border of thyroid cartilage, lateral wall of pharynx	XI via X	Pulls pharynx upward (in deglutition), lowers soft palate

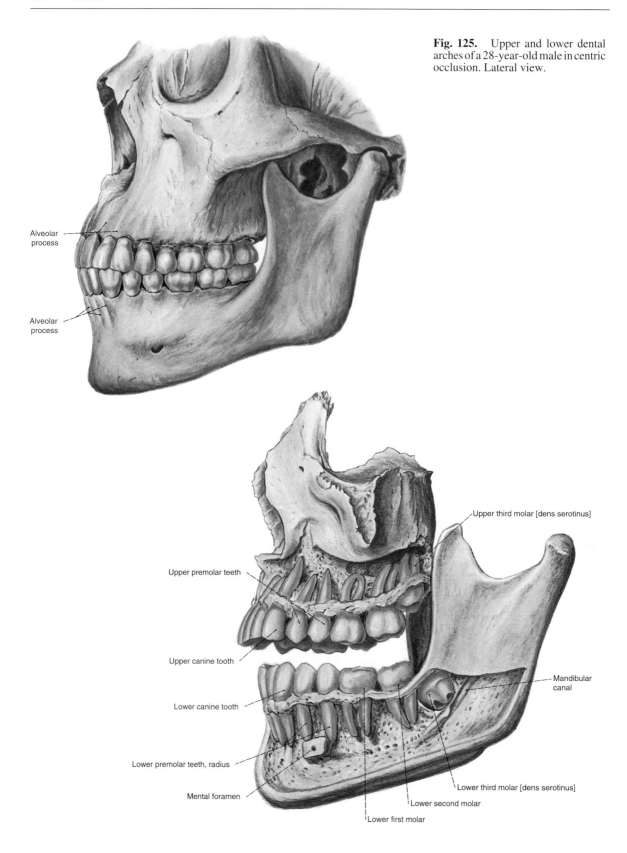

Fig. 125. Upper and lower dental arches of a 28-year-old male in centric occlusion. Lateral view.

Alveolar process

Alveolar process

Upper third molar [dens serotinus]

Upper premolar teeth

Upper canine tooth

Lower canine tooth

Lower premolar teeth, radius

Mental foramen

Mandibular canal

Lower third molar [dens serotinus]

Lower second molar

Lower first molar

Fig. 126. Maxilla and mandible of a 20-year-old male. All the permanent teeth, except the third lower molar, have erupted. Removal of the alveolar walls has exposed the roots of the teeth. The lingual roots of the upper molars are not visible in this lateral view.

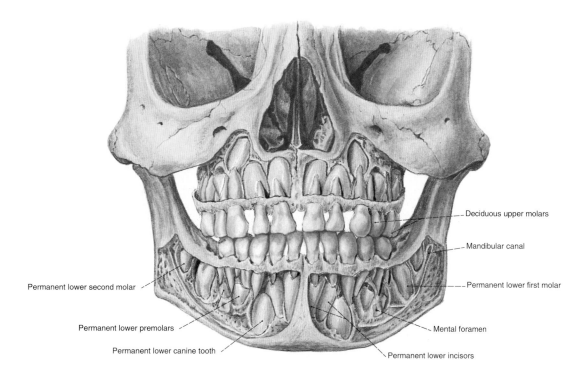

Deciduous upper molars

Mandibular canal

Permanent lower first molar

Permanent lower second molar

Permanent lower premolars

Permanent lower canine tooth

Mental foramen

Permanent lower incisors

Fig. 127.   Facial skeleton of a 5-year-old child with the deciduous teeth and rudiments of the permanent teeth (light blue), viewed from the front.

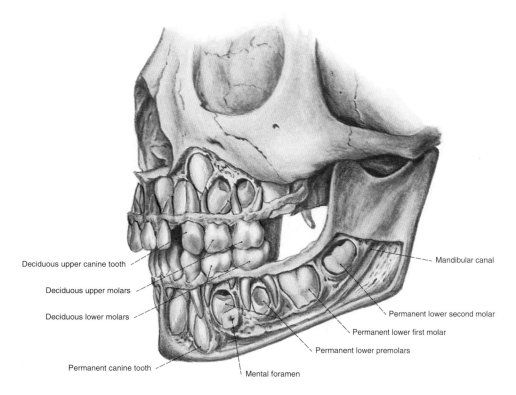

Deciduous upper canine tooth

Deciduous upper molars

Deciduous lower molars

Permanent canine tooth

Mental foramen

Mandibular canal

Permanent lower second molar

Permanent lower first molar

Permanent lower premolars

Fig. 128.   The same specimen as in Fig. 127, viewed from the side.

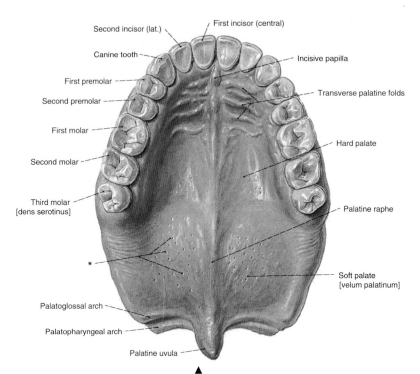

Fig. 129. The hard and soft palates and the upper dental arch in the maxilla, viewed from below.
* Openings of palatine glands

Fig. 131. The right half of a mandible with the permanent ▶ teeth (b). Section through the alveolar and root region of the same specimen to demonstrate the position of the dental roots in the bone (a). (From: Prof. Dr. H. FERNER, Vienna.)

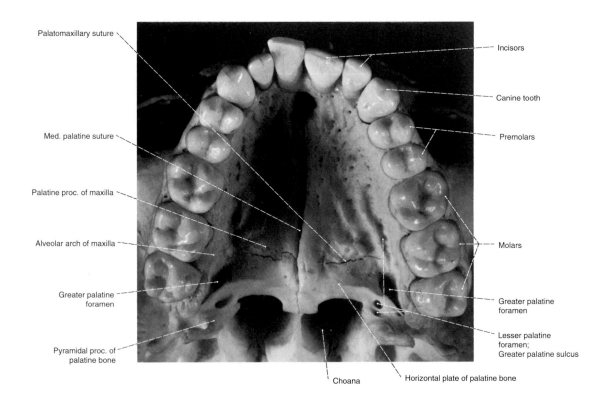

▲

Fig. 130. The bony hard palate and dental arch of the maxilla of an adult. (From: Prof. Dr. H. FERNER, Vienna.)

Fig. 132. Mandible with inferior dental arch and portions of ▶ the gingiva and oral mucosa.

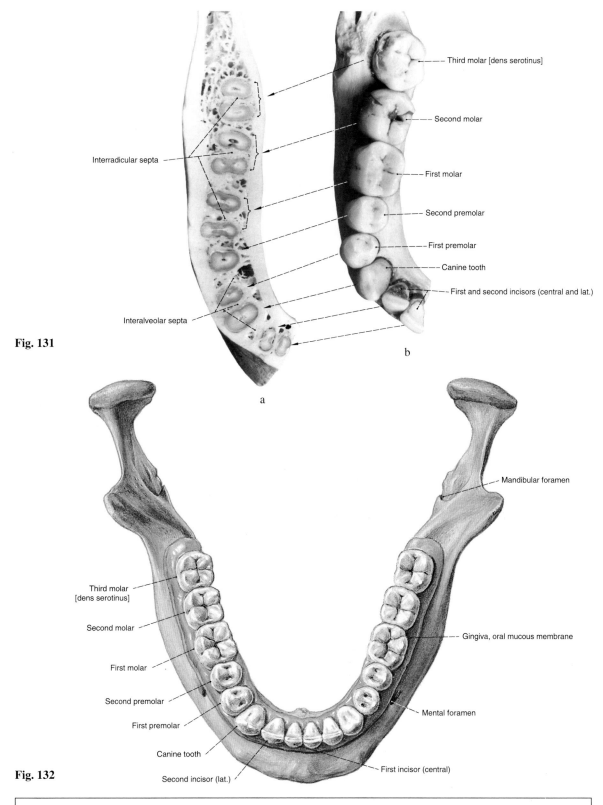

Fig. 131

Interradicular septa

Interalveolar septa

a

b

Third molar [dens serotinus]

Second molar

First molar

Second premolar

First premolar

Canine tooth

First and second incisors (central and lat.)

Fig. 132

Third molar [dens serotinus]

Second molar

First molar

Second premolar

First premolar

Canine tooth

Second incisor (lat.)

First incisor (central)

Mental foramen

Gingiva, oral mucous membrane

Mandibular foramen

Note:
The dentition of the lower jaw is arranged so that the front teeth are closer to the vestibular side of the alveoli while the molars increasingly approach the inner or lingual side of the mandible. This is especially noticeable in the section through the root region. The dental arch of the mandible resembles a parabola. The lower molars have two roots: an anterior (mesial), nearly vertical, and a posterior (distal), directed obliquely backward. The roots are surrounded by a thin shell of alveolar bone and are embedded between the strong cortical shells of the mandibular body.

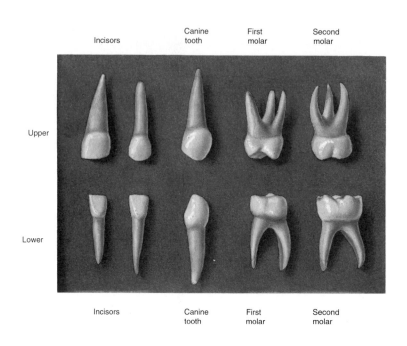

Incisors   Canine tooth   First molar   Second molar

Upper

Lower

Fig. 133a

Incisors   Canine tooth   First molar   Second molar

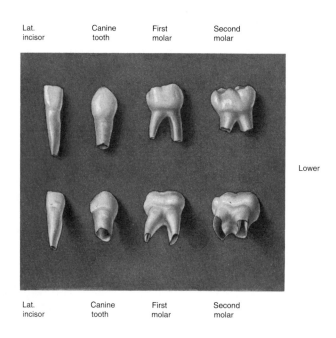

Lat. incisor   Canine tooth   First molar   Second molar

Lower

Fig. 133b

Lat. incisor   Canine tooth   First molar   Second molar

Fig. 133. a) Deciduous teeth of a 3-year-old child, viewed from the buccal surface. b) Deciduous teeth of a 2-year-old child. Upper row: viewed from the buccal surface. Lower row: viewed obliquely from below. Note the incomplete calcification of the dental roots and the various developmental stages of the illustrated deciduous teeth.

Fig. 136. Right lower permanent canine tooth viewed from ▶ the labial surface. The lower and upper canines have the longest roots of all the teeth. The root of the lower canines is occasionally bipartite.

Fig. 137. Left lower permanent second molar tooth (second ▶ left deciduous), viewed from the buccal surface. The occlusal surface usually has two vestibular (buccal) and two lingual tubercles or cusps. The two sturdy roots of this molar are bent slightly posteriorly (a typical feature of the root).

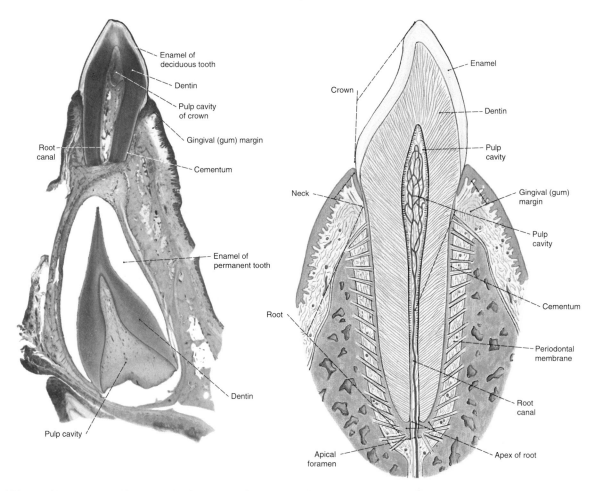

Fig. 134. Deciduous tooth with resorbed apical root region. Resorption of the root of a deciduous tooth progresses toward the neck until only a hollow crown remains. Below this lies the permanent tooth, of which only the crown, but not the root, has developed. (From: Prof. Dr. H. FERNER, Vienna).

Fig. 135. Schematic longitudinal section through an incisor tooth in its alveolar bony socket.

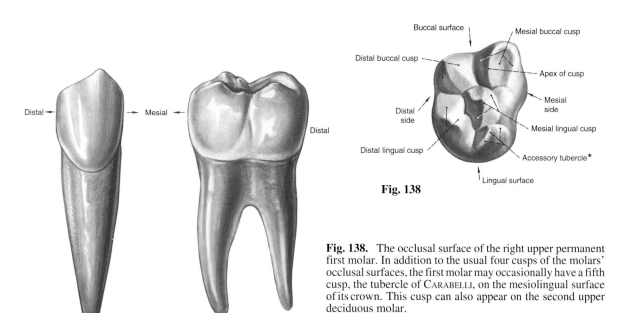

Fig. 138

Fig. 136   **Fig. 137**

Fig. 138. The occlusal surface of the right upper permanent first molar. In addition to the usual four cusps of the molars' occlusal surfaces, the first molar may occasionally have a fifth cusp, the tubercle of CARABELLI, on the mesiolingual surface of its crown. This cusp can also appear on the second upper deciduous molar.

* Clinical eponym: Tubercle of CARABELLI

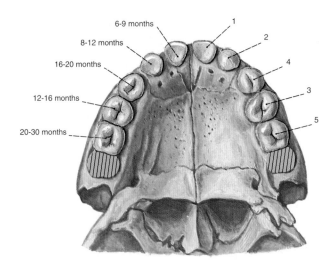

Fig. 139. Eruption schedule of the deciduous teeth in the maxilla. The numbers on the left in the figure indicate the time of eruption of each tooth in months, while the numbers on the right indicate the sequential order of appearance for the erupted deciduous teeth. The hatched lines outline the position of the later-appearing permanent third molars.

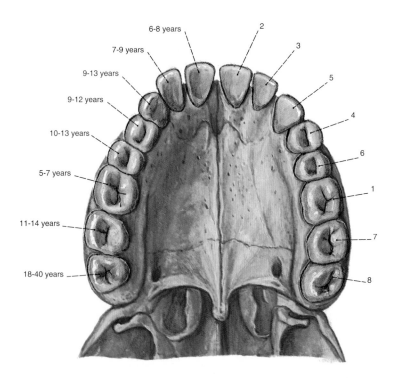

Fig. 140. Eruption schedule of the permanent teeth in the maxilla. The numbers on the left in the figure indicate the time of eruption of each tooth in years, while the numbers on the right indicate the sequential order of appearance of the erupted permanent teeth.

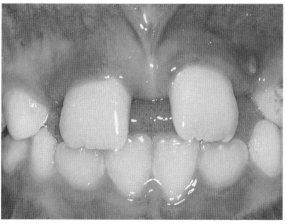

a

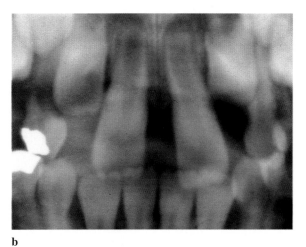

b

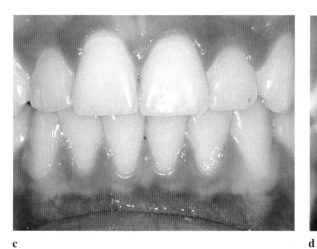

c

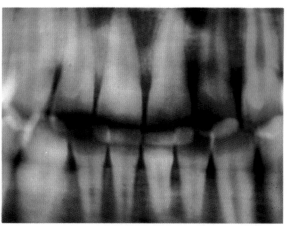

d

Fig. 141a–d.   Orthodontic correction of a diastema in a 9-year-old child. a,b = before initiation of treatment; c,d = after completion of treatment (approximately 12 months' duration). (From: Prof. Dr. Th. RAKOSI, Orthodontic Division, University Dental Clinic, Freiburg i. Br.).

Orthodontics
The branch of dentistry that is concerned with the correction and prevention of irregularities and malocclusion of the teeth (Stedman's Medical Dictionary, 24th Edition, Williams and Wilkins, Baltimore).

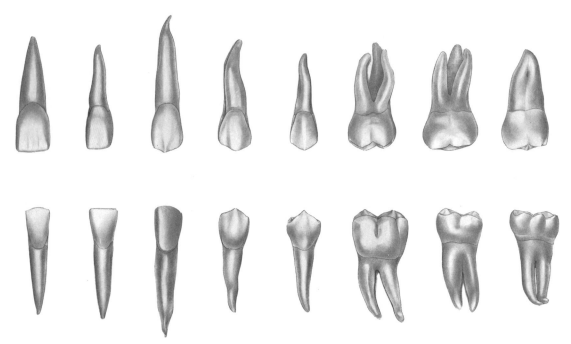

Fig. 142. Permanent teeth of the left maxilla and mandible. View from the vestibular (labial and buccal surfaces). The teeth in order of progression from left to right: 2 incisors, 1 canine, 2 premolars and 3 molars.

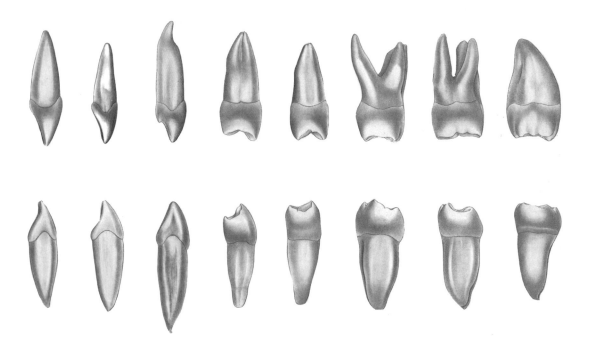

Fig. 143. Permanent teeth of the left maxilla and mandible. View from the mesial side. Order of teeth the same as in Fig. 142.

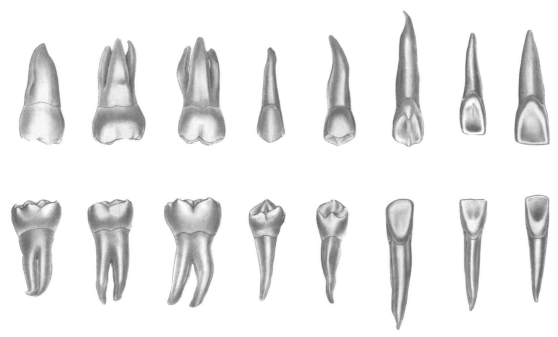

Fig. 144. Permanent teeth of the left maxilla and mandible. View from the oral (lingual) surface. The teeth in order of progression from right to left: 2 incisors, 1 canine, 2 premolars and 3 molars.

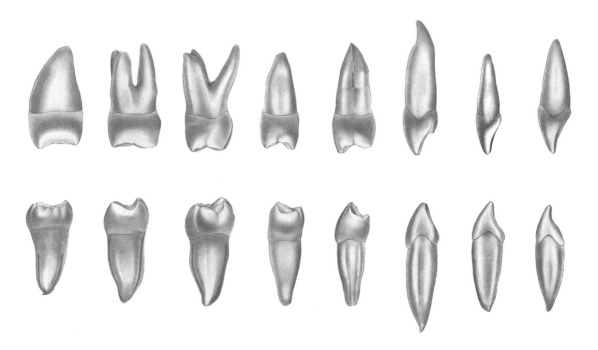

Fig. 145. Permanent teeth of the left maxilla and mandible. View from the distal side. Order of teeth the same as in Fig. 144.

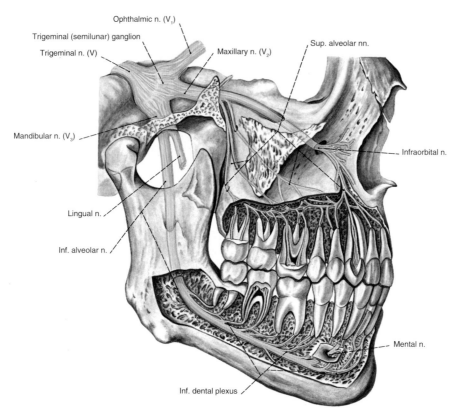

Ophthalmic n. (V₁)

Trigeminal (semilunar) ganglion

Trigeminal n. (V)

Maxillary n. (V₂)

Sup. alveolar nn.

Mandibular n. (V₃)

Infraorbital n.

Lingual n.

Inf. alveolar n.

Mental n.

Inf. dental plexus

▲

Fig. 146. Innervation of the teeth by the maxillary and mandibular nerves from the trigeminal nerve (V). Parts of the maxilla and mandible have been removed and the mandibular canal has been exposed.

Fig. 147. Maxillary nerve, pterygopalatine ganglion, facial nerve (VII). The orbit has been opened laterally and its contents have been removed. The tympanic cavity has been opened, the temporal bone and mastoid process have been sectioned obliquely and the trigeminal ganglion has been retracted upward.

▼

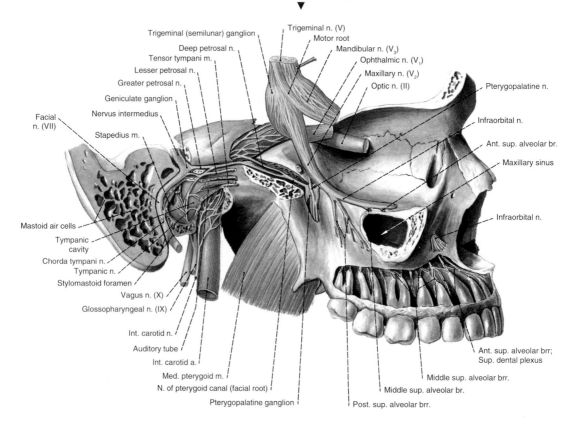

Trigeminal (semilunar) ganglion

Deep petrosal n.

Tensor tympani m.

Lesser petrosal n.

Greater petrosal n.

Geniculate ganglion

Nervus intermedius

Facial n. (VII)

Stapedius m.

Trigeminal n. (V)

Motor root

Mandibular n. (V₃)

Ophthalmic n. (V₁)

Maxillary n. (V₂)

Optic n. (II)

Pterygopalatine n.

Infraorbital n.

Ant. sup. alveolar br.

Maxillary sinus

Infraorbital n.

Mastoid air cells

Tympanic cavity

Chorda tympani n.

Tympanic n.

Stylomastoid foramen

Vagus n. (X)

Glossopharyngeal n. (IX)

Int. carotid n.

Auditory tube

Int. carotid a.

Med. pterygoid m.

N. of pterygoid canal (facial root)

Pterygopalatine ganglion

Ant. sup. alveolar brr;
Sup. dental plexus

Middle sup. alveolar brr.

Middle sup. alveolar br.

Post. sup. alveolar brr.

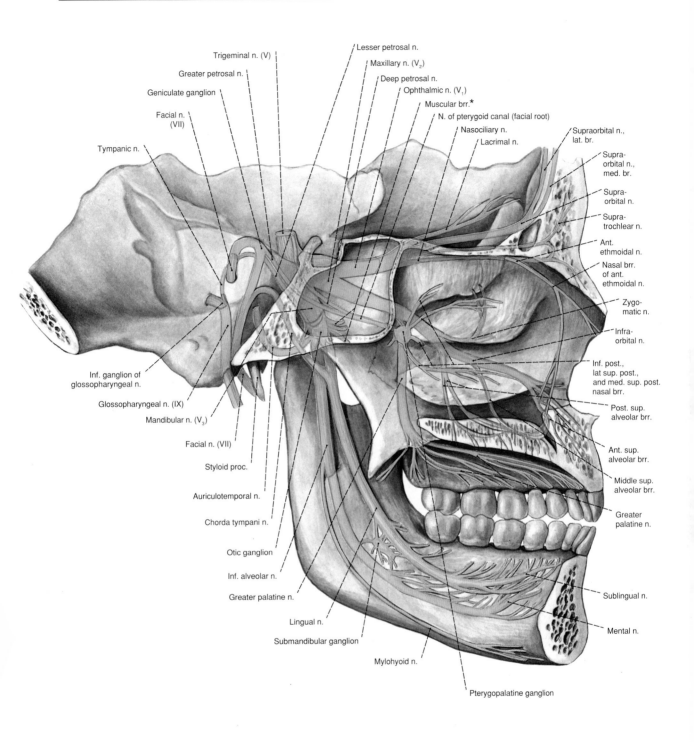

Trigeminal n. (V)

Greater petrosal n.

Geniculate ganglion

Facial n. (VII)

Tympanic n.

Lesser petrosal n.

Maxillary n. (V₂)

Deep petrosal n.

Ophthalmic n. (V₁)

Muscular brr.*

N. of pterygoid canal (facial root)

Nasociliary n.

Lacrimal n.

Supraorbital n., lat. br.

Supra-orbital n., med. br.

Supra-orbital n.

Supra-trochlear n.

Ant. ethmoidal n.

Nasal brr. of ant. ethmoidal n.

Zygo-matic n.

Infra-orbital n.

Inf. post., lat sup. post., and med. sup. post. nasal brr.

Post. sup. alveolar brr.

Ant. sup. alveolar brr.

Middle sup. alveolar brr.

Greater palatine n.

Sublingual n.

Mental n.

Inf. ganglion of glossopharyngeal n.

Glossopharyngeal n. (IX)

Mandibular n. (V₃)

Facial n. (VII)

Styloid proc.

Auriculotemporal n.

Chorda tympani n.

Otic ganglion

Inf. alveolar n.

Greater palatine n.

Lingual n.

Submandibular ganglion

Mylohyoid n.

Pterygopalatine ganglion

Fig. 148. Schematic representation of the branches of the trigeminal nerve (V) and their connections with the facial nerve [nervus intermedius] (VII) and the glossopharyngeal nerve (IX), projected onto a paramedian section of the skull. Medial view. The exposed portions of nerves are colored yellow, and those portions covered by bone, grayish yellow.

* Branches to muscles of mastication

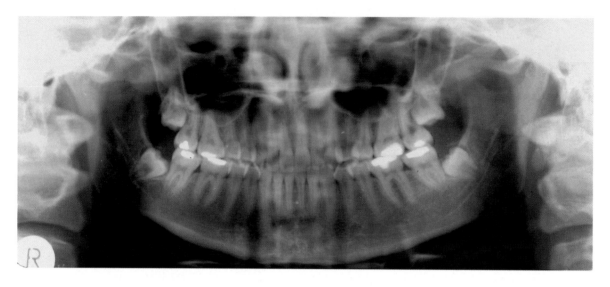

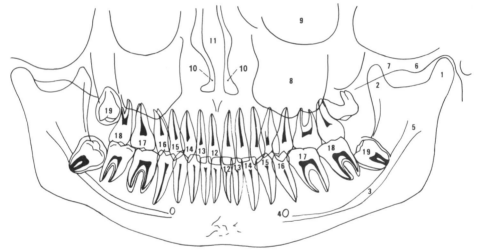

Fig. 149

1 Condylar proc.	7 Zygomatic arch	14 Permanent canine tooth
2 Coronoid proc.	8 Maxillary sinus	15 First permanent premolar
3 Mandibular canal	9 Orbit	16 Second permanent premolar
4 Mental foramen	10 Nasal cavity	17 First permanent molar
5 Mandibular foramen	11 Nasal septum	18 Second permanent molar
6 Articular tubercle	12 First permanent incisor	19 Third molar [dens serotinus]
	13 Second permanent incisor	

▲

Fig. 149. Panoramic radiograph of the jaws. Complete dentition of an 18-year-old female with impacted third molars (wisdom teeth) and molars with amalgam restorations. (From: Prof. Dr. J. DÜKER, University Dental Clinic, Freiburg i. Br.).

Fig. 150. Muscular floor of the oral cavity (clinically, the ▶ diaphragm of the mouth), viewed anteriorly from below.

* The origin of the mylohyoid muscle on the inner surface of the mandible (= mylohyoid line)
** Region of insertion of the anterior belly of the digastric muscle into the digastric fossa on the inner surface of the mandible.

Fig. 151. Muscular floor of the oral cavity viewed from ▶ above. Mandible, hyoid bone, with the mandibular rami of both sides sectioned at the level of the mandibular foramen.

Fig. 152. The floor of the oral cavity viewed from above. The ▶ tongue was extensively removed and stumps of the genioglossus and hyoglossus muscles were retained. The mandibular rami were sectioned horizontally and the oral mucosa removed almost to the gums.

* The portion of the submandibular gland overlying the mylohyoid muscle.

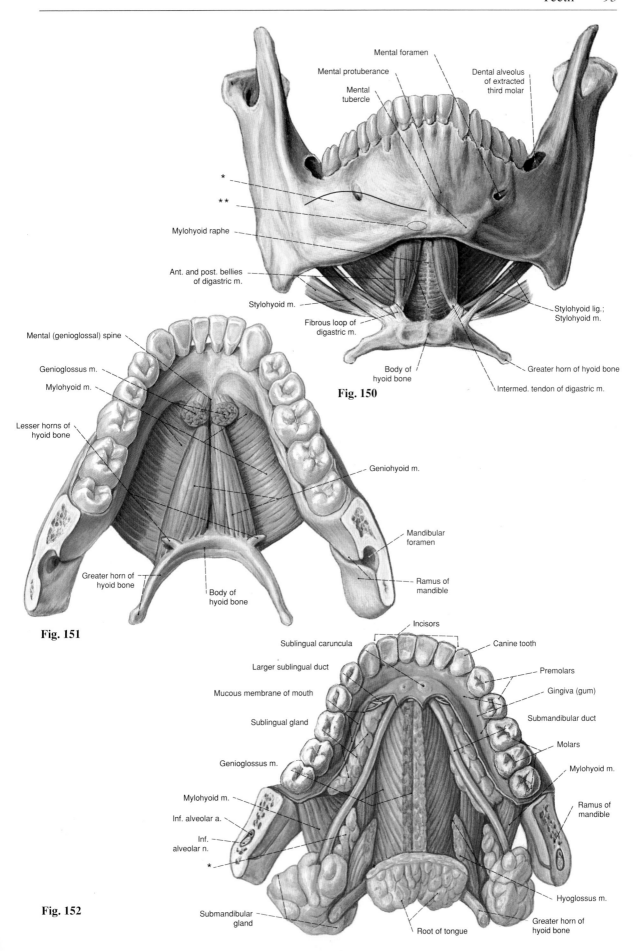

Mental foramen

Mental protuberance

Mental tubercle

Dental alveolus of extracted third molar

*

**

Mylohyoid raphe

Ant. and post. bellies of digastric m.

Stylohyoid m.

Fibrous loop of digastric m.

Body of hyoid bone

Stylohyoid lig.; Stylohyoid m.

Greater horn of hyoid bone

Intermed. tendon of digastric m.

Fig. 150

Mental (genioglossal) spine

Genioglossus m.

Mylohyoid m.

Lesser horns of hyoid bone

Geniohyoid m.

Mandibular foramen

Greater horn of hyoid bone

Body of hyoid bone

Ramus of mandible

Fig. 151

Incisors

Sublingual caruncula

Larger sublingual duct

Mucous membrane of mouth

Sublingual gland

Genioglossus m.

Mylohyoid m.

Inf. alveolar a.

Inf. alveolar n.

*

Canine tooth

Premolars

Gingiva (gum)

Submandibular duct

Molars

Mylohyoid m.

Ramus of mandible

Hyoglossus m.

Greater horn of hyoid bone

Submandibular gland

Root of tongue

Fig. 152

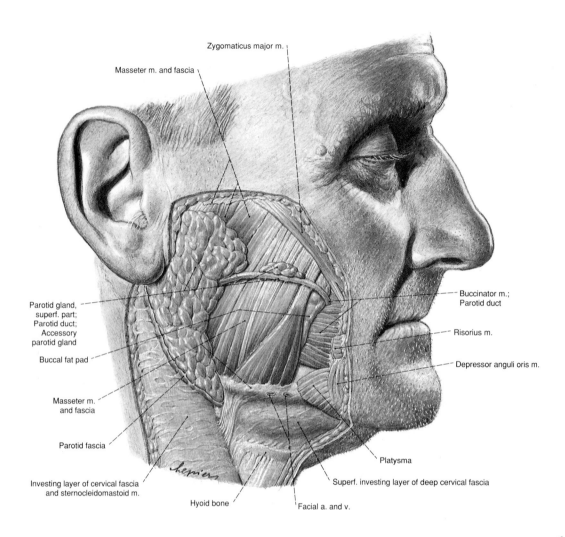

Zygomaticus major m.

Masseter m. and fascia

Parotid gland, superf. part; Parotid duct; Accessory parotid gland

Buccal fat pad

Masseter m. and fascia

Parotid fascia

Investing layer of cervical fascia and sternocleidomastoid m.

Hyoid bone

Buccinator m.; Parotid duct

Risorius m.

Depressor anguli oris m.

Platysma

Superf. investing layer of deep cervical fascia

Facial a. and v.

Fig. 153. The parotid gland and duct. Lateral view. The parotid duct is located one finger's breadth below the zygomatic arch, crosses the masseter muscle and, at the anterior border of this muscle, turns medially inward as it passes through the buccal fat pad and pierces the buccinator muscle, to open into the oral vestibule opposite the second upper molar. Note the accessory parotid gland along the parotid duct.

Fig. 154. The submandibular gland in the left submandibular triangle. Lateral view from below. The submandibular triangle is bounded below, front and back, by both bellies of the digastric muscle and cranially by the lower border of the body of the mandible. Note the relationships of the submandibular gland, as it overlies the mylohyoid muscle, to the facial vein and artery and to the submandibular lymph nodes along the lower border of the body of the mandible.

The parotid gland has a superficial part which lies **over** the facial nerve (VII) and a deep surface **under** the facial nerve (Fig. 81). The branches of the facial nerve emerge from the borders of the gland. The main portion of the gland extends vertically, nearly to but not above the zygomatic arch. The remainder of the gland extends deeply inward toward the pharyngeal wall. The gland's shape may vary from that of an inverted triangle (56% of cases) to an oval (31% of cases). In about 20% of cases, an accessory part of the gland may wrap over the parotid duct (cf. Figs. 80, 84, 153). The posterior border of the mandible lies deep beneath the gland. Glandular processes project to the mastoid process, the styloid process, the styloid group of muscles and the posterior belly of the digastric muscle. The posterior superficial portion of the gland is indented by the cartilaginous part of the external acoustic meatus. The size of the gland may vary considerably.

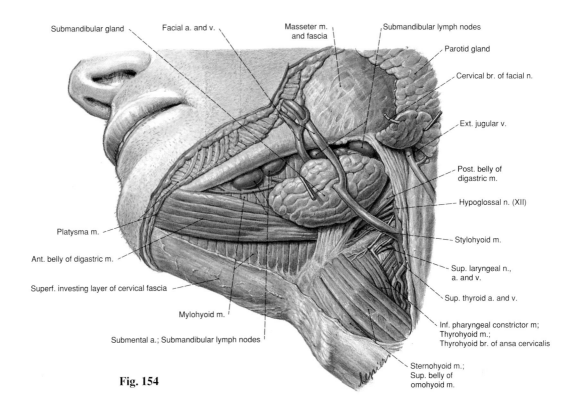

Submandibular gland — Facial a. and v. — Masseter m. and fascia — Submandibular lymph nodes — Parotid gland — Cervical br. of facial n. — Ext. jugular v. — Post. belly of digastric m. — Hypoglossal n. (XII) — Stylohyoid m. — Sup. laryngeal n., a. and v. — Sup. thyroid a. and v. — Inf. pharyngeal constrictor m; Thyrohyoid m.; Thyrohyoid br. of ansa cervicalis — Sternohyoid m.; Sup. belly of omohyoid m. — Platysma m. — Ant. belly of digastric m. — Superf. investing layer of cervical fascia — Mylohyoid m. — Submental a.; Submandibular lymph nodes

Fig. 154

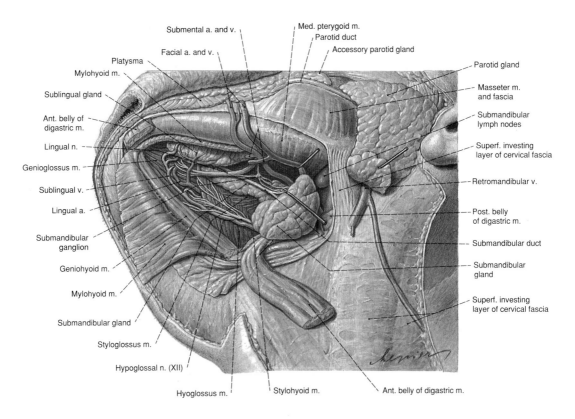

Submental a. and v. — Med. pterygoid m. — Parotid duct — Accessory parotid gland — Facial a. and v. — Platysma — Mylohyoid m. — Sublingual gland — Ant. belly of digastric m. — Lingual n. — Genioglossus m. — Sublingual v. — Lingual a. — Submandibular ganglion — Geniohyoid m. — Mylohyoid m. — Submandibular gland — Styloglossus m. — Hypoglossal n. (XII) — Hyoglossus m. — Stylohyoid m. — Ant. belly of digastric m. — Parotid gland — Masseter m. and fascia — Submandibular lymph nodes — Superf. investing layer of cervical fascia — Retromandibular v. — Post. belly of digastric m. — Submandibular duct — Submandibular gland — Superf. investing layer of cervical fascia

Fig. 155. Position of the three major salivary glands after reflection of the mylohyoid muscle and the anterior belly of the digastric muscle. View from below and slightly lateral. Exposed are the submandibular duct, the hypoglossal nerve (XII), the sublingual vein, the lingual artery, and the lingual nerve with the submandibular ganglion (a parasympathetic ganglion in which preganglionic fibers of the chorda tympani synapse with postganglionic neurons whose fibers innervate the submandibular and sublingual glands).

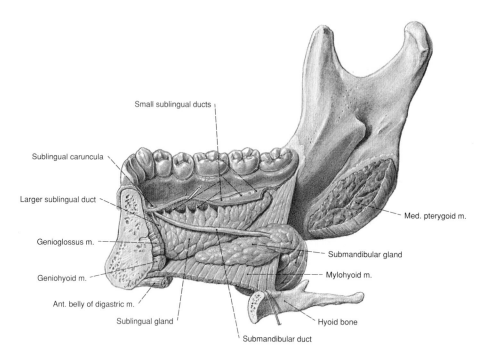

Small sublingual ducts

Sublingual caruncula

Larger sublingual duct

Genioglossus m.

Geniohyoid m.

Ant. belly of digastric m.

Sublingual gland

Submandibular duct

Med. pterygoid m.

Submandibular gland

Mylohyoid m.

Hyoid bone

Fig. 156.   Position of the submandibular and sublingual glands. Medial view. Mandible and hyoid bone are sectioned in the midline.

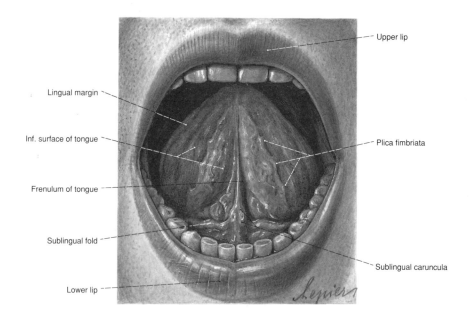

Upper lip

Lingual margin

Inf. surface of tongue

Frenulum of tongue

Sublingual fold

Lower lip

Plica fimbriata

Sublingual caruncula

Fig. 157. Anterior sublingual region of the oral cavity. The sublingual caruncula is the orifice for the duct of the submandibular gland and the larger duct of the sublingual gland.

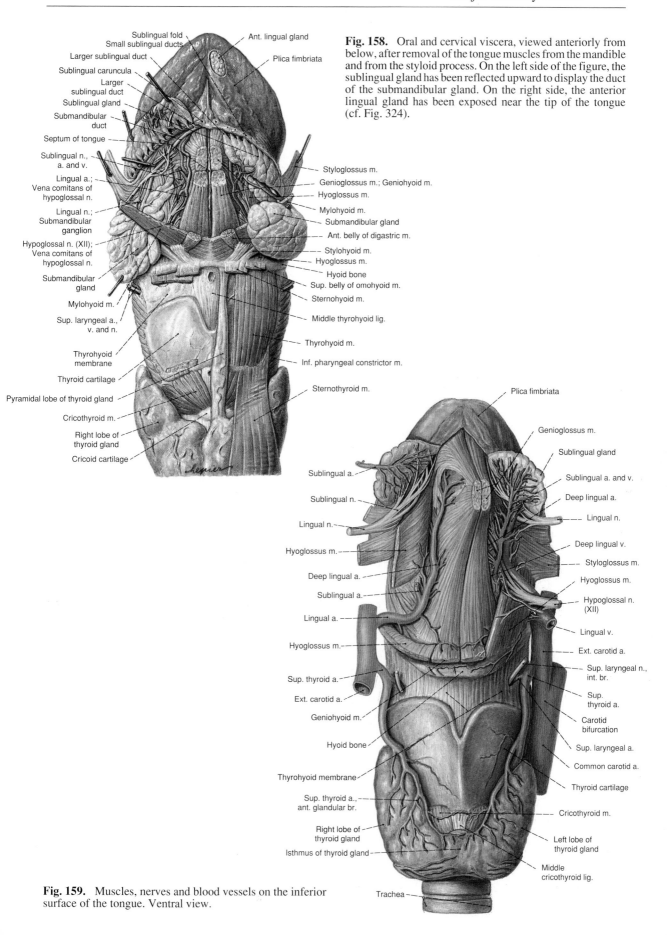

Sublingual fold
Small sublingual ducts
Ant. lingual gland
Larger sublingual duct
Sublingual caruncula
Plica fimbriata
Larger sublingual duct
Sublingual gland
Submandibular duct
Septum of tongue
Sublingual n., a. and v.
Styloglossus m.
Lingual a.; Vena comitans of hypoglossal n.
Geniglossus m.; Geniohyoid m.
Hyoglossus m.
Lingual n.; Submandibular ganglion
Mylohyoid m.
Submandibular gland
Ant. belly of digastric m.
Hypoglossal n. (XII); Vena comitans of hypoglossal n.
Stylohyoid m.
Hyoglossus m.
Hyoid bone
Submandibular gland
Sup. belly of omohyoid m.
Mylohyoid m.
Sternohyoid m.
Sup. laryngeal a., v. and n.
Middle thyrohyoid lig.
Thyrohyoid membrane
Thyrohyoid m.
Thyroid cartilage
Inf. pharyngeal constrictor m.
Pyramidal lobe of thyroid gland
Sternothyroid m.
Cricothyroid m.
Right lobe of thyroid gland
Cricoid cartilage

Fig. 158. Oral and cervical viscera, viewed anteriorly from below, after removal of the tongue muscles from the mandible and from the styloid process. On the left side of the figure, the sublingual gland has been reflected upward to display the duct of the submandibular gland. On the right side, the anterior lingual gland has been exposed near the tip of the tongue (cf. Fig. 324).

Plica fimbriata
Genioglossus m.
Sublingual gland
Sublingual a. and v.
Sublingual a.
Deep lingual a.
Sublingual n.
Lingual n.
Lingual n.
Deep lingual v.
Hyoglossus m.
Styloglossus m.
Deep lingual a.
Hyoglossus m.
Sublingual a.
Hypoglossal n. (XII)
Lingual a.
Lingual v.
Hyoglossus m.
Ext. carotid a.
Sup. thyroid a.
Sup. laryngeal n., int. br.
Ext. carotid a.
Sup. thyroid a.
Geniohyoid m.
Carotid bifurcation
Hyoid bone
Sup. laryngeal a.
Common carotid a.
Thyrohyoid membrane
Thyroid cartilage
Sup. thyroid a., ant. glandular br.
Cricothyroid m.
Right lobe of thyroid gland
Left lobe of thyroid gland
Isthmus of thyroid gland
Middle cricothyroid lig.
Trachea

Fig. 159. Muscles, nerves and blood vessels on the inferior surface of the tongue. Ventral view.

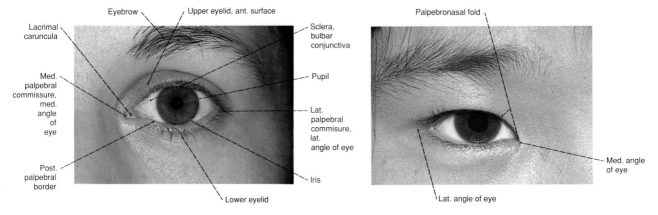

Fig. 160a.   The left eye and eyelids.

Fig. 160b.   The right eye with palpebronasal fold. The fold draws the upper eyelid over the medial angle (canthus) toward the nose and covers portions of the anterior surface of the upper lid. This feature is characteristic of Orientals.

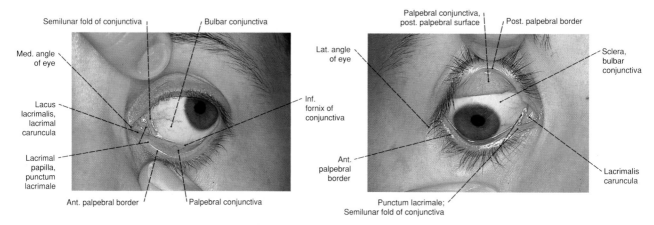

Fig. 161.   Left palpebral fissure with eyelids pulled apart. The gaze is oriented superiorly and laterally.

Fig. 162.   Eversion of the upper lid of the right eye. This is made possible through the stiffness of the superior tarsus. Tarsal glands are visible through the conjunctiva. Position of eyelids for removal of foreign bodies from the conjunctiva.

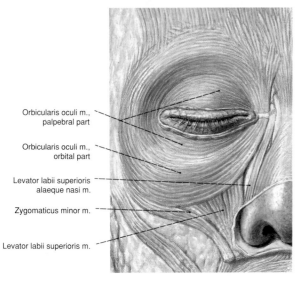

Fig. 163.   The muscles of facial expression surrounding the eye.

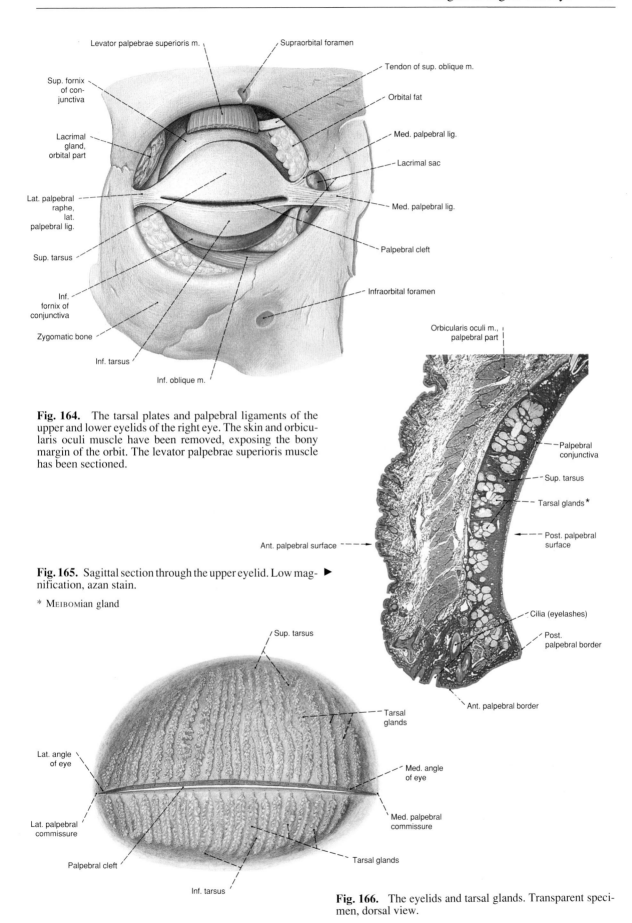

Fig. 164. The tarsal plates and palpebral ligaments of the upper and lower eyelids of the right eye. The skin and orbicularis oculi muscle have been removed, exposing the bony margin of the orbit. The levator palpebrae superioris muscle has been sectioned.

Fig. 165. Sagittal section through the upper eyelid. Low magnification, azan stain. ▶

* MEIBOMian gland

Fig. 166. The eyelids and tarsal glands. Transparent specimen, dorsal view.

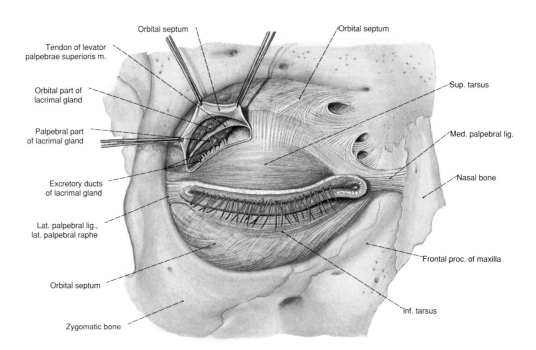

Fig. 167. The right eye viewed from anterior after removal of the skin, superficial fascia and orbicularis oculi muscle to expose the orbital septum. In the lateral part of the upper lid the orbital septum is incised and reflected to expose the lacrimal gland and the tendon of the levator palpebrae superioris muscle.

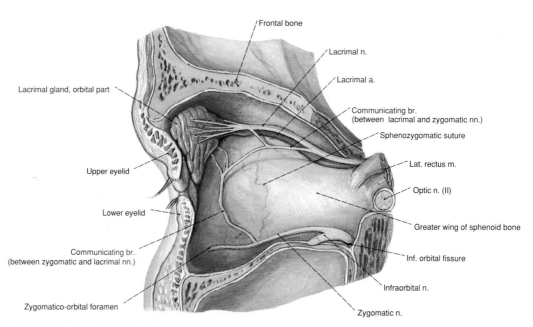

Fig. 168. The lateral wall of the right orbital cavity (contents removed) with the lacrimal gland and its arterial supply and innervation.

The Nasolacrimal Canal

The nasolacrimal canal is a passageway 2 cm in length and approximately 0.5 cm wide. It begins in the lacrimal fossa which lodges the lacrimal sac and opens into the inferior meatus under the inferior nasal concha. Its ventro-lateral wall is formed by the lacrimal sulcus of the maxilla, its dorsomedial boundary by the upper (orbital) portion of the lacrimal bone, and its inferior (nasal) part by the inferior nasal concha. The nasolacrimal canal transmits the nasolacrimal duct.

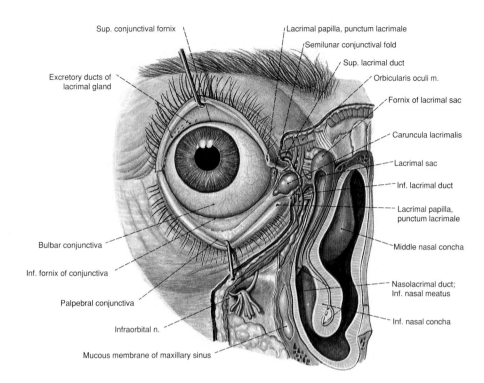

Fig. 169. The lacrimal apparatus. The eyelids have been pulled away from the eyeball to expose the conjunctiva. At the medial angle (canthus), the lacrimal ducts and their entry into the lacrimal sac are visible. Small muscle bundles of the orbicularis oculi muscle acting as sphincters surround the lacrimal ducts. The lacrimal sac and the nasolacrimal duct have been opened showing the orifice of the latter beneath the inferior nasal concha.

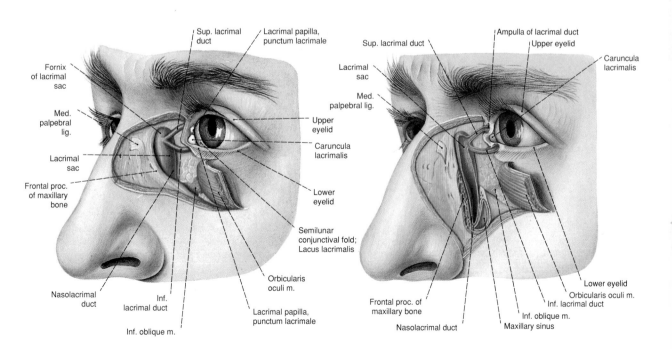

Fig. 170. The lacrimal apparatus, lacrimal ducts and lacrimal sac. Left side, superficial dissection.

Fig. 171. The lacrimal apparatus. Deep dissection of the lacrimal ducts, lacrimal sac and nasolacrimal duct in the nasolacrimal canal on the left side.

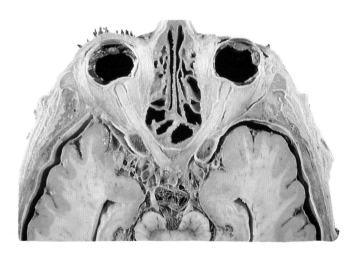
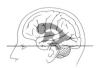

Fig. 172a. A horizontal section through both orbits at the level indicated in the adjacent diagram.

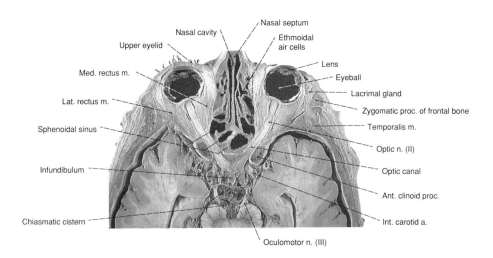

Nasal septum

Nasal cavity

Ethmoidal
air cells

Upper eyelid

Lens

Med. rectus m.

Eyeball

Lacrimal gland

Lat. rectus m.

Zygomatic proc. of frontal bone

Sphenoidal sinus

Temporalis m.

Optic n. (II)

Infundibulum

Optic canal

Ant. clinoid proc.

Chiasmatic cistern

Int. carotid a.

Oculomotor n. (III)

Fig. 172b. The same section as in Fig. 172a, with structures labeled. (From: Dr. R. UNSÖLD, University Eye Clinic, Düsseldorf.)

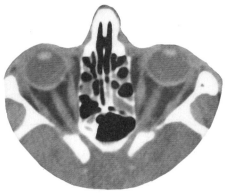

Refer to text on page 174, Volume 2
for an explanation of
computed axial tomography.

Fig. 172c. Computed tomographic reconstruction of the section shown in Figs. 172a, b. (From: Dr. R. UNSÖLD, University Eye Clinic, Düsseldorf.)

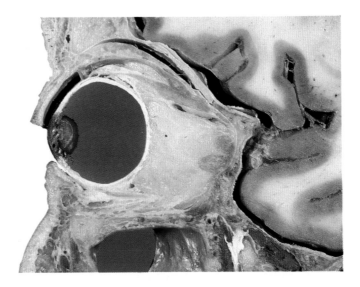

Fig. 173a. A sagittal section through the middle of the orbit and adjacent regions of the brain in the anterior cranial fossa.

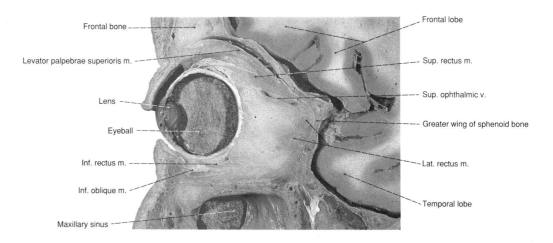

Frontal bone

Levator palpebrae superioris m.

Lens

Eyeball

Inf. rectus m.

Inf. oblique m.

Maxillary sinus

Frontal lobe

Sup. rectus m.

Sup. ophthalmic v.

Greater wing of sphenoid bone

Lat. rectus m.

Temporal lobe

Fig. 173b. The same section as in Fig. 173a, with structures labeled.

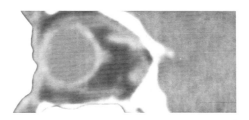

Fig. 173c. Computed tomographic reconstruction of the section shown in Figs. 173a, b. (From: Dr. R. Unsöld, University Eye Clinic, Düsseldorf.)

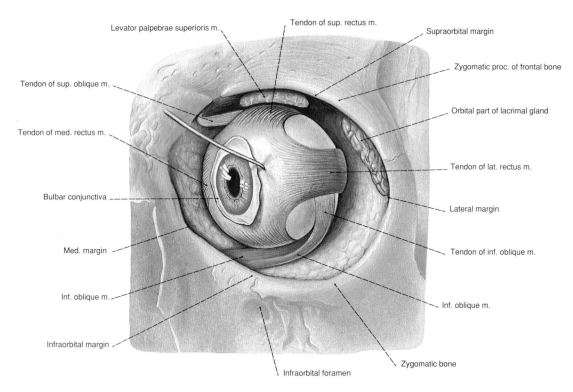

Levator palpebrae superioris m.

Tendon of sup. rectus m.

Supraorbital margin

Zygomatic proc. of frontal bone

Tendon of sup. oblique m.

Orbital part of lacrimal gland

Tendon of med. rectus m.

Tendon of lat. rectus m.

Bulbar conjunctiva

Lateral margin

Med. margin

Tendon of inf. oblique m.

Inf. oblique m.

Inf. oblique m.

Infraorbital margin

Zygomatic bone

Infraorbital foramen

Fig. 174. The ocular muscles of the left eye. Anterolateral view. The skin, eyelids and fascia have been removed and the eyeball has been pulled medially (by a hook).

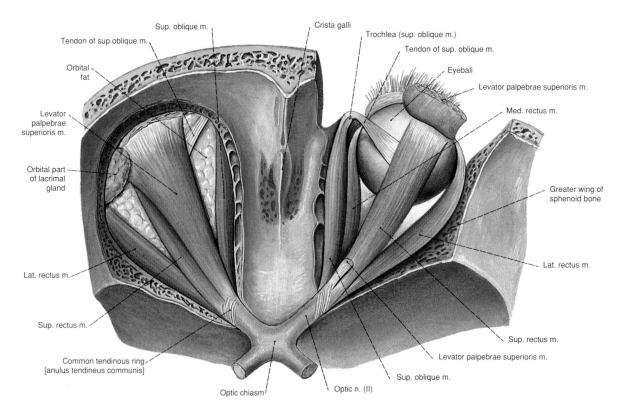

Sup. oblique m.

Crista galli

Trochlea (sup. oblique m.)

Tendon of sup.oblique m.

Tendon of sup. oblique m.

Orbital fat

Eyeball

Levator palpebrae superioris m.

Levator palpebrae superioris m.

Med. rectus m.

Orbital part of lacrimal gland

Greater wing of sphenoid bone

Lat. rectus m.

Lat. rectus m.

Sup. rectus m.

Sup. rectus m.

Common tendinous ring [anulus tendineus communis]

Levator palpebrae superioris m.

Sup. oblique m.

Optic chiasm

Optic n. (II)

Fig. 175. The extrinsic ocular muscles, their position and direction within the orbital cavity. On the right side the levator palpebrae superioris muscle has been resected and the orbital fat has been removed. Cranial view.

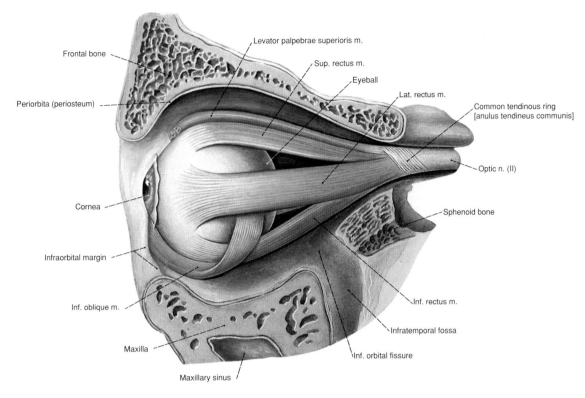

Frontal bone

Levator palpebrae superioris m.

Sup. rectus m.

Eyeball

Lat. rectus m.

Periorbita (periosteum)

Common tendinous ring
[anulus tendineus communis]

Optic n. (II)

Cornea

Sphenoid bone

Infraorbital margin

Inf. oblique m.

Inf. rectus m.

Maxilla

Infratemporal fossa

Maxillary sinus

Inf. orbital fissure

Fig. 176. The ocular muscles, left lateral view. The lateral wall of the left orbit has been removed.

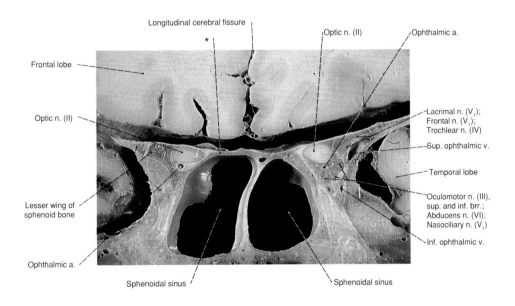

Longitudinal cerebral fissure

*

Optic n. (II)

Ophthalmic a.

Frontal lobe

Lacrimal n. (V₁);
Frontal n. (V₁);
Trochlear n. (IV)

Optic n. (II)

Sup. ophthalmic v.

Temporal lobe

Lesser wing of
sphenoid bone

Oculomotor n. (III),
sup. and inf. brr.;
Abducens n. (VI);
Nasociliary n. (V₁)

Inf. ophthalmic v.

Ophthalmic a.

Sphenoidal sinus

Sphenoidal sinus

Fig. 177. An oblique frontal section through the apex of the orbit with the left side more posterior through the intracranial end of the optic canal and the right side through the orbital end of the canal. (From: Dr. R. UNSÖLD, University Eye Clinic, Düsseldorf.)

* Clinically: Sphenoidal plane

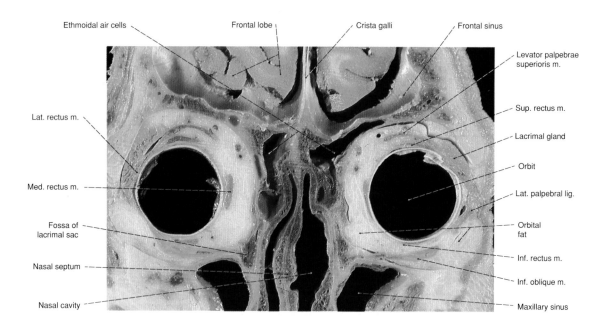

Ethmoidal air cells

Frontal lobe

Crista galli

Frontal sinus

Levator palpebrae
superioris m.

Lat. rectus m.

Sup. rectus m.

Lacrimal gland

Med. rectus m.

Orbit

Lat. palpebral lig.

Fossa of
lacrimal sac

Orbital
fat

Nasal septum

Inf. rectus m.

Inf. oblique m.

Nasal cavity

Maxillary sinus

Fig. 178a. A frontal section through the anterior cranial fossa, orbit and nasal cavity at the level of the eyeball.

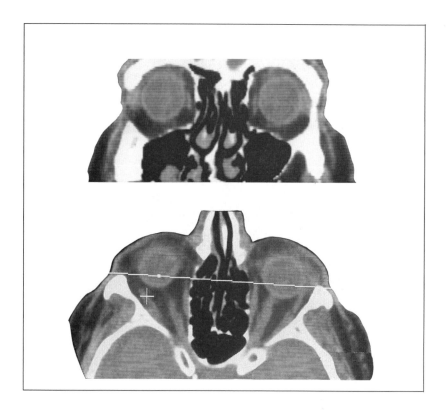

Fig. 178b. Top: computed tomographic reconstruction of the section shown in Fig. 178a; bottom: transverse CT-scan through the middle of the orbit indicating the level of the reconstruction pictured above. (From: Dr. R. Unsöld, University Eye Clinic, Düsseldorf.)

Refer to text on page 174, Volume 2 for an explanation of computed axial tomography.

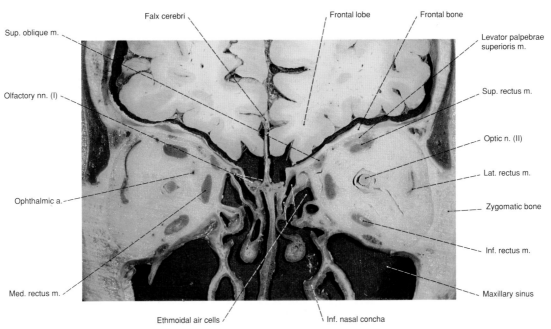

Falx cerebri

Frontal lobe

Frontal bone

Sup. oblique m.

Levator palpebrae superioris m.

Olfactory nn. (I)

Sup. rectus m.

Optic n. (II)

Lat. rectus m.

Ophthalmic a.

Zygomatic bone

Inf. rectus m.

Med. rectus m.

Maxillary sinus

Ethmoidal air cells

Inf. nasal concha

Fig. 178c. Retrobulbar frontal section through the anterior cranial fossa, orbit and nasal cavity.

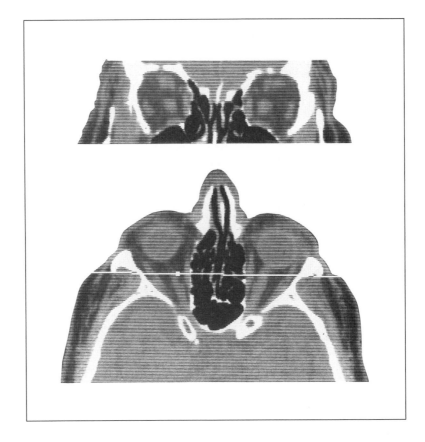

Fig. 178d. Top: computed tomographic reconstruction of the section shown in Fig. 178c; bottom: transverse CT-scan through the retrobulbar orbit indicating the level of the reconstruction pictured above. (From: Dr. R. UNSÖLD, University Eye Clinic, Düsseldorf.)

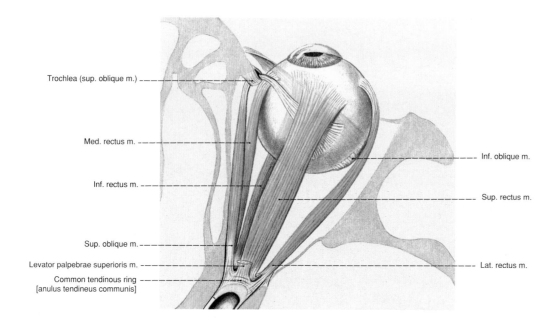

Fig. 179. Diagram of the muscles of the right eye viewed from above. Note the angle between the axis of the eyeball and the direction of the muscles. (From PERNKOPF Anatomy: Atlas of Topographic and Applied Human Anatomy, Vol. 1, 3rd edition W. PLATZER, Ed., Urban & Schwarzenberg, Baltimore–Munich, 1989.)

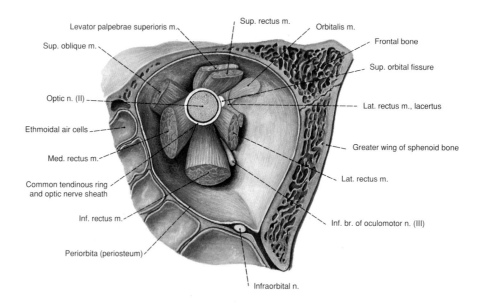

Fig. 180. Origins of the ocular muscles from the common tendinous ring. Anterior view of a frontal section through the apex of the left orbit. The optic nerve (II) has been sectioned, stumps of the encircling ocular muscles are retained; of the intraorbital nerves, only the inferior branch of the oculomotor nerve (III) is visible.

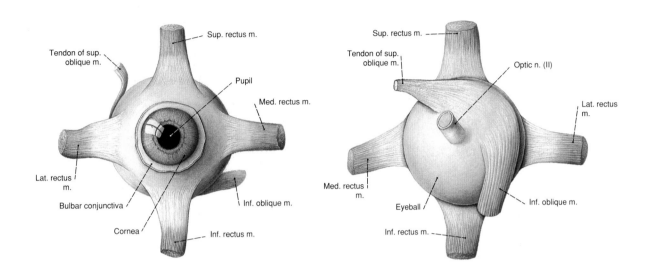

Fig. 181. Right eyeball with insertions of the ocular muscles, viewed from the front.

Fig. 182. Right eyeball with insertions of the ocular muscles, viewed from behind and below.

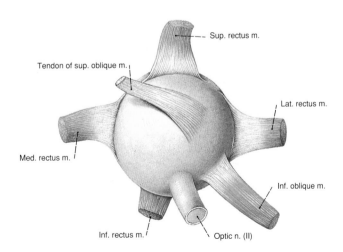

Fig. 183. Right eyeball with insertions of the ocular muscles, viewed from behind and above. The inferior oblique muscle has been deflected from the eyeball.

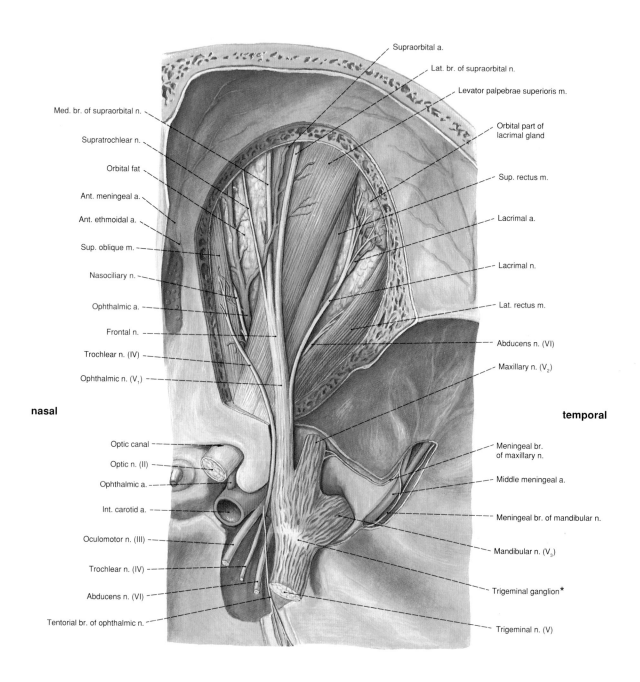

Supraorbital a.

Lat. br. of supraorbital n.

Levator palpebrae superioris m.

Orbital part of lacrimal gland

Sup. rectus m.

Lacrimal a.

Lacrimal n.

Lat. rectus m.

Abducens n. (VI)

Maxillary n. (V$_2$)

Meningeal br. of maxillary n.

Middle meningeal a.

Meningeal br. of mandibular n.

Mandibular n. (V$_3$)

Trigeminal ganglion*

Trigeminal n. (V)

Med. br. of supraorbital n.

Supratrochlear n.

Orbital fat

Ant. meningeal a.

Ant. ethmoidal a.

Sup. oblique m.

Nasociliary n.

Ophthalmic a.

Frontal n.

Trochlear n. (IV)

Ophthalmic n. (V$_1$)

nasal

temporal

Optic canal

Optic n. (II)

Ophthalmic a.

Int. carotid a.

Oculomotor n. (III)

Trochlear n. (IV)

Abducens n. (VI)

Tentorial br. of ophthalmic n.

Fig. 184.   Contents of the right orbit after removal of the
orbital plate of the frontal bone. The dura mater has been split
along the middle meningeal artery and has been removed over
the trigeminal ganglion and over the nerves to the ocular muscles. Superior view, superficial layer.

* Semilunar or GASSERIAN ganglion

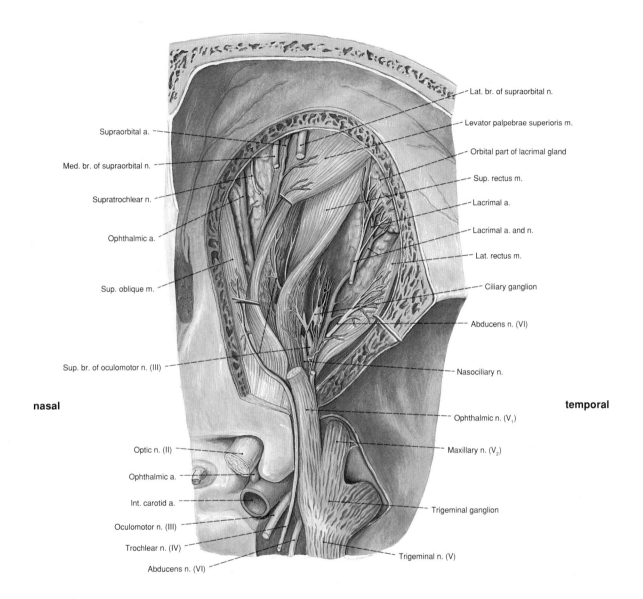

Supraorbital a.

Med. br. of supraorbital n.

Supratrochlear n.

Ophthalmic a.

Sup. oblique m.

Sup. br. of oculomotor n. (III)

nasal

Optic n. (II)

Ophthalmic a.

Int. carotid a.

Oculomotor n. (III)

Trochlear n. (IV)

Abducens n. (VI)

Lat. br. of supraorbital n.

Levator palpebrae superioris m.

Orbital part of lacrimal gland

Sup. rectus m.

Lacrimal a.

Lacrimal a. and n.

Lat. rectus m.

Ciliary ganglion

Abducens n. (VI)

Nasociliary n.

temporal

Ophthalmic n. (V$_1$)

Maxillary n. (V$_2$)

Trigeminal ganglion

Trigeminal n. (V)

Fig. 185. Contents of the right orbit after removal of the orbital plate of the frontal bone. Specimen similar to the one in Fig. 184. The ophthalmic division of the trigeminal nerve (V) and its lacrimal, supratrochlear and frontal branches have been cut. The levator palpebrae superioris and superior rectus muscles have been retracted medially. Note the ciliary ganglion between the optic nerve (II) and the lateral rectus muscle. Superior view.

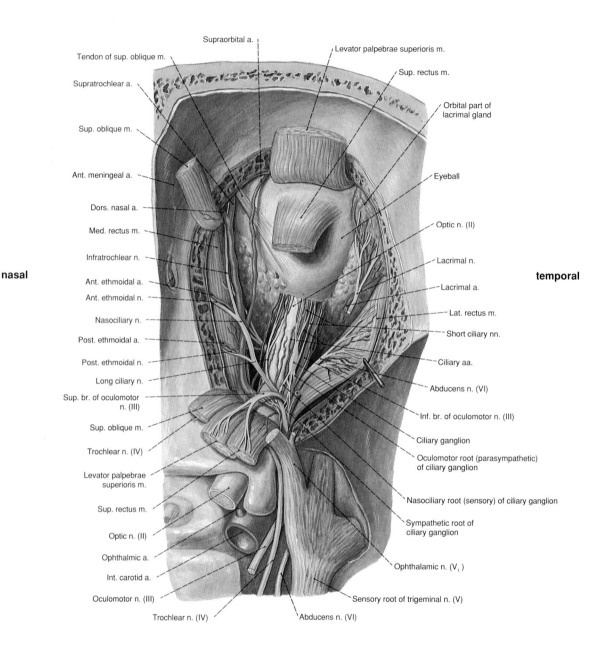

Supraorbital a.

Tendon of sup. oblique m.

Supratrochlear a.

Sup. oblique m.

Ant. meningeal a.

Dors. nasal a.

Med. rectus m.

Infratrochlear n.

Ant. ethmoidal a.
Ant. ethmoidal n.

Nasociliary n.

Post. ethmoidal a.

Post. ethmoidal n.

Long ciliary n.

Sup. br. of oculomotor
n. (III)

Sup. oblique m.

Trochlear n. (IV)

Levator palpebrae
superioris m.

Sup. rectus m.

Optic n. (II)

Ophthalmic a.

Int. carotid a.

Oculomotor n. (III)

Trochlear n. (IV)

nasal

Levator palpebrae superioris m.

Sup. rectus m.

Orbital part of
lacrimal gland

Eyeball

Optic n. (II)

Lacrimal n.

Lacrimal a.

Lat. rectus m.

Short ciliary nn.

Ciliary aa.

Abducens n. (VI)

Inf. br. of oculomotor n. (III)

Ciliary ganglion

Oculomotor root (parasympathetic)
of ciliary ganglion

Nasociliary root (sensory) of ciliary ganglion

Sympathetic root of
ciliary ganglion

Ophthalmic n. (V₁)

Sensory root of trigeminal n. (V)

Abducens n. (VI)

temporal

Fig. 186.   Contents of the right orbit after removal of the orbital plate of the frontal bone. Specimen similar to those in Figs. 184 and 185. Severed additionally were: levator palpebrae superioris muscle, superior rectus muscle and superior oblique muscle. Note the branches to and from the ciliary ganglion. Superior view.

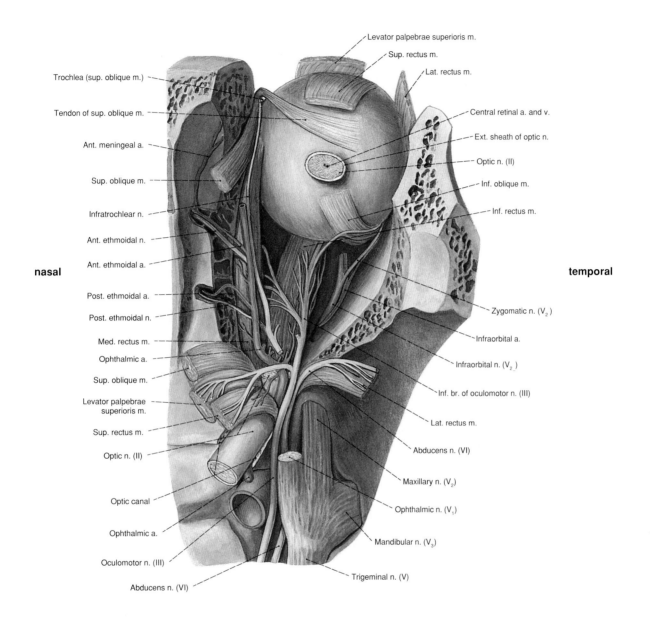

Levator palpebrae superioris m.

Sup. rectus m.

Lat. rectus m.

Trochlea (sup. oblique m.)

Tendon of sup. oblique m.

Central retinal a. and v.

Ant. meningeal a.

Ext. sheath of optic n.

Sup. oblique m.

Optic n. (II)

Inf. oblique m.

Infratrochlear n.

Inf. rectus m.

Ant. ethmoidal n.

Ant. ethmoidal a.

nasal

temporal

Post. ethmoidal a.

Post. ethmoidal n.

Zygomatic n. (V$_2$)

Med. rectus m.

Infraorbital a.

Ophthalmic a.

Infraorbital n. (V$_2$)

Sup. oblique m.

Inf. br. of oculomotor n. (III)

Levator palpebrae
superioris m.

Lat. rectus m.

Sup. rectus m.

Abducens n. (VI)

Optic n. (II)

Maxillary n. (V$_2$)

Optic canal

Ophthalmic n. (V$_1$)

Ophthalmic a.

Mandibular n. (V$_3$)

Oculomotor n. (III)

Trigeminal n. (V)

Abducens n. (VI)

Fig. 187. Muscles, nerves and arteries of the right orbit. The optic nerve (II) and parts of the ocular muscles have been removed. Note the course and branches of the oculomotor nerve (III). Superior view.

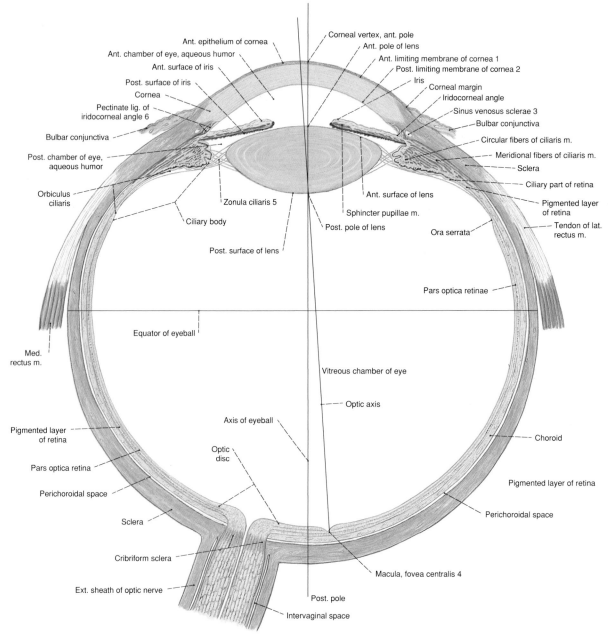

Fig. 188. Schematic horizontal section through the right eyeball at the level of the optic nerve and disc. View from above.

1 = BOWMAN's membrane
2 = membrane of DESCEMET
3 = canal of SCHLEMM
4 = yellow spot
5 = zonule of ZINN
6 = pectinate ligament or spongy iridocornealis = spongy layer at the junction of cornea and iris. The NOMINA HISTOLOGICA (2nd edition, 1980) differentiate for the iridocorneal angle
  - Pectinate ligament
  - Space of the iridocorneal angle
  - Trabecular network

Measurements of the eyeball (average values from the anatomical and ophthalmological literature):
outer diameter (axial) = 24.27 mm, inner diameter (axial) = 21.74 mm, diameter at equator = 24.32 mm, vertical diameter = 23.60 mm, radius of scleral curvature = 12.70 mm, radius of corneal curvature = 7.75 mm, depth of anterior chamber = 3 mm, sagittal axis of lens = 4 mm, diameter of lens at equator = 9–10 mm, distance of lens from retina = 16,7 mm, range of pupillary movement, left to right = 56–70 mm, horizontal diameter of cornea = 11.9 mm, vertical diameter of cornea = 11.0 mm, diameter of pupil = 1.5–8 mm.

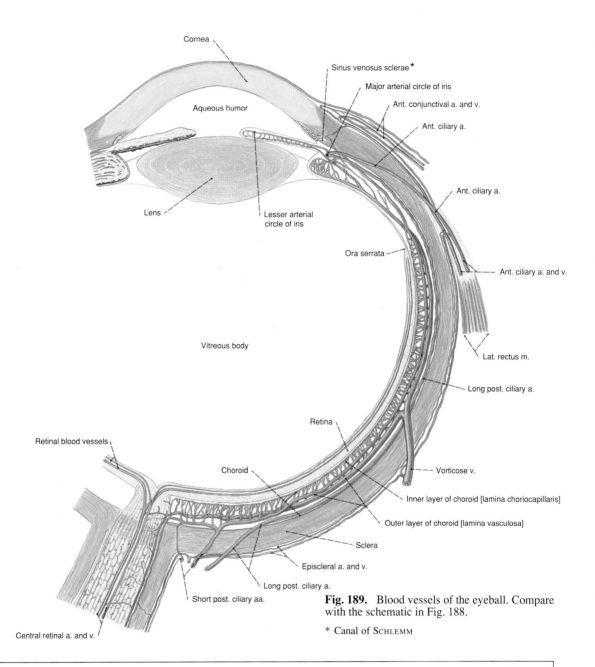

Cornea

Sinus venosus sclerae*

Major arterial circle of iris

Ant. conjunctival a. and v.

Ant. ciliary a.

Aqueous humor

Ant. ciliary a.

Lens

Lesser arterial circle of iris

Ora serrata

Ant. ciliary a. and v.

Vitreous body

Lat. rectus m.

Long post. ciliary a.

Retina

Retinal blood vessels

Vorticose v.

Choroid

Inner layer of choroid [lamina choriocapillaris]

Outer layer of choroid [lamina vasculosa]

Sclera

Episcleral a. and v.

Long post. ciliary a.

Short post. ciliary aa.

Central retinal a. and v.

Fig. 189. Blood vessels of the eyeball. Compare with the schematic in Fig. 188.

* Canal of SCHLEMM

Tunics of the Eyeball

1. Outer, fibrous tunic
a) Cornea, anterior smaller portion (1/5), pronounced curvature, translucent.
b) Sclera, whitish, posterior larger portion (4/5), lesser curvature, bluish-white in infancy, yellowish-white in senescence (yellow in icteric conditions).

2. Middle, vascular tunic
a) Iris with central round opening, the pupil.
b) Ciliary body with ciliary muscle, ciliary processes, ciliary zonule with suspensory ligament of the lens and the spatia zonularia (clinical eponym: zonule of ZINN).
c) Choroid.

3. Internal tunic, retina (inner layer of the optic cup)
a) Anterior non-neural part, from pupillary margin of iris to ora serrata.
  1. Pars iridica retinae, single layered, heavily pigmented epithelium.
  2. Pars ciliaris retinae, single layered, unpigmented epithelium.
b) Posterior neural part; stratified. Three neuron system: 1st neuron, situated toward the outside, bordering on the pigmented epithelium, neuroepithelium, rod cells, cone cells, avascularized. 2nd and 3rd neurons form the layers of the brain and carry blood vessels. Specialized areas in the posterior segment of the eyeball: macula (yellow spot) with fovea centralis (site of most acute vision) and the optic disc with porus opticus (blind spot) which marks the point of exit of the optic nerve (II) and entrance of the central retinal artery.

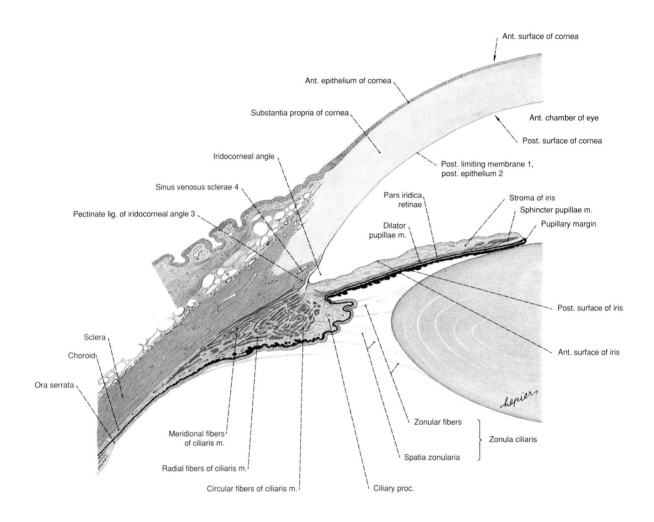

Fig. 190. Horizontal section through the anterior part of the eyeball. (From PERNKOPF Anatomy: Atlas of Topographic and Applied Human Anatomy, Vol. 1, 3rd edition, W. PLATZER, editor, Urban & Schwarzenberg, Baltimore-Munich, 1989.)

1 = Membrane of DESCEMET
2 = NOMINA ANATOMICA (5th edition, 1983) designates the posterior corneal **endothelium** as posterior **epithelium**
3 = Terminology according to NOMINA HISTOLOGICA (2nd Edition, 1980), compare with footnote 6 for Fig. 188
4 = Canal of SCHLEMM

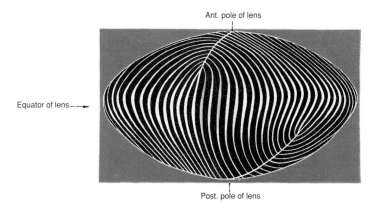

Fig. 191. Ocular lens of a newborn. Equatorial view. Course of the lens fibers. Note that they begin and end at the septa of the anterior and posterior surfaces, respectively. The free edges of the septa correspond with the lines radiating from the anterior and posterior poles toward the equator. In the adult there are more radiations than in the newborn (cf. Fig. 195).

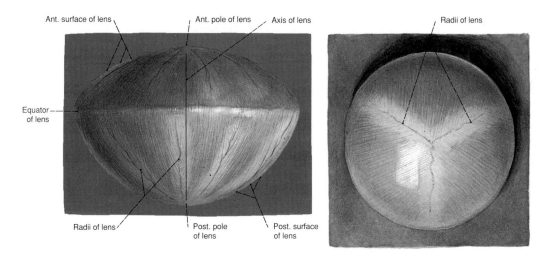

Fig. 192. Ocular lens viewed from the equator.

Fig. 193. Ocular lens of a child viewed from anterior (lens "star" with three radiations).

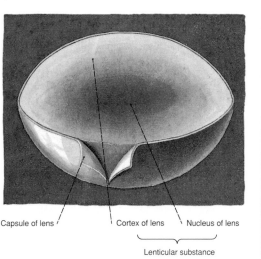

Fig. 194. Ocular lens halved through the equator, with its capsule partially elevated.

Fig. 195. Ocular lens of an adult viewed from anterior (lens "star" with multiple radiations).

Fig. 196. Suspensory apparatus of the ocular lens viewed from anterior. Cornea and iris have been removed. The anterior ends of the ciliary processes can be seen. Between them are the zonular fibers of the suspensory ligament which insert at the equator of the lens as well as at the anterior and posterior surfaces of the lens capsule. The anterior pole of the lens and its radiations lines are also visible.   ▶

* see Fig. 188

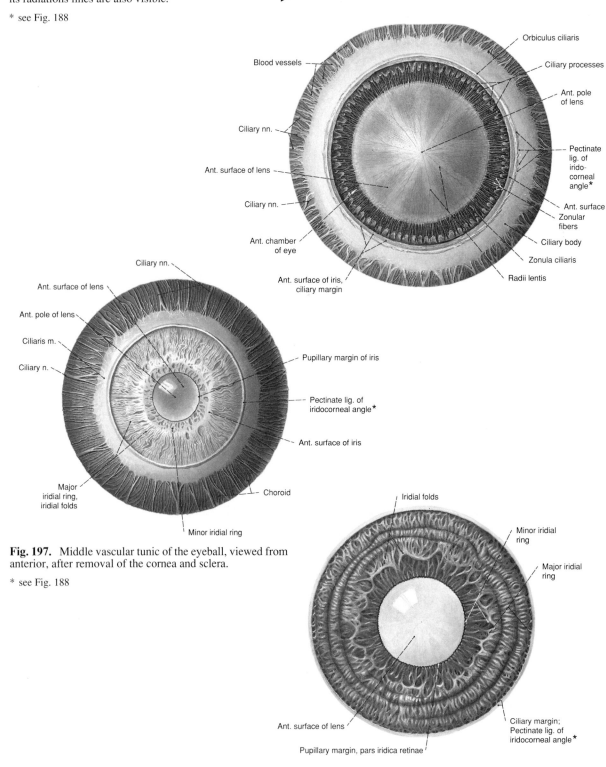

Fig. 197. Middle vascular tunic of the eyeball, viewed from anterior, after removal of the cornea and sclera.

* see Fig. 188

Fig. 198. Anterior surface of the iris.

* see Fig. 188

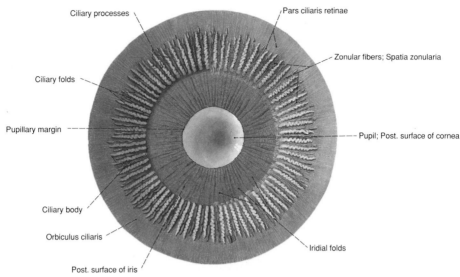

Ciliary processes

Pars ciliaris retinae

Zonular fibers; Spatia zonularia

Ciliary folds

Pupillary margin

Pupil; Post. surface of cornea

Ciliary body

Orbiculus ciliaris

Post. surface of iris

Iridial folds

Fig. 199. Posterior surface of the iris and the ciliary body after removal of the lens. The zonular fibers on the left side have been removed, and on the right side they are sectioned close to the ciliary body. The posterior surface of the cornea can be seen through the pupil.

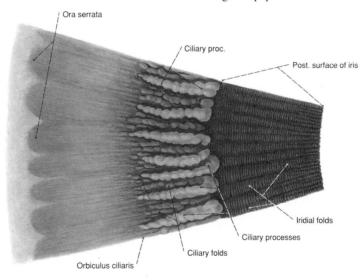

Ora serrata

Ciliary proc.

Post. surface of iris

Iridial folds

Ciliary processes

Ciliary folds

Orbiculus ciliaris

Fig. 200. Enlarged section of Fig. 199. From left to right: ora serrata, orbicularis ciliaris, ciliary processes, posterior surface of iris.

Note: The ring-shaped suspensory ligament of the lens, zonula ciliaris (clinically: zonule of ZINN) consists of meridionally arranged, very delicate, yet rigid fibrils, the zonular fibers, and their interstices, spatia zonularia. They originate from the entire stretch of the orbiculus ciliaris and from the valleys between the ciliary processes. These fine fibers are grouped into dense bundles. They course in the valleys between the ciliary processes toward the lens where, in the area of the equator, they partially cross before inserting in the lens capsule (Fig. 190).

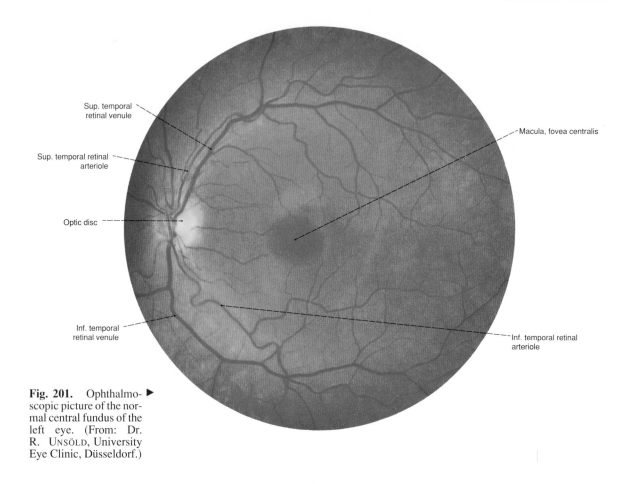

Sup. temporal
retinal venule

Sup. temporal retinal
arteriole

Optic disc

Inf. temporal
retinal venule

Macula, fovea centralis

Inf. temporal retinal
arteriole

Fig. 201. Ophthalmo- ▶
scopic picture of the normal central fundus of the
left eye. (From: Dr.
R. Unsöld, University
Eye Clinic, Düsseldorf.)

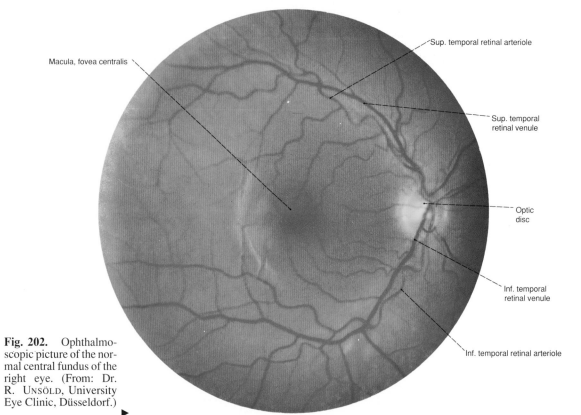

Macula, fovea centralis

Sup. temporal retinal arteriole

Sup. temporal
retinal venule

Optic
disc

Inf. temporal
retinal venule

Inf. temporal retinal arteriole

Fig. 202. Ophthalmo-
scopic picture of the normal central fundus of the
right eye. (From: Dr.
R. Unsöld, University
Eye Clinic, Düsseldorf.)

▶

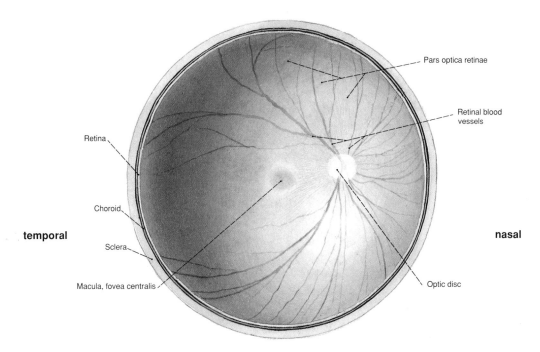

Pars optica retinae

Retinal blood vessels

Retina

Choroid

Sclera

Macula, fovea centralis

temporal

nasal

Optic disc

Fig. 203. Posterior half of an equatorially sectioned right eyeball. Anterior view of the optic disc (white "blind spot") and the fovea centralis (gray). The vitreous body has been removed.

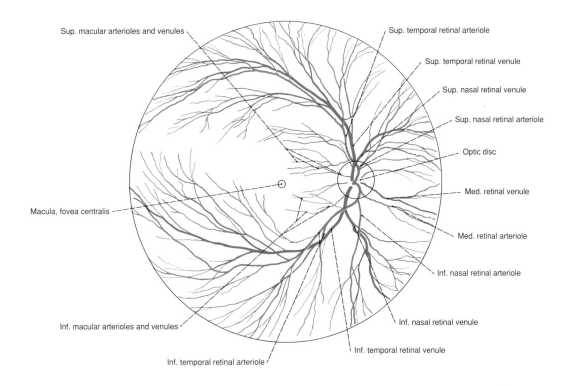

Sup. macular arterioles and venules

Sup. temporal retinal arteriole

Sup. temporal retinal venule

Sup. nasal retinal venule

Sup. nasal retinal arteriole

Optic disc

Med. retinal venule

Macula, fovea centralis

Med. retinal arteriole

Inf. nasal retinal arteriole

Inf. macular arterioles and venules

Inf. nasal retinal venule

Inf. temporal retinal venule

Inf. temporal retinal arteriole

Fig. 204. Retinal blood vessels of the right eye. Schematic view onto the fundus of the eye.

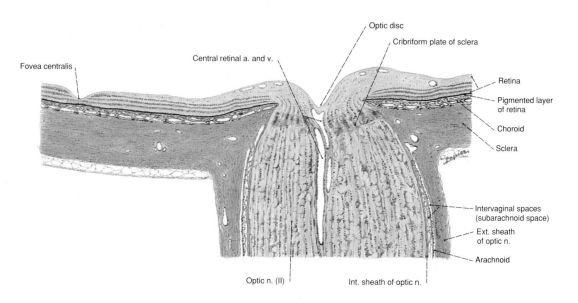

Fovea centralis

Central retinal a. and v.

Optic disc

Cribriform plate of sclera

Retina

Pigmented layer
of retina

Choroid

Sclera

Intervaginal spaces
(subarachnoid space)

Ext. sheath
of optic n.

Arachnoid

Optic n. (II)

Int. sheath of optic n.

Fig. 205. Horizontal section through the point of exit of the optic nerve (II) from the eyeball. (From PERNKOPF Anatomy: Atlas of Topographic and Applied Human Anatomy, Vol. 1, 3rd edition, W. PLATZER, Editor, Urban & Schwarzenberg, Baltimore–Munich, 1989.)

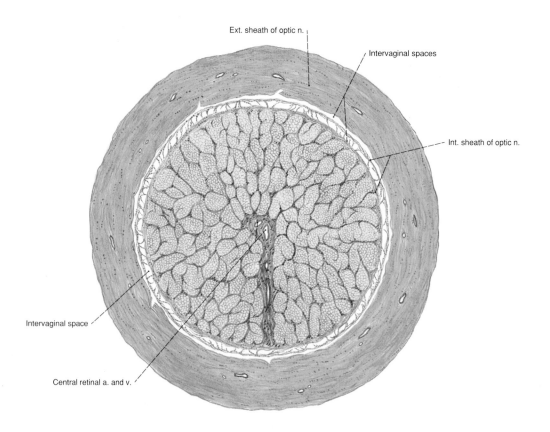

Ext. sheath of optic n.

Intervaginal spaces

Int. sheath of optic n.

Intervaginal space

Central retinal a. and v.

Fig. 206. Cross section through the optic nerve (II) near the eyeball.

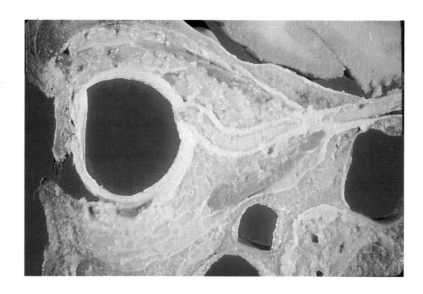

Fig. 207a. Oblique sagittal section through the eyeball and the optic nerve (II) in situ; note the tortuous course of the optic nerve.

Frontal bone

Optic disc

Bulb of eye

Upper eyelid

Lower eyelid

Inf. oblique m.

Sup. rectus m.

Frontal lobe

Optic n. (II)

Optic canal;
Lesser wing of sphenoid bone

Inf. rectus m.

Sphenoidal sinus

Maxillary sinus

Fig. 207b. The same section as in Fig. 207a with structures labeled.

Refer to the text in Volume 2, page 174 for an explanation of computed axial tomography.

Fig. 207c. Computed tomographic reconstruction corresponding to Figs. 207 a and b. (From: Dr. R. UNSÖLD, University Eye Clinic, Düsseldorf.)

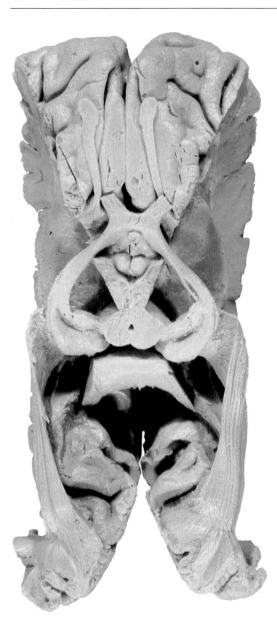

Fig. 208a. Course of the visual pathways dissected within the brain. (From: Dr. S. ZULEGER, Anatomical Institute, University of Freiburg.)

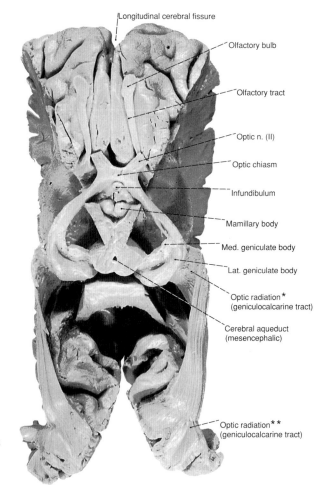

Longitudinal cerebral fissure

Olfactory bulb

Olfactory tract

Optic n. (II)

Optic chiasm

Infundibulum

Mamillary body

Med. geniculate body

Lat. geniculate body

Optic radiation *
(geniculocalcarine tract)

Cerebral aqueduct
(mesencephalic)

Optic radiation * *
(geniculocalcarine tract)

Fig. 208b. The same specimen as in Fig. 208a with structures labeled.

 * Temporal component of optic radiation
** Occipital component of optic radiation

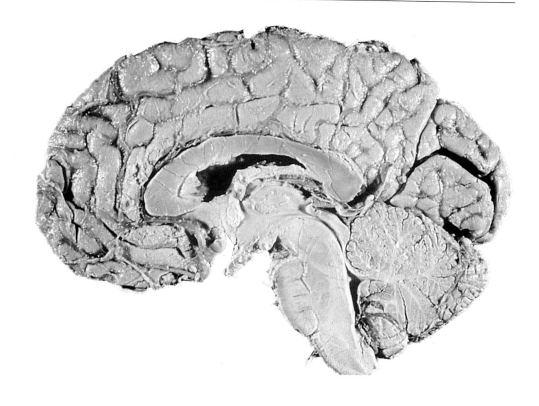

Fig. 208c. Right half of the brain, viewed from medial. The parieto-occipital sulcus separates the occipital lobe from the parietal lobe. The calcarine sulcus is artificially expanded.

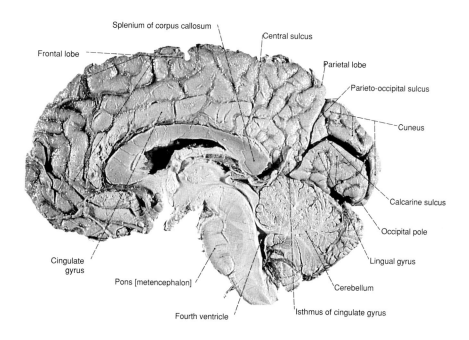

Splenium of corpus callosum

Central sulcus

Frontal lobe

Parietal lobe

Parieto-occipital sulcus

Cuneus

Calcarine sulcus

Occipital pole

Lingual gyrus

Cingulate gyrus

Pons [metencephalon]

Cerebellum

Isthmus of cingulate gyrus

Fourth ventricle

Fig. 208d. The same specimen as in Fig. 208c with structures labeled. (From: Dr. R. UNSÖLD, University Eye Clinic, Düsseldorf.)

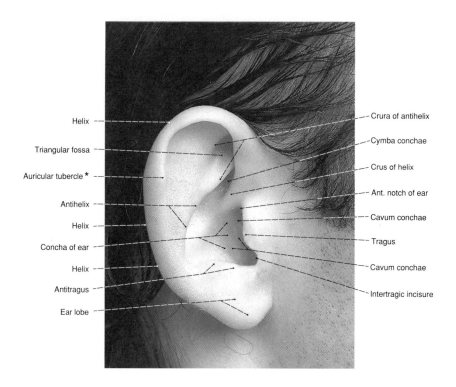

Helix
Triangular fossa
Auricular tubercle *
Antihelix
Helix
Concha of ear
Helix
Antitragus
Ear lobe

Crura of antihelix
Cymba conchae
Crus of helix
Ant. notch of ear
Cavum conchae
Tragus
Cavum conchae
Intertragic incisure

Fig. 209. The right external ear, lateral view.

* Tubercle of DARWIN

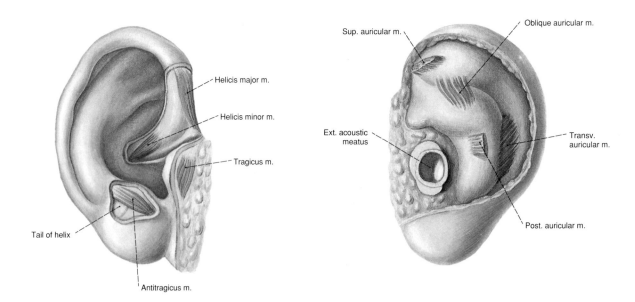

Helicis major m.
Helicis minor m.
Tragicus m.
Tail of helix
Antitragicus m.

Sup. auricular m.
Oblique auricular m.
Ext. acoustic meatus
Transv. auricular m.
Post. auricular m.

Fig. 210. Muscles of the lateral surface of the right external ear.

Fig. 211. Muscles of the medial surface of the right external ear. Ligaments with the same names may be present instead of the superior and posterior auricular muscles.

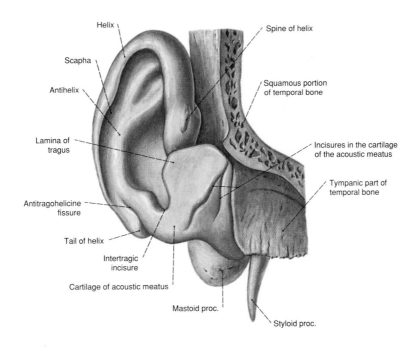

Fig. 212. The cartilage of the right external ear, viewed from the front.

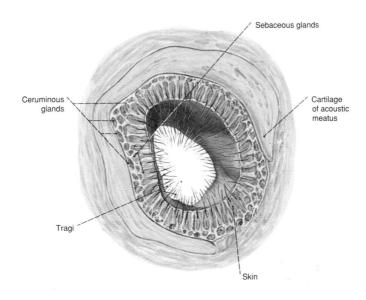

Fig. 213. Cross section through the cartilaginous portion of the external acoustic meatus.

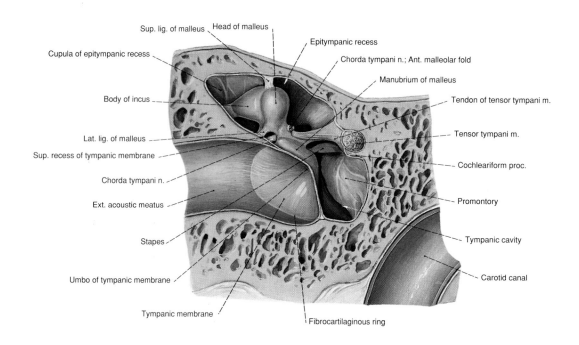

Sup. lig. of malleus · Head of malleus · Epitympanic recess · Chorda tympani n.; Ant. malleolar fold · Manubrium of malleus · Cupula of epitympanic recess · Tendon of tensor tympani m. · Body of incus · Tensor tympani m. · Lat. lig. of malleus · Cochleariform proc. · Sup. recess of tympanic membrane · Chorda tympani n. · Promontory · Ext. acoustic meatus · Stapes · Tympanic cavity · Umbo of tympanic membrane · Carotid canal · Tympanic membrane · Fibrocartilaginous ring

Boundaries of the Tympanic Cavity and Their Clinical Significance

Name	Constituents	Neighboring organs	Pecularities	Clinical complications
Roof (tegmental wall)	Epitympanic recess, tegmen tympani of temporal bone, petrosquamous suture	Middle cranial fossa, meninges, temporal lobe	Vascular channels in the roof and suture: route for infections	Meningitis, abscess of temporal lobe
Floor (jugular wall)	Styloid prominence	Jugular fossa, superior bulb of jugular vein	Variable form and size of air cells, bony lamina may be partially absent	Septic thrombosis of internal jugular vein (pyemia)
Medial (labyrinthine wall)	Promontory, fenestrae cochleae and vestibuli, prominence of the facial canal	Membranous labyrinth, facial nerve (VII) [nervus inter mediofacialis]		Infections of the labyrinth (deafness), facial paresis
Lateral (membranous wall)	Tympanic membrane, manubrium mallei (chorda tympani)	External acoustic meatus, temporo-mandibular joint		Perforation of tympanic membrane (e.g., by careless cleaning)
Posterior (mastoid wall)	Mastoid antrum, mastoid air cells, prominence of lat. semicircular canal, prominence of facial canal	Facial nerve (VII) [nervus intermedio-facialis], sigmoid sinus, post. cranial fossa, cerebellum	Variable pneuma-tization of the mastoid process	Mastoiditis, sinus thrombosis, menin-gitis, cerebellar abs-cess, facial paresis
Anterior (carotid wall)	Tympanic opening of auditory tube, musculotubular canal	Carotid canal, cavernous sinus, abducens nerve (VI), trigeminal ganglion	Apical pneumatization of the pyramid	Auditory tube as route for infection of apical air cells, abducens paresis, otitis of middle ear

Note:	The tympanic cavity can be divided into three levels* (Figs. 214, 226):	**Middle level:**	area of the tympanic membrane and the tympanic opening of the auditory tube.
Upper level:	epitympanic recess with mastoid an-trum extending to the neck of the malleus and the anterior and posterior malleolar folds.	**Lower level:**	extending to the floor or jugular wall of the tympanic cavity.
		* (clinically: epi-, meso-, hypotympanum)	

◄ **Fig. 214.** Frontal section through the right external acoustic meatus, the tympanic membrane and the tympanic cavity. Note the narrowest part of the tympanic cavity between the umbo of the tympanic membrane and the promontory. The cupula of the epitympanic recess is the apex of the tympanic cavity. Surgical access to the upper level of the tympanic cavity from the external acoustic meatus is by way of that portion of the temporal squama which, as the "wall of Loguette", reaches to the flaccid portion of the tympanic membrane (cf. Fig. 226).

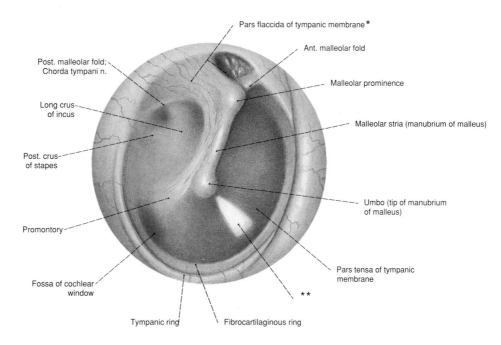

Fig. 215. Right tympanic membrane as seen through an otoscope in a living person. Magnification approximately 6-fold.

* SHRAPNELL's membrane

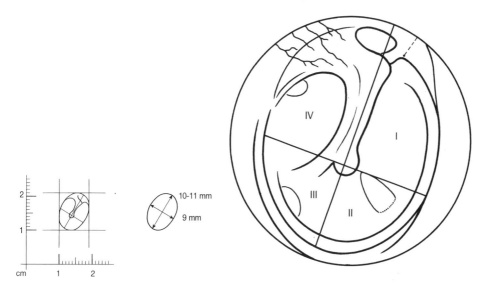

Natural size of the tympanic membrane in an adult

Fig. 216. Labeled line diagram of Fig. 215. For practical reasons, the surface of the tympanic membrane is divided into quadrants (I-IV). In the otoscopic picture of the normal tympanic membrane a glistening triangular reflection of light occurs in front of the umbo in the region of quadrant II **.

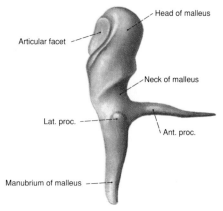

Fig. 217. Right malleus, lateral view.

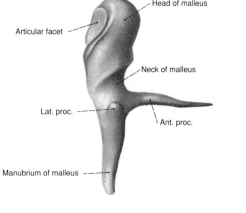

Fig. 218. Right malleus, anterior view.

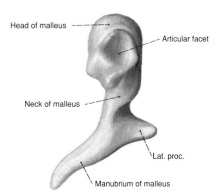

Fig. 219. Right malleus, dorsal view.

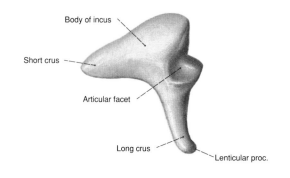

Fig. 220. Right incus, lateral view.

Fig. 221. Right incus, medial view.

Fig. 223. Right stapes, cranial view.

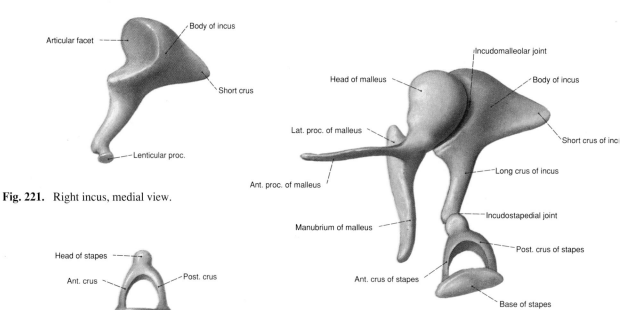

Fig. 222. The auditory ossicles of the right ear in their natural relationship. Medial, cranial view.

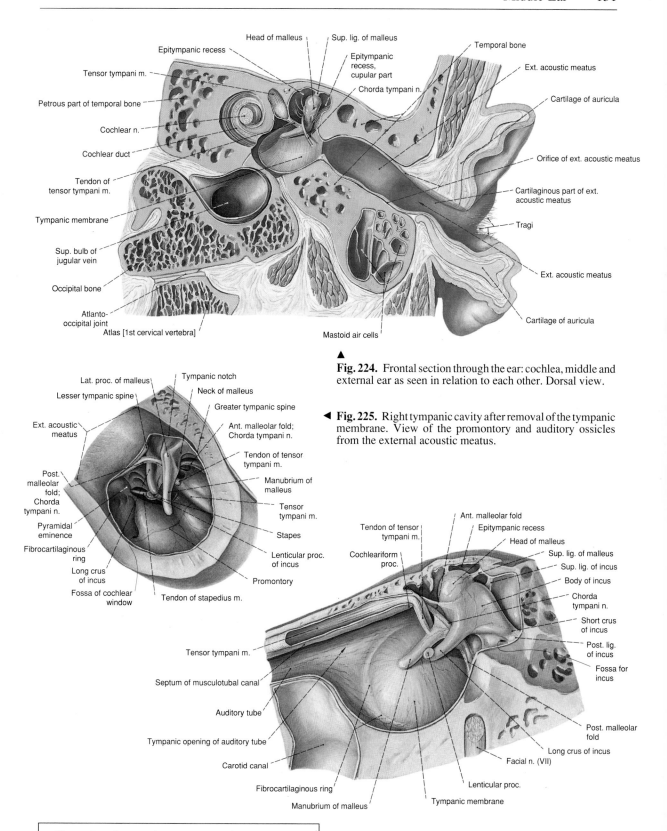

Head of malleus
Sup. lig. of malleus
Epitympanic recess
Tensor tympani m.
Epitympanic recess, cupular part
Temporal bone
Chorda tympani n.
Ext. acoustic meatus
Petrous part of temporal bone
Cochlear n.
Cochlear duct
Cartilage of auricula
Tendon of tensor tympani m.
Orifice of ext. acoustic meatus
Tympanic membrane
Cartilaginous part of ext. acoustic meatus
Sup. bulb of jugular vein
Tragi
Occipital bone
Ext. acoustic meatus
Atlanto-occipital joint
Atlas [1st cervical vertebra]
Cartilage of auricula
Mastoid air cells

▲
Fig. 224. Frontal section through the ear: cochlea, middle and external ear as seen in relation to each other. Dorsal view.

◀ **Fig. 225.** Right tympanic cavity after removal of the tympanic membrane. View of the promontory and auditory ossicles from the external acoustic meatus.

Lat. proc. of malleus
Tympanic notch
Lesser tympanic spine
Neck of malleus
Greater tympanic spine
Ext. acoustic meatus
Ant. malleolar fold; Chorda tympani n.
Tendon of tensor tympani m.
Manubrium of malleus
Post. malleolar fold; Chorda tympani n.
Tensor tympani m.
Pyramidal eminence
Stapes
Fibrocartilaginous ring
Lenticular proc. of incus
Long crus of incus
Promontory
Fossa of cochlear window
Tendon of stapedius m.

Ant. malleolar fold
Epitympanic recess
Tendon of tensor tympani m.
Head of malleus
Cochleariform proc.
Sup. lig. of malleus
Sup. lig. of incus
Body of incus
Chorda tympani n.
Short crus of incus
Tensor tympani m.
Post. lig. of incus
Septum of musculotubal canal
Fossa for incus
Auditory tube
Tympanic opening of auditory tube
Post. malleolar fold
Carotid canal
Long crus of incus
Facial n. (VII)
Fibrocartilaginous ring
Lenticular proc.
Manubrium of malleus
Tympanic membrane

Note: Curtailment of the mobility of the chain of auditory ossicles leads to hearing impairment (conductive deafness). In stapedial ankylosis, the mobility of the base of the stapes within the oval window is impaired or completely blocked by newly formed bone (otosclerosis).

Fig. 226. Lateral (membranous) wall of the right tympanic cavity, medial view. Removal of most of the septum of the musculotubal canal up to the cochleariform process has exposed the tensor tympani muscle. Note the insertion of its tendon into the manubrium of the malleus. The dense fascial cover of the tensor tympani muscle has been partially retained.

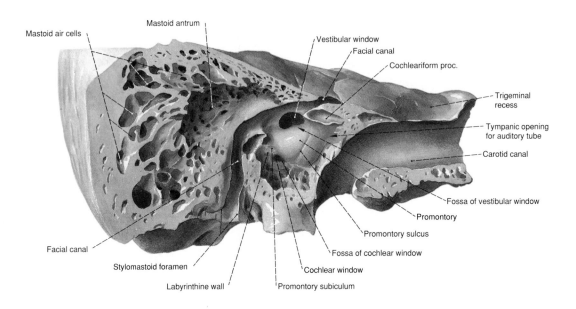

Mastoid antrum
Mastoid air cells
Vestibular window
Facial canal
Cochleariform proc.
Trigeminal recess
Tympanic opening for auditory tube
Carotid canal
Fossa of vestibular window
Promontory
Promontory sulcus
Fossa of cochlear window
Cochlear window
Promontory subiculum
Labyrinthine wall
Stylomastoid foramen
Facial canal

Fig. 227. Parasagittal section through the petrous part of the right temporal bone. Anterolateral view. In the center are the labyrinthine (medial) wall and the vestibular (oval) window. The facial and the carotid canals have been opened. Also visible are the mastoid air cells as they communicate via the mastoid antrum with the epitympanic recess of the tympanic cavity. This section is deeper than the dissection shown in Fig. 228.

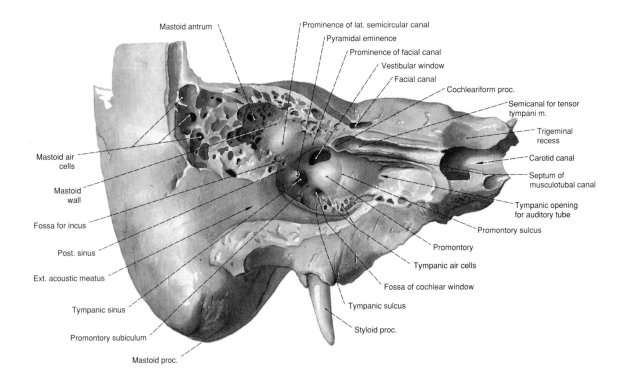

Mastoid antrum
Prominence of lat. semicircular canal
Pyramidal eminence
Prominence of facial canal
Vestibular window
Facial canal
Cochleariform proc.
Semicanal for tensor tympani m.
Trigeminal recess
Carotid canal
Septum of musculotubal canal
Tympanic opening for auditory tube
Promontory sulcus
Promontory
Tympanic air cells
Fossa of cochlear window
Tympanic sulcus
Styloid proc.
Mastoid air cells
Mastoid wall
Fossa for incus
Post. sinus
Ext. acoustic meatus
Tympanic sinus
Promontory subiculum
Mastoid proc.

Fig. 228. Right tympanic cavity after removal of the lateral wall and the adjacent portions of the anterior and superior walls. Anterolateral view.

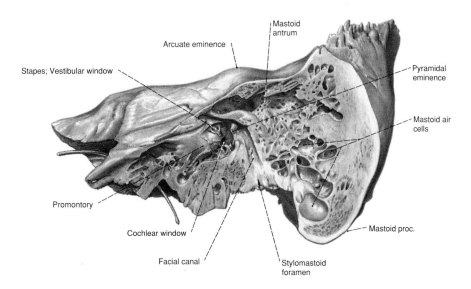

Mastoid antrum

Arcuate eminence

Stapes; Vestibular window

Pyramidal eminence

Mastoid air cells

Promontory

Cochlear window

Facial canal

Stylomastoid foramen

Mastoid proc.

Fig. 229. Parasagittal section through the petrous part of the left temporal bone. Anterolateral view of half of the labyrinthine (medial) wall and promontory. A probe is in the carotid canal. Note the pneumatization of the mastoid process and surrounding bone.

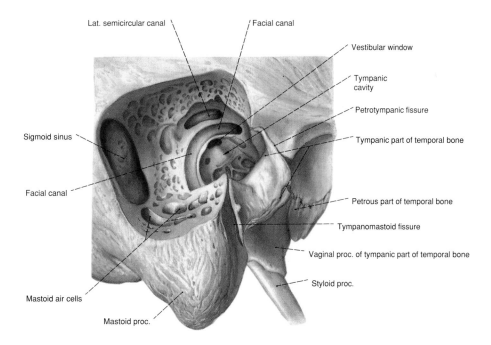

Lat. semicircular canal

Facial canal

Vestibular window

Tympanic cavity

Petrotympanic fissure

Sigmoid sinus

Tympanic part of temporal bone

Facial canal

Petrous part of temporal bone

Tympanomastoid fissure

Vaginal proc. of tympanic part of temporal bone

Styloid proc.

Mastoid air cells

Mastoid proc.

Fig. 230. The right temporal bone opened from the outside. Lateral view demonstrating the medial wall of the tympanic cavity and the relative positions of the sigmoid sinus, the facial canal and the organ of hearing and equilibration (vestibulocochlear organ).

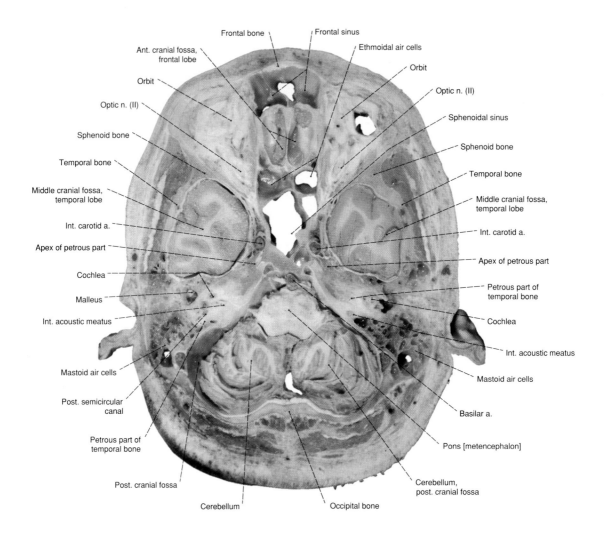

Fig. 231. Horizontal section through the head and brain of a 26-year-old woman parallel to and about 1 cm cranial to the canthomeatal plane (plane through the canthus of the eye and the midpoint of the external acoustic meatus). Postmortem injection of red dye into the cranial arteries. (From: Dr. G. RILLING, Institute of Anatomy and Special Embryology, Fribourg, Switzerland.)

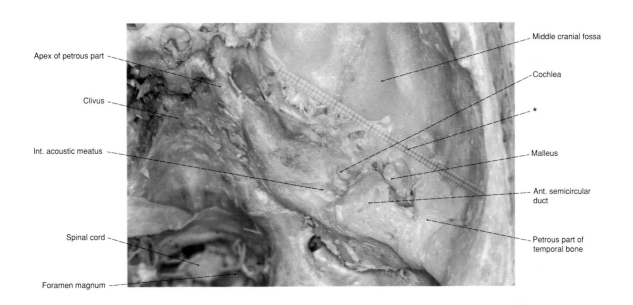

Apex of petrous part

Clivus

Int. acoustic meatus

Spinal cord

Foramen magnum

Middle cranial fossa

Cochlea

*

Malleus

Ant. semicircular duct

Petrous part of temporal bone

Fig. 232. Dissection of the petrous part of the right temporal bone, viewed from above. (From: Dr. G. RILLING, Institute of Anatomy and Special Embryology, Fribourg, Switzerland.)

* Millimeter scale

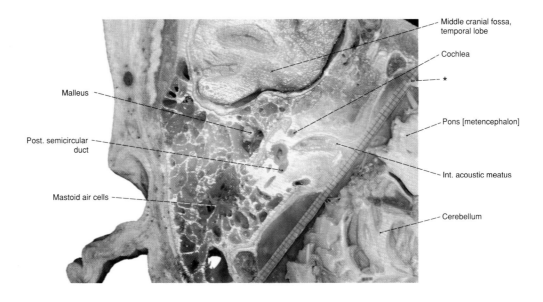

Malleus

Post. semicircular duct

Mastoid air cells

Middle cranial fossa, temporal lobe

Cochlea

*

Pons [metencephalon]

Int. acoustic meatus

Cerebellum

Fig. 233. Enlargement of a portion of Fig. 231: View of the opened petrous portion of the left temporal bone. (From: Dr. G. RILLING, Institute of Anatomy and Special Embryology, Fribourg, Switzerland.)

* Millimeter scale

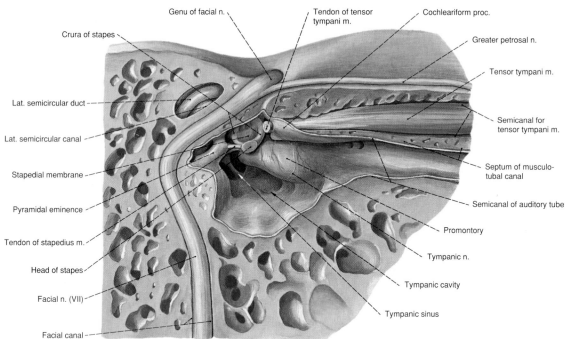

Genu of facial n.
Tendon of tensor tympani m.
Cochleariform proc.
Crura of stapes
Greater petrosal n.
Tensor tympani m.
Lat. semicircular duct
Lat. semicircular canal
Semicanal for tensor tympani m.
Stapedial membrane
Septum of musculo-tubal canal
Pyramidal eminence
Semicanal of auditory tube
Tendon of stapedius m.
Promontory
Tympanic n.
Head of stapes
Tympanic cavity
Facial n. (VII)
Tympanic sinus
Facial canal

Fig. 234. Medial wall of the right tympanic cavity, lateral view, with stapes and the course of the middle and the lower segments of the facial nerve (VII) as they pass close to the middle ear. The tympanic cavity has been divided by a parasagittal section of the petrous part of the temporal bone; the lateral half of the tympanic membrane, malleus and incus have been removed. The tendon of the tensor tympani muscle has been cut at the cochleariform process, the musculotubal canal has been opened and the facial canal exposed from its hiatus to the region of the stylomastoid foramen.

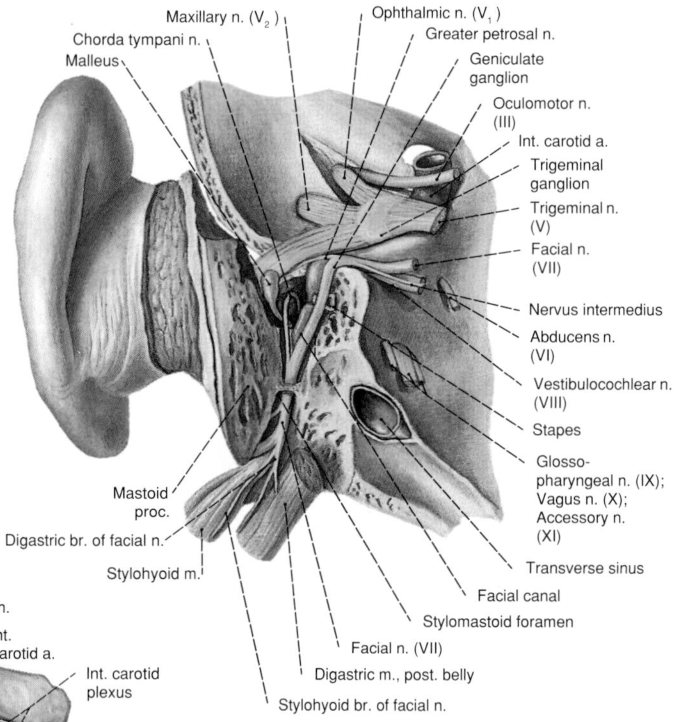

Maxillary n. (V₂)
Ophthalmic n. (V₁)
Chorda tympani n.
Greater petrosal n.
Malleus
Geniculate ganglion
Oculomotor n. (III)
Int. carotid a.
Trigeminal ganglion
Trigeminal n. (V)
Facial n. (VII)
Nervus intermedius
Abducens n. (VI)
Vestibulocochlear n. (VIII)
Stapes
Glosso-pharyngeal n. (IX); Vagus n. (X); Accessory n. (XI)
Mastoid proc.
Digastric br. of facial n.
Stylohyoid m.
Transverse sinus
Facial canal
Stylomastoid foramen
Facial n. (VII)
Digastric m., post. belly
Stylohyoid br. of facial n.

Fig. 235

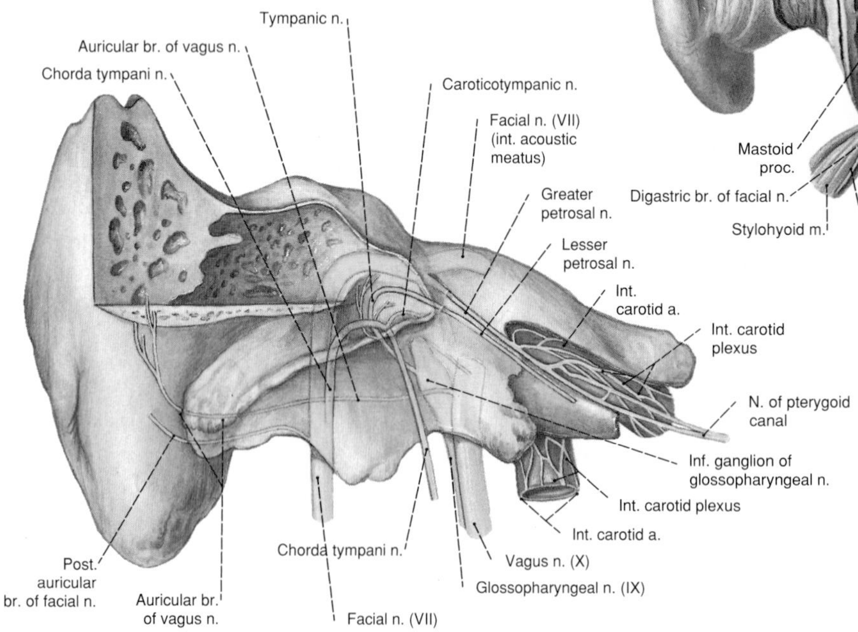

Tympanic n.
Auricular br. of vagus n.
Chorda tympani n.
Caroticotympanic n.
Facial n. (VII) (int. acoustic meatus)
Greater petrosal n.
Lesser petrosal n.
Int. carotid a.
Int. carotid plexus
N. of pterygoid canal
Inf. ganglion of glossopharyngeal n.
Int. carotid plexus
Int. carotid a.
Vagus n. (X)
Glossopharyngeal n. (IX)
Chorda tympani n.
Facial n. (VII)
Post. auricular br. of facial n.
Auricular br. of vagus n.

Fig. 236. The intracranial course of the facial nerve (VII) and its connections schematically projected onto the right temporal bone.

Fig. 235. The intracranial course of the facial nerve (VII), viewed from behind. The facial canal and tympanic cavity have been opened from the mastoid process.

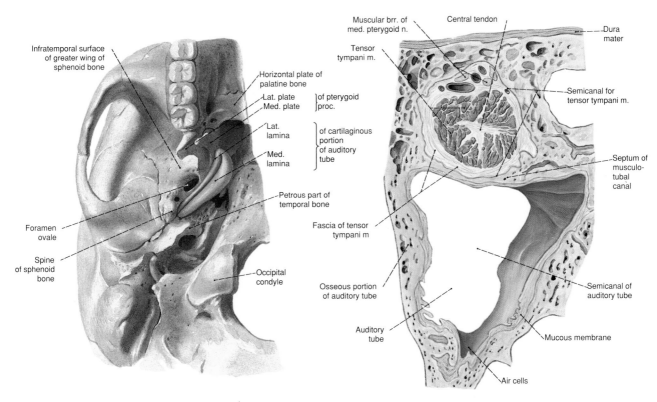

Fig. 237. Cartilage of the right auditory tube (blue) in its normal position on the external aspect of the base of the skull.

Fig. 238. Transverse section through the musculotubal canal showing the osseous portion of the auditory tube and the tensor tympani muscle (magnification approximately 15-fold).

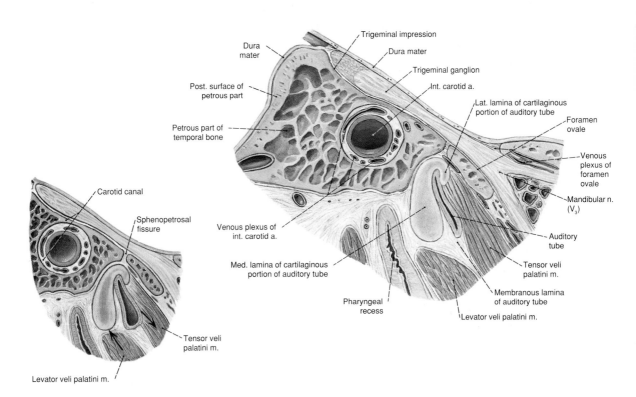

Fig. 239. Transverse section through the cartilaginous portion of the left auditory tube, the levator veli palatini muscle and the insertion of the tensor veli palatini muscle. The tube is open.

Fig. 240. Transverse section through the cartilaginous portion of the left auditory tube near its pharyngeal opening. The tube is closed.

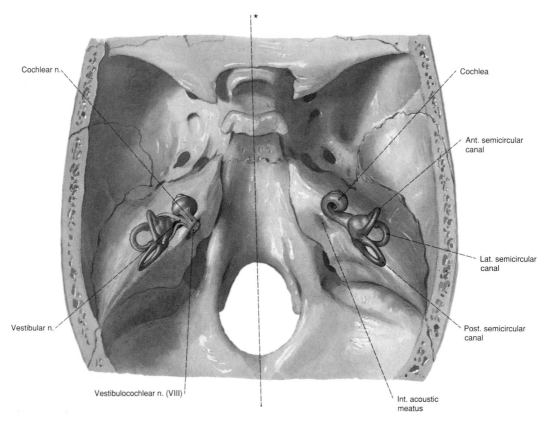

Cochlear n.

Cochlea

Ant. semicircular canal

Lat. semicircular canal

Vestibular n.

Post. semicircular canal

Vestibulocochlear n. (VIII)

Int. acoustic meatus

Fig. 241. Phantom projection of the osseous labyrinths of both sides onto the internal aspect of the base of the skull with the semicircular canals and the cochlea in their natural position. On the left the nerves are also depicted. Note the oblique orientation of the cochlear axis from posteromedially above to anterolaterally below. Its apex (cupula) is directed anterolaterally and inferiorly.
* Median line

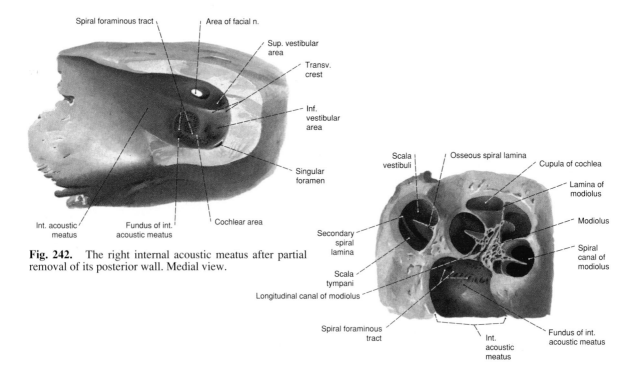

Spiral foraminous tract

Area of facial n.

Sup. vestibular area

Transv. crest

Inf. vestibular area

Singular foramen

Int. acoustic meatus

Fundus of int. acoustic meatus

Cochlear area

Fig. 242. The right internal acoustic meatus after partial removal of its posterior wall. Medial view.

Scala vestibuli

Osseous spiral lamina

Cupula of cochlea

Lamina of modiolus

Modiolus

Secondary spiral lamina

Spiral canal of modiolus

Scala tympani

Longitudinal canal of modiolus

Spiral foraminous tract

Int. acoustic meatus

Fundus of int. acoustic meatus

Fig. 243. The left bony cochlea sectioned along the axis of the modiolus.

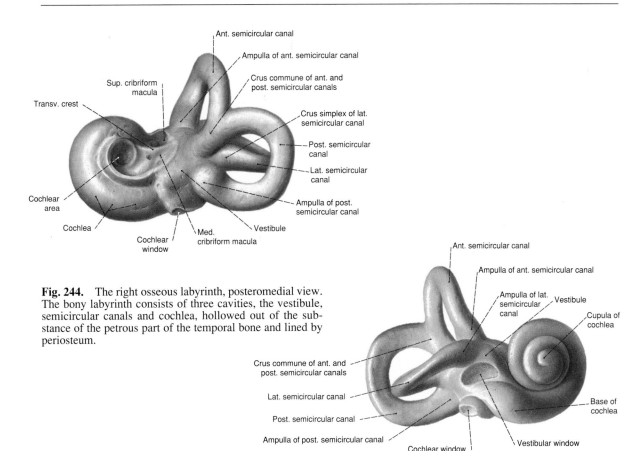

Fig. 244. The right osseous labyrinth, posteromedial view. The bony labyrinth consists of three cavities, the vestibule, semicircular canals and cochlea, hollowed out of the substance of the petrous part of the temporal bone and lined by periosteum.

Fig. 245. The osseous labyrinth hollowed out of the petrous part of the right temporal bone, anterolateral view.

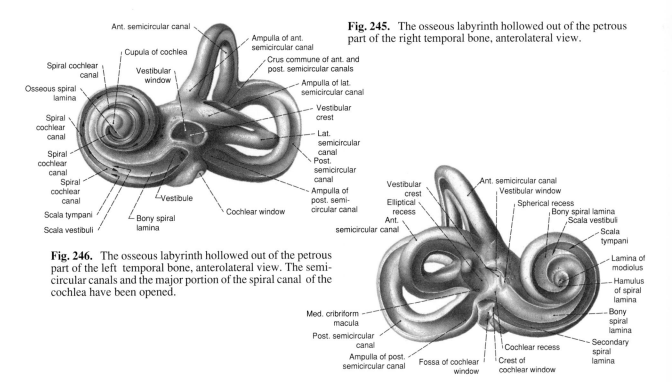

Fig. 246. The osseous labyrinth hollowed out of the petrous part of the left temporal bone, anterolateral view. The semicircular canals and the major portion of the spiral canal of the cochlea have been opened.

Fig. 247. The osseous labyrinth hollowed out of the petrous part of the right temporal bone. The dissection is the same as in Fig. 245, but additionally the vestibule and the cochlea have been opened to the cupula.

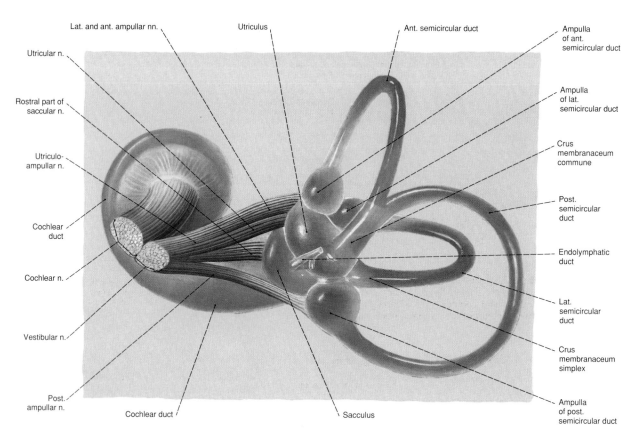

Lat. and ant. ampullar nn.

Utricular n.

Rostral part of saccular n.

Utriculo-ampullar n.

Cochlear duct

Cochlear n.

Vestibular n.

Post. ampullar n.

Cochlear duct

Utriculus

Sacculus

Ant. semicircular duct

Ampulla of ant. semicircular duct

Ampulla of lat. semicircular duct

Crus membranaceum commune

Post. semicircular duct

Endolymphatic duct

Lat. semicircular duct

Crus membranaceum simplex

Ampulla of post. semicircular duct

Fig. 248. The right membranous labyrinth, medial view. Branches of the vestibulocochlear nerve (VIII) to the membranous labyrinth which is lodged within the osseous labyrinth.

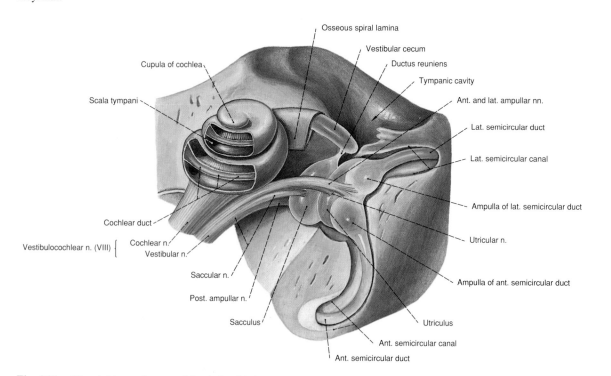

Cupula of cochlea

Scala tympani

Cochlear duct

Vestibulocochlear n. (VIII) {

Cochlear n.

Vestibular n.

Saccular n.

Post. ampullar n.

Sacculus

Osseous spiral lamina

Vestibular cecum

Ductus reuniens

Tympanic cavity

Ant. and lat. ampullar nn.

Lat. semicircular duct

Lat. semicircular canal

Ampulla of lat. semicircular duct

Utricular n.

Ampulla of ant. semicircular duct

Utriculus

Ant. semicircular canal

Ant. semicircular duct

Fig. 249. The right membranous labyrinth with its nerve supply partially exposed within the temporal bone. Cochlea opened laterally. The membranous labyrinth is in blue.

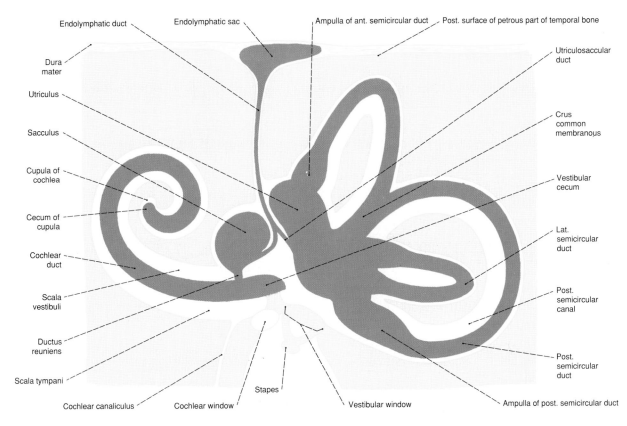

Endolymphatic duct
Endolymphatic sac
Ampulla of ant. semicircular duct
Post. surface of petrous part of temporal bone
Dura mater
Utriculosaccular duct
Utriculus
Sacculus
Crus common membranous
Cupula of cochlea
Cecum of cupula
Vestibular cecum
Cochlear duct
Lat. semicircular duct
Scala vestibuli
Post. semicircular canal
Ductus reuniens
Post. semicircular duct
Scala tympani
Cochlear canaliculus
Cochlear window
Stapes
Vestibular window
Ampulla of post. semicircular duct

Fig. 250. Diagram of the right membranous labyrinth. The endolymphatic spaces are shown in blue-gray, the bony structures are colored yellow and the perilymphatic spaces are diagrammed as white.

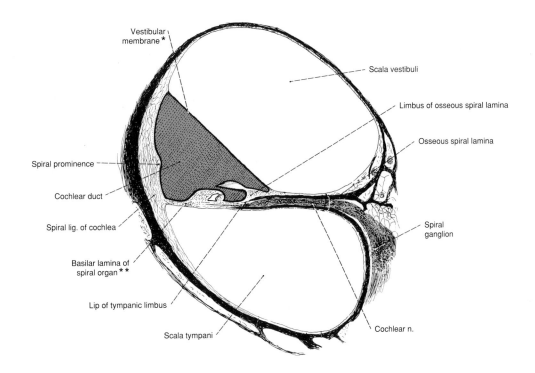

Vestibular membrane *
Scala vestibuli
Limbus of osseous spiral lamina
Osseous spiral lamina
Spiral prominence
Cochlear duct
Spiral lig. of cochlea
Spiral ganglion
Basilar lamina of spiral organ **
Lip of tympanic limbus
Scala tympani
Cochlear n.

* Reissner's membrane
** Organ of Corti

Fig. 251. Schematic section through the cochlear duct. The endolymphatic space is colored blue-gray.

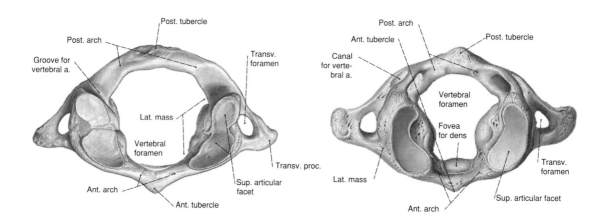

Fig. 252.　First cervical vertebra, atlas. View from above.

Fig. 253.　First cervical vertebra, atlas. View from above. The canal of the vertebral artery is an enclosed variation of the groove for the vertebral artery.

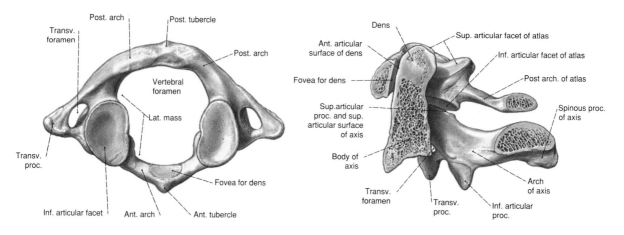

Fig. 254.　First cervical vertebra, atlas. Caudal view.

Fig. 255.　Median sagittal section through the 1st, atlas, and 2nd cervical vertebra, axis. Medial view.

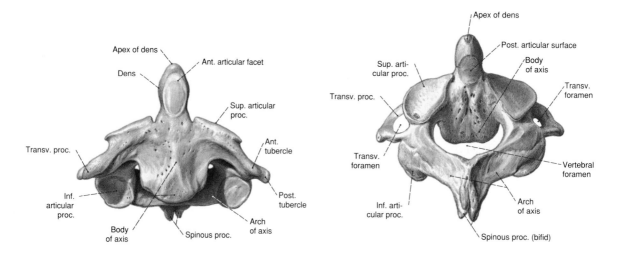

Fig. 256.　Second cervical vertebra, axis. Anterior view.

Fig. 257.　Second cervical vertebra, axis. Posterior view.

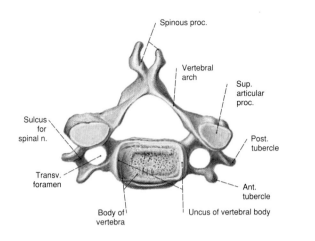

Fig. 258. Fifth cervical vertebra. View from above.

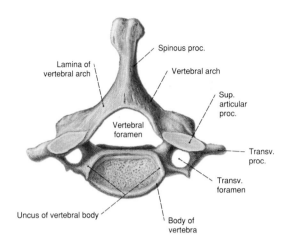

Fig. 259. Seventh cervical vertebra (vertebra prominens). View from above.

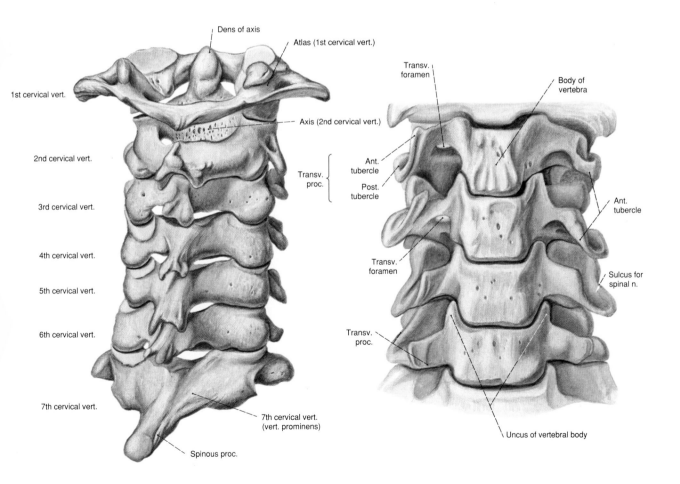

Fig. 260. The seven vertebrae of the cervical spinal column. Posterior view, somewhat from the right.

Fig. 261. Vertebrae of the lower cervical spinal column. Anterior view.

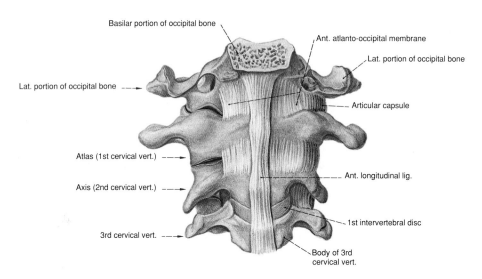

Fig. 262. Articulations of the occipital bone and the first three cervical vertebrae, anterior view. The anterior atlanto-occipital membrane and the rostral end of the anterior longitudinal ligament are visible. Articular capsules are removed on the right.

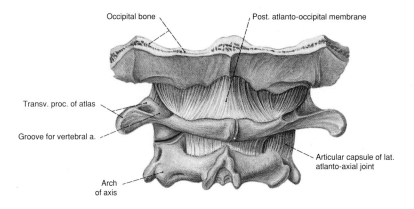

Fig. 263. Part of the occipital bone, first (atlas) and second (axis) cervical vertebrae with their ligaments and the posterior atlanto-occipital membrane. Posterior view. The articular capsule of the lateral atlanto-axial joint has been removed on the left.

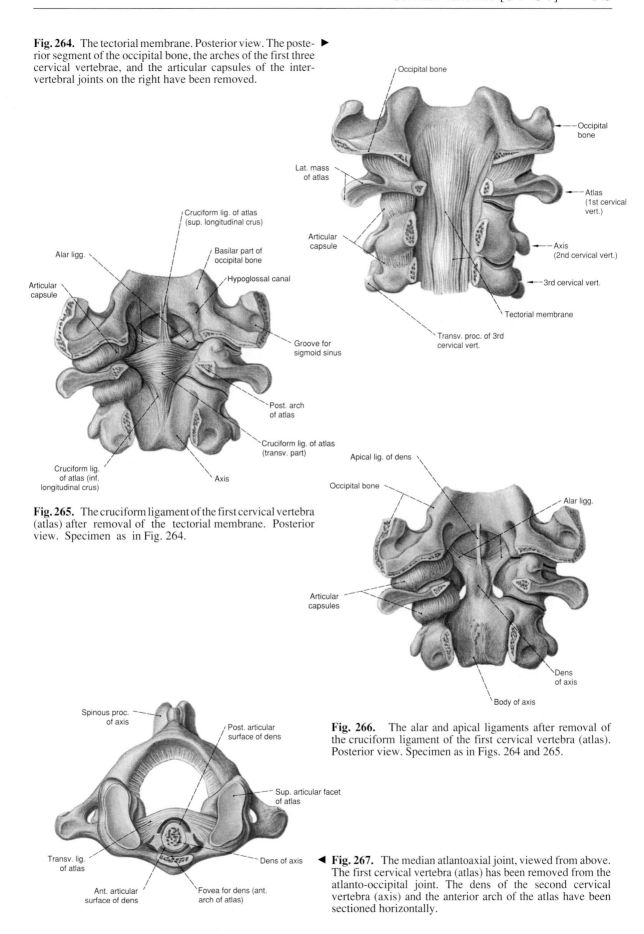

Fig. 264. The tectorial membrane. Posterior view. The posterior segment of the occipital bone, the arches of the first three cervical vertebrae, and the articular capsules of the intervertebral joints on the right have been removed. ▶

Occipital bone

Occipital bone

Lat. mass of atlas

Articular capsule

Atlas (1st cervical vert.)

Axis (2nd cervical vert.)

3rd cervical vert.

Tectorial membrane

Transv. proc. of 3rd cervical vert.

Cruciform lig. of atlas (sup. longitudinal crus)

Alar ligg.

Basilar part of occipital bone

Hypoglossal canal

Articular capsule

Groove for sigmoid sinus

Post. arch of atlas

Cruciform lig. of atlas (transv. part)

Cruciform lig. of atlas (inf. longitudinal crus)

Axis

Fig. 265. The cruciform ligament of the first cervical vertebra (atlas) after removal of the tectorial membrane. Posterior view. Specimen as in Fig. 264.

Apical lig. of dens

Occipital bone

Alar ligg.

Articular capsules

Dens of axis

Body of axis

Fig. 266. The alar and apical ligaments after removal of the cruciform ligament of the first cervical vertebra (atlas). Posterior view. Specimen as in Figs. 264 and 265.

Spinous proc. of axis

Post. articular surface of dens

Sup. articular facet of atlas

Transv. lig. of atlas

Dens of axis

Ant. articular surface of dens

Fovea for dens (ant. arch of atlas)

◀ **Fig. 267.** The median atlantoaxial joint, viewed from above. The first cervical vertebra (atlas) has been removed from the atlanto-occipital joint. The dens of the second cervical vertebra (axis) and the anterior arch of the atlas have been sectioned horizontally.

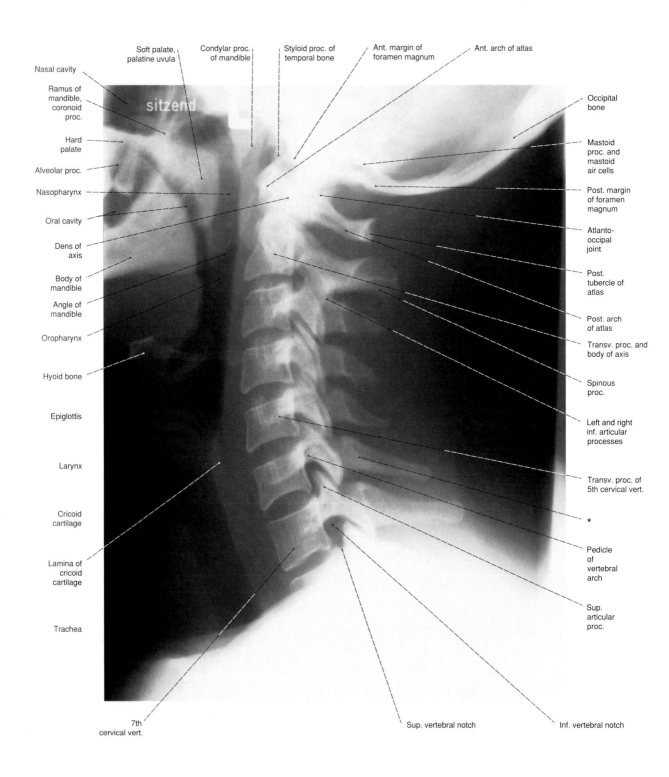

Nasal cavity

Ramus of
mandible,
coronoid
proc.

Hard
palate

Alveolar proc.

Nasopharynx

Oral cavity

Dens of
axis

Body of
mandible

Angle of
mandible

Oropharynx

Hyoid bone

Epiglottis

Larynx

Cricoid
cartilage

Lamina of
cricoid
cartilage

Trachea

Soft palate,
palatine uvula

Condylar proc.
of mandible

Styloid proc. of
temporal bone

Ant. margin of
foramen magnum

Ant. arch of atlas

sitzend

Occipital
bone

Mastoid
proc. and
mastoid
air cells

Post. margin
of foramen
magnum

Atlanto-
occipital
joint

Post.
tubercle of
atlas

Post. arch
of atlas

Transv. proc. and
body of axis

Spinous
proc.

Left and right
inf. articular
processes

Transv. proc. of
5th cervical vert.

*

Pedicle
of
vertebral
arch

Sup.
articular
proc.

7th
cervical vert.

Sup. vertebral notch

Inf. vertebral notch

Fig. 268.   Radiograph of the lateral aspect of the cervical
vertebral column. (From: Dr. G. Greeven, St. Elizabeth Hospital, Neuwied.)

* Spinolaminar line (= line marking the closure of the posterior arch)

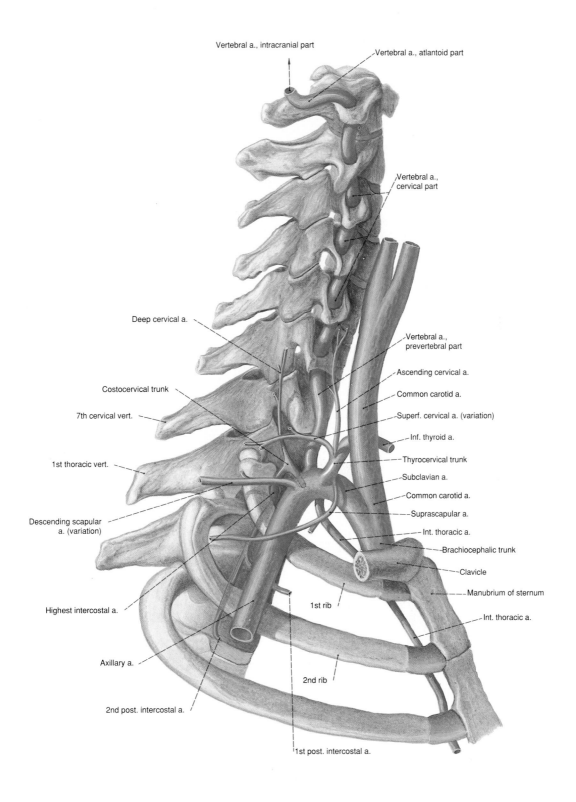

Vertebral a., intracranial part

Vertebral a., atlantoid part

Vertebral a., cervical part

Deep cervical a.

Vertebral a., prevertebral part

Ascending cervical a.

Costocervical trunk

Common carotid a.

Superf. cervical a. (variation)

7th cervical vert.

Inf. thyroid a.

Thyrocervical trunk

1st thoracic vert.

Subclavian a.

Common carotid a.

Suprascapular a.

Int. thoracic a.

Descending scapular a. (variation)

Brachiocephalic trunk

Clavicle

Manubrium of sternum

Highest intercostal a.

1st rib

Int. thoracic a.

Axillary a.

2nd rib

2nd post. intercostal a.

1st post. intercostal a.

Fig. 269. The right subclavian artery with its branches, in particular, the origin and course of the right vertebral artery.

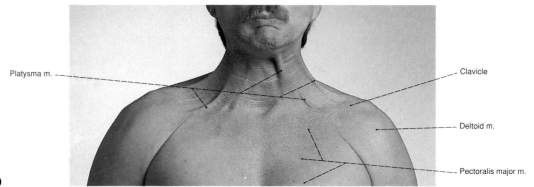

Platysma m.

Clavicle

Deltoid m.

Pectoralis major m.

Fig. 270

Fig. 270. Chest and neck regions of a male viewed from anterior. By inducing contraction (as in grimacing), the muscular fascicles of the platysma become visible underneath the skin of the neck and posteriorly to the clavicle.

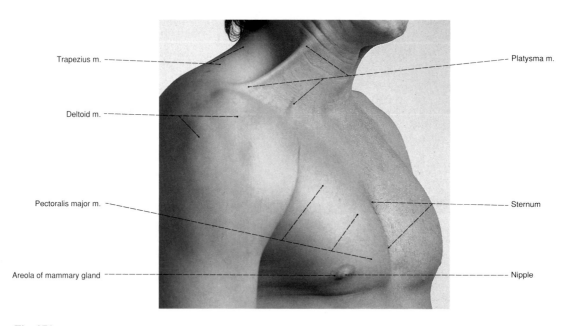

Trapezius m.

Platysma m.

Deltoid m.

Pectoralis major m.

Sternum

Areola of mammary gland

Nipple

Fig. 271

Fig. 271. Chest and neck regions of a male viewed anterolaterally. When the platysma muscle is induced to contract, its free posterolateral border produces a distinct ridge in the skin of the neck.

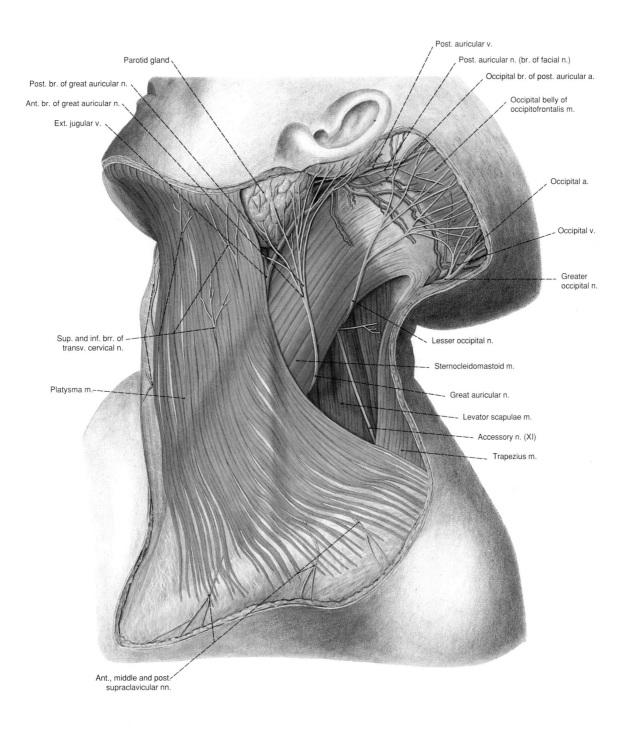

Post. auricular v.

Post. auricular n. (br. of facial n.)

Occipital br. of post. auricular a.

Occipital belly of occipitofrontalis m.

Parotid gland

Post. br. of great auricular n.

Ant. br. of great auricular n.

Ext. jugular v.

Occipital a.

Occipital v.

Greater occipital n.

Sup. and inf. brr. of transv. cervical n.

Lesser occipital n.

Sternocleidomastoid m.

Platysma m.

Great auricular n.

Levator scapulae m.

Accessory n. (XI)

Trapezius m.

Ant., middle and post. supraclavicular nn.

Fig. 272. Cutaneous branches of the cervical plexus, left side of the neck. The point on the posterior border of the sterno-cleidomastoid muscle at the level of the 3rd cervical vertebra where the cutaneous nerves of the cervical plexus exit is known as the punctum nervosum.

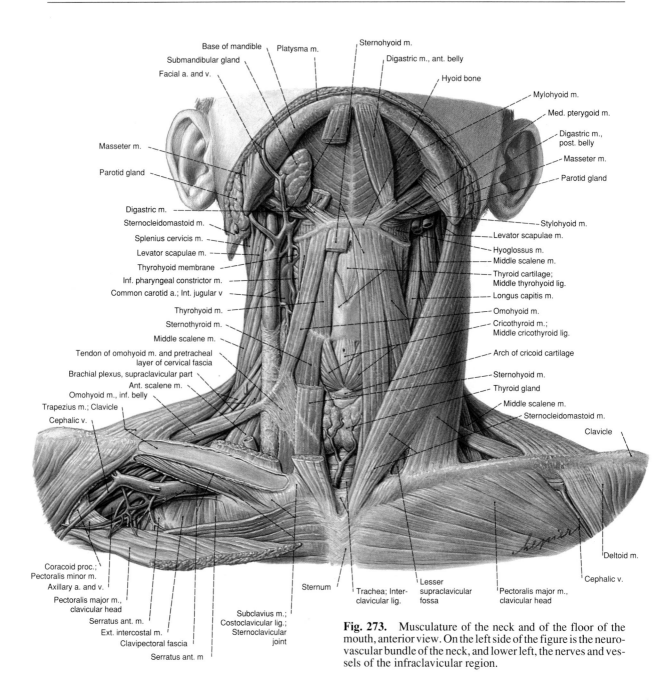

Base of mandible
Platysma m.
Sternohyoid m.
Submandibular gland
Digastric m., ant. belly
Facial a. and v.
Hyoid bone
Mylohyoid m.
Med. pterygoid m.
Masseter m.
Digastric m., post. belly
Masseter m.
Parotid gland
Parotid gland
Digastric m.
Stylohyoid m.
Sternocleidomastoid m.
Levator scapulae m.
Splenius cervicis m.
Hyoglossus m.
Levator scapulae m.
Middle scalene m.
Thyrohyoid membrane
Thyroid cartilage; Middle thyrohyoid lig.
Inf. pharyngeal constrictor m.
Common carotid a.; Int. jugular v
Longus capitis m.
Thyrohyoid m.
Omohyoid m.
Sternothyroid m.
Cricothyroid m.; Middle cricothyroid lig.
Middle scalene m.
Arch of cricoid cartilage
Tendon of omohyoid m. and pretracheal layer of cervical fascia
Sternohyoid m.
Brachial plexus, supraclavicular part
Thyroid gland
Ant. scalene m.
Middle scalene m.
Omohyoid m., inf. belly
Sternocleidomastoid m.
Trapezius m.; Clavicle
Clavicle
Cephalic v.
Deltoid m.
Coracoid proc.; Pectoralis minor m.
Cephalic v.
Axillary a. and v.
Sternum
Trachea; Inter-clavicular lig.
Lesser supraclavicular fossa
Pectoralis major m., clavicular head
Pectoralis major m., clavicular head
Serratus ant. m.
Subclavius m.; Costoclavicular lig.; Sternoclavicular joint
Ext. intercostal m.
Clavipectoral fascia
Serratus ant. m

Fig. 273. Musculature of the neck and of the floor of the mouth, anterior view. On the left side of the figure is the neurovascular bundle of the neck, and lower left, the nerves and vessels of the infraclavicular region.

Name	Origin	Insertion	Innervation	Function
Sternocleidomastoid muscle Sternal head	Ventral surface of the manubrium of the sternum	Lateral surface of mastoid process; lateral half of superior nuchal line of occipital bone	Spinal part of the accessory nerve (XI), cervical plexus	Both sides together hold the head, ant. fibers flex the head, post. fibers extend the head. One side alone, bends the head laterally, rotates the head toward the opposite side. If the head is fixed, the two sides assist in elevating the thorax during deep inspiration.
Clavicular head	Medial third of clavicle			

Suprahyoid Muscles

Name	Origin	Insertion	Innervation	Function
1. **Digastric muscle** Divided into two bellies by an intertendon which is connected to the side of the body and greater horn of the hyoid bone by a fibrous loop	Mastoid notch of temporal bone (posterior belly)	Digastric fossa of mandible (anterior belly)	Ant. belly: mylohyoid nerve (V_3); Post. belly: facial nerve (VII)	Opens the mouth by depressing mandible, elevates (fixes) the hyoid bone; assists 3
2. **Stylohyoid muscle** Is perforated by the intertendon of the digastric muscle	Styloid process of temporal bone	Lateral margin of body of hyoid bone near the greater horn	Facial nerve (VII)	Fixes hyoid bone, pulling it backward and upward during deglutition
3. **Mylohyoid muscle** (The muscles of both sides are joined by a midline raphe.)	Mylohyoid line of the mandible; both together form a plate across the mandibular arch	Mylohyoid raphe and body of hyoid bone	Mylohyoid nerve (V_3)	Raises the floor of the mouth in swallowing, depresses the mandible, elevates hyoid bone
4. **Geniohyoid muscle** Borders directly on the tongue muscles, especially the genioglossus	Inferior mental spine of the mandible, the muscles of both sides lie close together, with a midline septum between them	Anterior surface of body of hyoid bone	1st cervical nerve via hypoglossal nerve (XII)	Assists the mylohyoid muscle in elevating the tongue, elevates and fixes the hyoid bone, depresses the mandible

Infrahyoid Muscles

Name	Origin	Insertion	Innervation	Function
1. **Sternohyoid muscle** (often has a transverse tendinous inscription caudally)	Inner surface of manubrium sterni and sternoclavicular joint	Body of hyoid bone	Ansa cervicalis (cervical plexus)	Depress the larynx and hyoid after they have been elevated with the pharynx in deglutition (thyrohyoid muscle elevates the larynx); indirectly flex the head and cervical joints; omohyoid muscle tenses the cervical fascia through the connection of its central tendon to the carotid sheath; assist in respiration (pull sternum cranially in inspiration)
2. **Sternothyroid muscle** (Often has a transverse tendinous inscription caudally.)	Inner surface of 1st costal cartilage and manubrium sterni, caudal to **1**	Outer surface of thyroid cartilage (opposite the origin of thyrohyoid muscle)		
3. **Thyrohyoid muscle**	Outer surface of the thyroid cartilage	Inferior border of greater horn of hyoid bone		
4. **Omohyoid muscle** (two fleshy bellies united by a central tendon connected to the carotid sheath)	Superior margin of scapula between superior angle and scapular notch (inferior belly)	Lower border of body of hyoid bone (superior belly)		

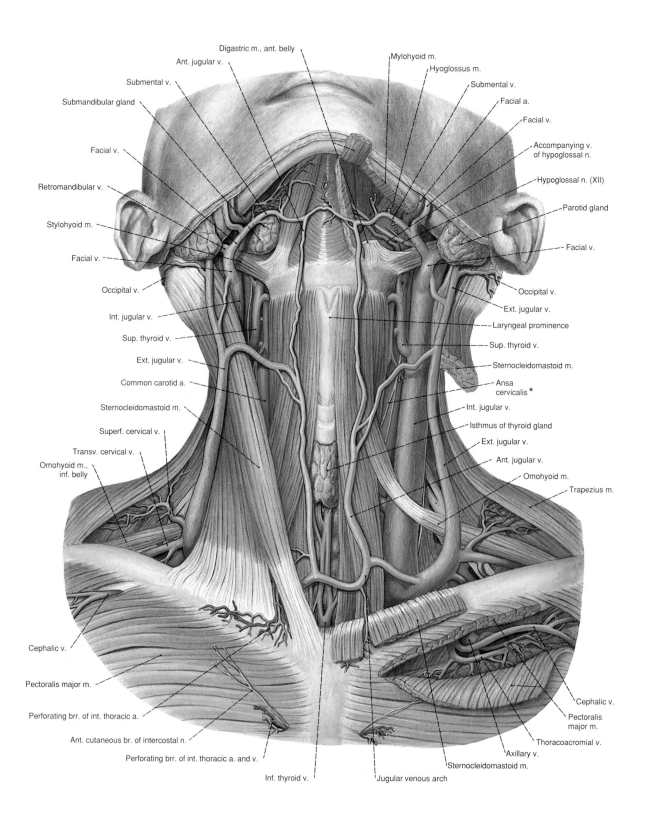

Fig. 274. Superficial veins of the anterior region of the neck and the veins of the infraclavicular region. The major portion of the left sternocleidomastoid muscle has been removed.

* Formerly: Ansa hypoglossi

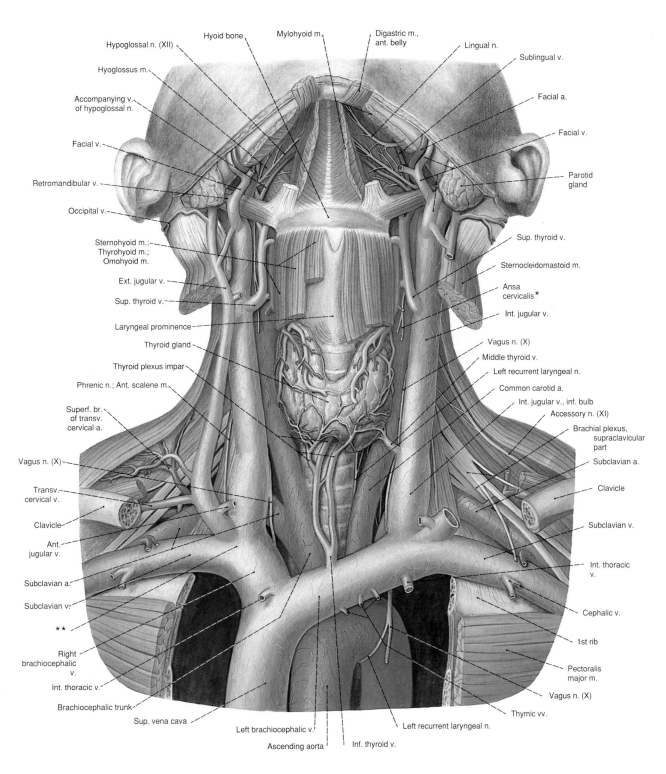

Fig. 275. Deep arteries and veins and nerves of the neck and the upper thoracic aperture. The sternocleidomastoid and infrahyoid muscles have been partially removed. The mylohyoid and anterior digastric muscles have been cut. The sternum, anterior portions of the first and second ribs and the clavicle have been resected.

* Formerly: Ansa hypoglossi
** The junction of the jugular and subclavian veins is frequently termed the venous angle.

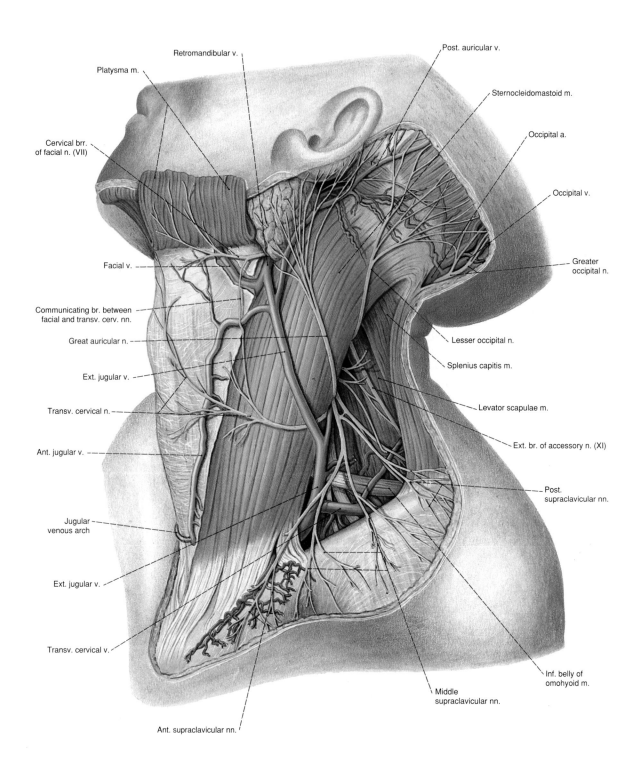

Retromandibular v.

Platysma m.

Post. auricular v.

Sternocleidomastoid m.

Cervical brr.
of facial n. (VII)

Occipital a.

Occipital v.

Facial v.

Greater
occipital n.

Communicating br. between
facial and transv. cerv. nn.

Great auricular n.

Lesser occipital n.

Ext. jugular v.

Splenius capitis m.

Transv. cervical n.

Levator scapulae m.

Ant. jugular v.

Ext. br. of accessory n. (XI)

Post.
supraclavicular nn.

Jugular
venous arch

Ext. jugular v.

Transv. cervical v.

Inf. belly of
omohyoid m.

Middle
supraclavicular nn.

Ant. supraclavicular nn.

Fig. 276. Cutaneous and muscular branches of the cervical
plexus and the superficial blood vessels of the left side of the
neck. The cranial portion of the platysma has been reflected
upwards. The investing layer of the cervical fascia is retained
anterior to the sternocleidomastoid muscle and split along the
facial vein.

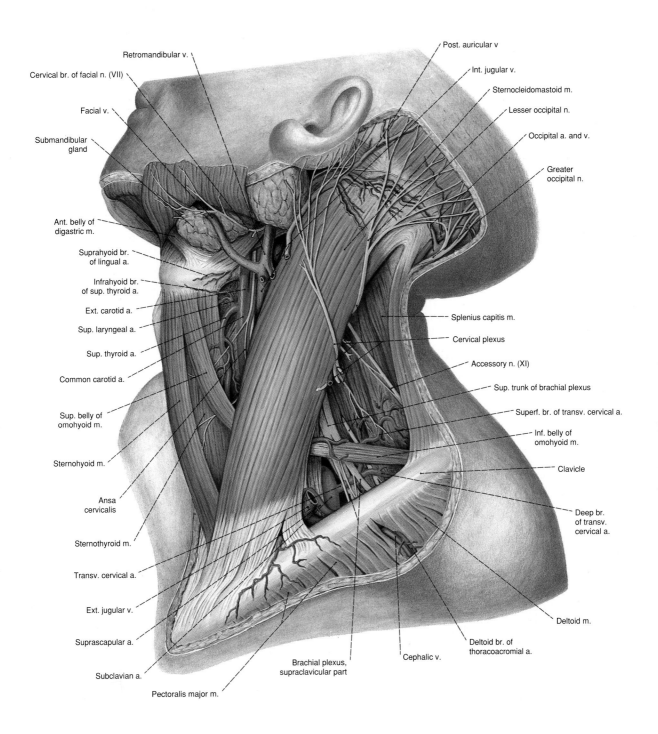

Fig. 277. The anterior and posterior triangles of the neck. Blood vessels and nerves after removal of the investing layer of the cervical fascia.

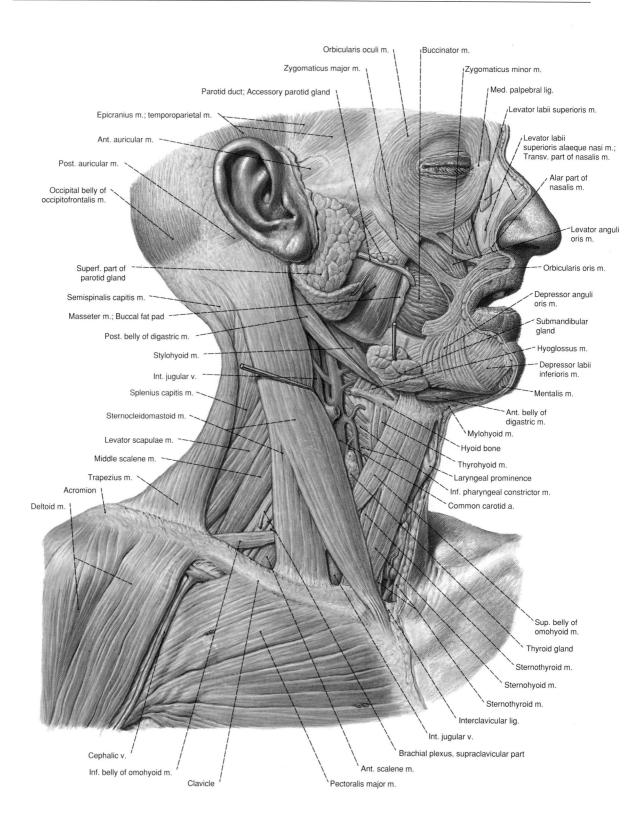

Orbicularis oculi m.

Buccinator m.

Zygomaticus major m.

Zygomaticus minor m.

Parotid duct; Accessory parotid gland

Med. palpebral lig.

Epicranius m.; temporoparietal m.

Levator labii superioris m.

Ant. auricular m.

Levator labii superioris alaeque nasi m.; Transv. part of nasalis m.

Post. auricular m.

Alar part of nasalis m.

Occipital belly of occipitofrontalis m.

Levator anguli oris m.

Superf. part of parotid gland

Orbicularis oris m.

Semispinalis capitis m.

Depressor anguli oris m.

Masseter m.; Buccal fat pad

Submandibular gland

Post. belly of digastric m.

Hyoglossus m.

Stylohyoid m.

Depressor labii inferioris m.

Int. jugular v.

Mentalis m.

Splenius capitis m.

Ant. belly of digastric m.

Sternocleidomastoid m.

Mylohyoid m.

Levator scapulae m.

Hyoid bone

Middle scalene m.

Thyrohyoid m.

Trapezius m.

Laryngeal prominence

Acromion

Inf. pharyngeal constrictor m.

Deltoid m.

Common carotid a.

Sup. belly of omohyoid m.

Thyroid gland

Sternothyroid m.

Sternohyoid m.

Sternothyroid m.

Interclavicular lig.

Int. jugular v.

Cephalic v.

Brachial plexus, supraclavicular part

Inf. belly of omohyoid m.

Ant. scalene m.

Clavicle

Pectoralis major m.

Fig. 278. Head, neck and upper thorax. Right side. Superficial layer.

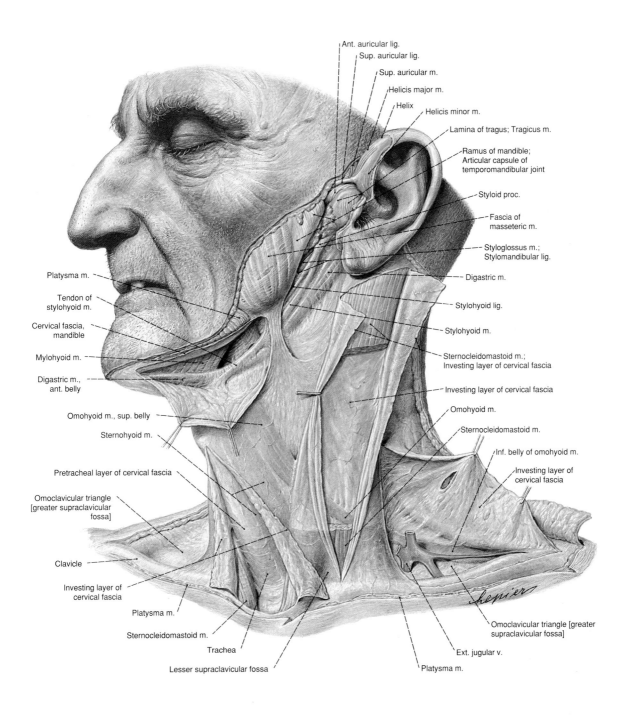

Ant. auricular lig.

Sup. auricular lig.

Sup. auricular m.

Helicis major m.

Helix

Helicis minor m.

Lamina of tragus; Tragicus m.

Ramus of mandible;
Articular capsule of
temporomandibular joint

Styloid proc.

Fascia of
masseteric m.

Styloglossus m.;
Stylomandibular lig.

Digastric m.

Stylohyoid lig.

Stylohyoid m.

Sternocleidomastoid m.;
Investing layer of cervical fascia

Investing layer of cervical fascia

Omohyoid m.

Sternocleidomastoid m.

Inf. belly of omohyoid m.

Investing layer of
cervical fascia

Omoclavicular triangle [greater
supraclavicular fossa]

Ext. jugular v.

Platysma m.

Platysma m.

Tendon of
stylohyoid m.

Cervical fascia,
mandible

Mylohyoid m.

Digastric m.,
ant. belly

Omohyoid m., sup. belly

Sternohyoid m.

Pretracheal layer of cervical fascia

Omoclavicular triangle
[greater supraclavicular
fossa]

Clavicle

Investing layer of
cervical fascia

Platysma m.

Sternocleidomastoid m.

Trachea

Lesser supraclavicular fossa

Fig. 279.   The fasciae of the neck: the investing and
pretracheal layers of the cervical fascia.

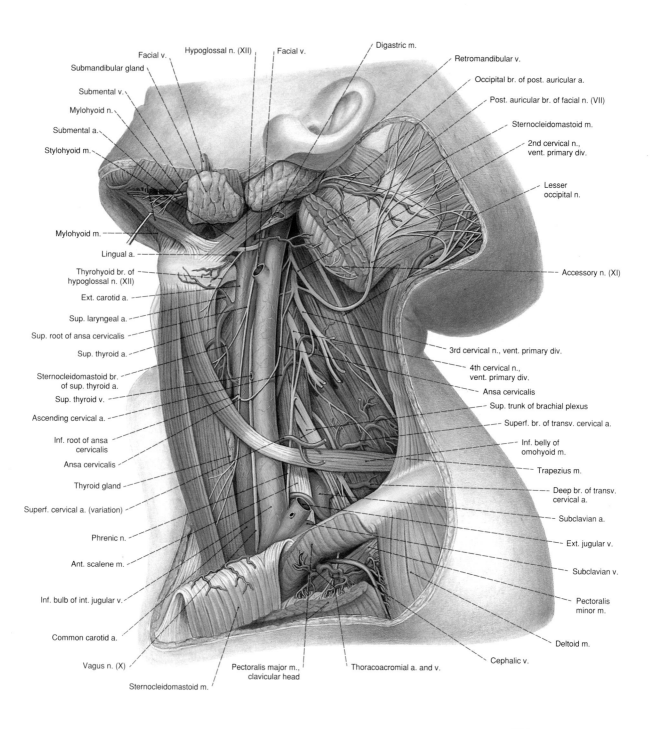

Hypoglossal n. (XII)

Facial v.

Submandibular gland

Facial v.

Digastric m.

Submental v.

Retromandibular v.

Mylohyoid n.

Occipital br. of post. auricular a.

Submental a.

Post. auricular br. of facial n. (VII)

Stylohyoid m.

Sternocleidomastoid m.

2nd cervical n.,
vent. primary div.

Mylohyoid m.

Lesser
occipital n.

Lingual a.

Thyrohyoid br. of
hypoglossal n. (XII)

Accessory n. (XI)

Ext. carotid a.

Sup. laryngeal a.

Sup. root of ansa cervicalis

3rd cervical n., vent. primary div.

Sup. thyroid a.

4th cervical n.,
vent. primary div.

Sternocleidomastoid br.
of sup. thyroid a.

Ansa cervicalis

Sup. thyroid v.

Sup. trunk of brachial plexus

Ascending cervical a.

Superf. br. of transv. cervical a.

Inf. root of ansa
cervicalis

Inf. belly of
omohyoid m.

Ansa cervicalis

Trapezius m.

Thyroid gland

Deep br. of transv.
cervical a.

Superf. cervical a. (variation)

Subclavian a.

Phrenic n.

Ext. jugular v.

Ant. scalene m.

Subclavian v.

Inf. bulb of int. jugular v.

Pectoralis
minor m.

Common carotid a.

Deltoid m.

Vagus n. (X)

Cephalic v.

Sternocleidomastoid m.

Pectoralis major m.,
clavicular head

Thoracoacromial a. and v.

Fig. 280.   The left cervical region after removal of the
sternocleidomastoid muscle and the deep cervical fascia. Dissection of the hiatus of the scalene muscles, the cervical plexus
and the brachial plexus.

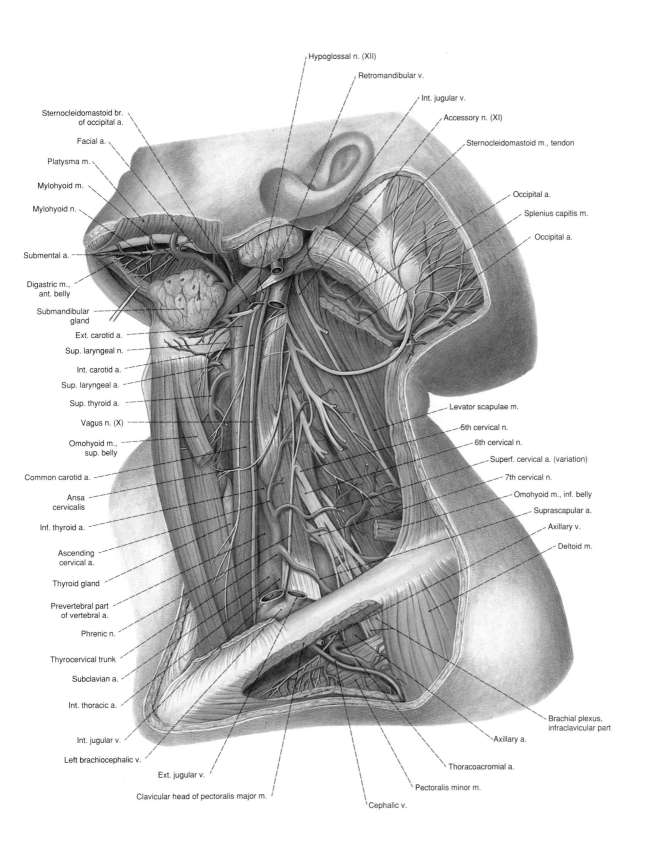

Hypoglossal n. (XII)

Retromandibular v.

Int. jugular v.

Accessory n. (XI)

Sternocleidomastoid br. of occipital a.

Facial a.

Platysma m.

Mylohyoid m.

Mylohyoid n.

Submental a.

Digastric m., ant. belly

Submandibular gland

Ext. carotid a.

Sup. laryngeal n.

Int. carotid a.

Sup. laryngeal a.

Sup. thyroid a.

Vagus n. (X)

Omohyoid m., sup. belly

Common carotid a.

Ansa cervicalis

Inf. thyroid a.

Ascending cervical a.

Thyroid gland

Prevertebral part of vertebral a.

Phrenic n.

Thyrocervical trunk

Subclavian a.

Int. thoracic a.

Int. jugular v.

Left brachiocephalic v.

Ext. jugular v.

Clavicular head of pectoralis major m.

Cephalic v.

Sternocleidomastoid m., tendon

Occipital a.

Splenius capitis m.

Occipital a.

Levator scapulae m.

5th cervical n.

6th cervical n.

Superf. cervical a. (variation)

7th cervical n.

Omohyoid m., inf. belly

Suprascapular a.

Axillary v.

Deltoid m.

Brachial plexus, infraclavicular part

Axillary a.

Thoracoacromial a.

Pectoralis minor m.

Fig. 281. The deeper layer of nerves and arteries of the left side of the neck and the deltopectoral triangle opened medially.

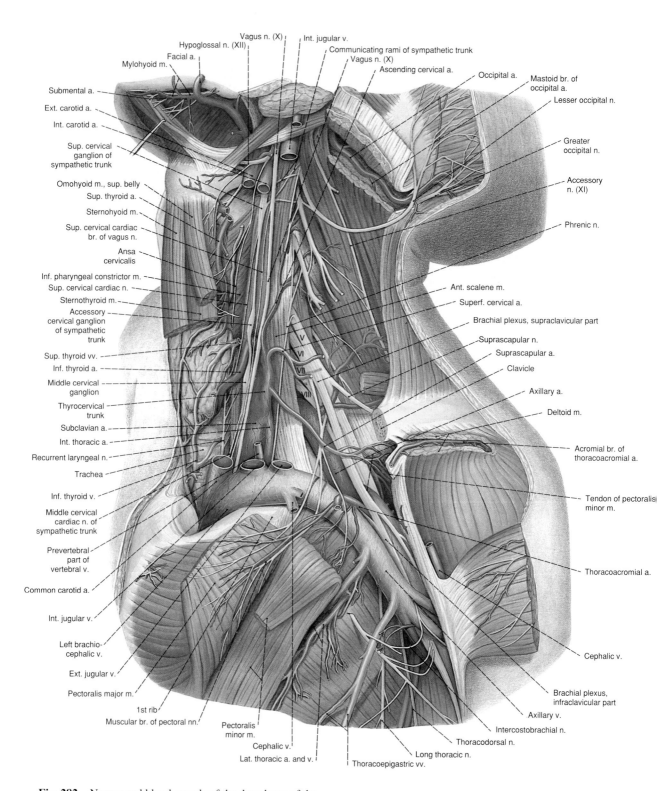

Vagus n. (X)
Hypoglossal n. (XII)
Facial a.
Mylohyoid m.
Int. jugular v.
Communicating rami of sympathetic trunk
Vagus n. (X)
Ascending cervical a.
Occipital a.
Mastoid br. of occipital a.
Submental a.
Ext. carotid a.
Int. carotid a.
Sup. cervical ganglion of sympathetic trunk
Omohyoid m., sup. belly
Sup. thyroid a.
Sternohyoid m.
Sup. cervical cardiac br. of vagus n.
Ansa cervicalis
Inf. pharyngeal constrictor m.
Sup. cervical cardiac n.
Sternothyroid m.
Accessory cervical ganglion of sympathetic trunk
Sup. thyroid vv.
Inf. thyroid a.
Middle cervical ganglion
Thyrocervical trunk
Subclavian a.
Int. thoracic a.
Recurrent laryngeal n.
Trachea
Inf. thyroid v.
Middle cervical cardiac n. of sympathetic trunk
Prevertebral part of vertebral v.
Common carotid a.
Int. jugular v.
Left brachio-cephalic v.
Ext. jugular v.
Pectoralis major m.
1st rib
Muscular br. of pectoral nn.
Pectoralis minor m.
Cephalic v.
Lat. thoracic a. and v.
Thoracoepigastric vv.
Long thoracic n.
Thoracodorsal n.
Intercostobrachial n.
Axillary v.
Brachial plexus, infraclavicular part
Cephalic v.
Thoracoacromial a.
Tendon of pectoralis minor m.
Acromial br. of thoracoacromial a.
Deltoid m.
Axillary a.
Clavicle
Suprascapular a.
Suprascapular n.
Brachial plexus, supraclavicular part
Superf. cervical a.
Ant. scalene m.
Phrenic n.
Accessory n. (XI)
Greater occipital n.
Lesser occipital n.
Mastoid br. of occipital a.

V
VI
VII
VIII

Fig. 282. Nerves and blood vessels of the deep layer of the neck and the axilla. The clavicle has been removed from the sternoclavicular joint and sectioned medial to the insertion of the trapezius muscle. The infrahyoid muscles, sternocleido-mastoid muscle and major cervical blood vessels have been mostly removed. Roman numerals V–VIII = ventral primary divisions of the fifth to the eighth cervical nerves.

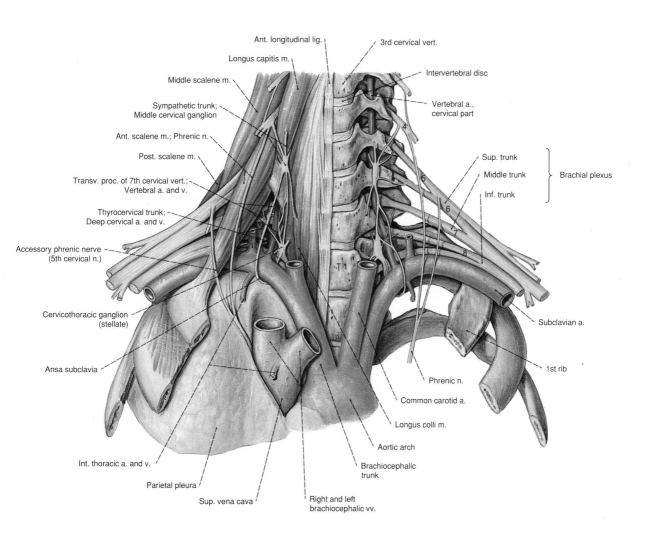

Ant. longitudinal lig.

3rd cervical vert.

Longus capitis m.

Intervertebral disc

Middle scalene m.

Vertebral a., cervical part

Sympathetic trunk; Middle cervical ganglion

Ant. scalene m.; Phrenic n.

Post. scalene m.

Sup. trunk

Middle trunk

Brachial plexus

Transv. proc. of 7th cervical vert.; Vertebral a. and v.

Inf. trunk

Thyrocervical trunk; Deep cervical a. and v.

Accessory phrenic nerve (5th cervical n.)

Cervicothoracic ganglion (stellate)

Subclavian a.

Ansa subclavia

1st rib

Phrenic n.

Common carotid a.

Int. thoracic a. and v.

Longus colli m.

Parietal pleura

Aortic arch

Brachiocephalic trunk

Sup. vena cava

Right and left brachiocephalic vv.

Organization of the Brachial Plexus

BRACHIAL PLEXUS
Roots

Trunks
  Superior trunk
  Middle trunk
  Inferior trunk

Divisions
  Anterior division
  Posterior division

Cords (named in relation to axillary artery)
  Lateral: Ant. divisions of sup. and middle trunks
  Medial: Ant. division of inf. trunk
  Posterior: Post. divisions of sup., middle and inf. trunks

Nerves

Fig. 283. Brachial and cervical plexuses and their relationships to the vertebral column (right side of figure), the surrounding muscles (left side of figure), the structures of the upper thoracic aperture and the sympathetic trunk. Anterior view. 4-8 = roots of the brachial plexus from the anterior primary divisions of the fourth to eighth cervical nerves. T1 = 1st thoracic vertebra.

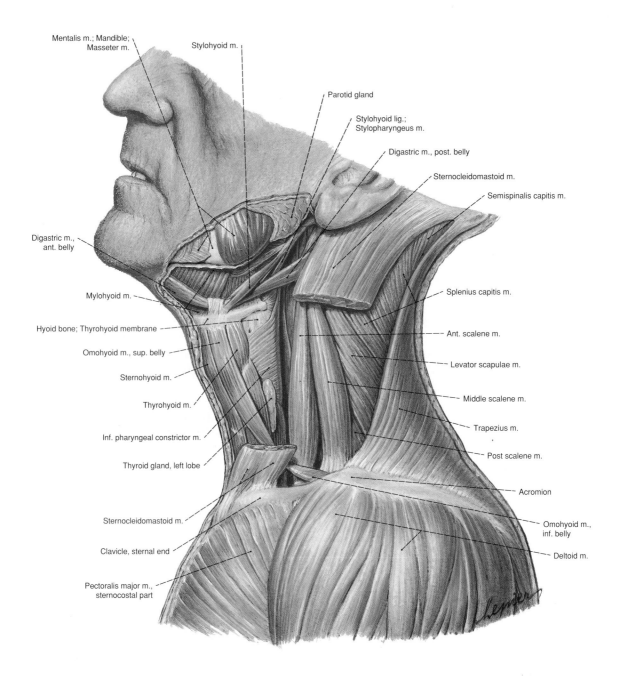

Mentalis m.; Mandible; Masseter m.

Stylohyoid m.

Parotid gland

Stylohyoid lig.; Stylopharyngeus m.

Digastric m., post. belly

Sternocleidomastoid m.

Semispinalis capitis m.

Digastric m., ant. belly

Mylohyoid m.

Hyoid bone; Thyrohyoid membrane

Omohyoid m., sup. belly

Sternohyoid m.

Thyrohyoid m.

Inf. pharyngeal constrictor m.

Thyroid gland, left lobe

Sternocleidomastoid m.

Clavicle, sternal end

Pectoralis major m., sternocostal part

Splenius capitis m.

Ant. scalene m.

Levator scapulae m.

Middle scalene m.

Trapezius m.

Post scalene m.

Acromion

Omohyoid m., inf. belly

Deltoid m.

Fig. 284. Muscles of the left side of the neck after removal of the midportion of the sternocleidomastoid muscle.

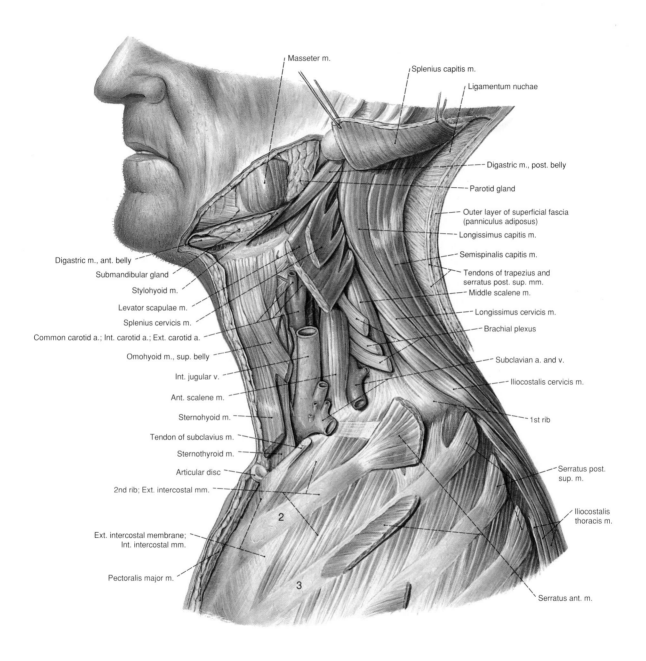

Masseter m.

Splenius capitis m.

Ligamentum nuchae

Digastric m., post. belly

Parotid gland

Outer layer of superficial fascia
(panniculus adiposus)

Longissimus capitis m.

Semispinalis capitis m.

Tendons of trapezius and
serratus post. sup. mm.

Middle scalene m.

Longissimus cervicis m.

Brachial plexus

Subclavian a. and v.

Iliocostalis cervicis m.

1st rib

Serratus post.
sup. m.

Iliocostalis
thoracis m.

Serratus ant. m.

Digastric m., ant. belly

Submandibular gland

Stylohyoid m.

Levator scapulae m.

Splenius cervicis m.

Common carotid a.; Int. carotid a.; Ext. carotid a.

Omohyoid m., sup. belly

Int. jugular v.

Ant. scalene m.

Sternohyoid m.

Tendon of subclavius m.

Sternothyroid m.

Articular disc

2nd rib; Ext. intercostal mm.

Ext. intercostal membrane;
Int. intercostal mm.

Pectoralis major m.

2

3

Fig. 285. Cervical and upper thoracic regions, deeper layers.
The levator scapulae and splenius cervicis muscles have been
cut and reflected anteriorly. Note the subclavian artery and
brachial plexus emerging between the anterior and middle
scalene muscles.

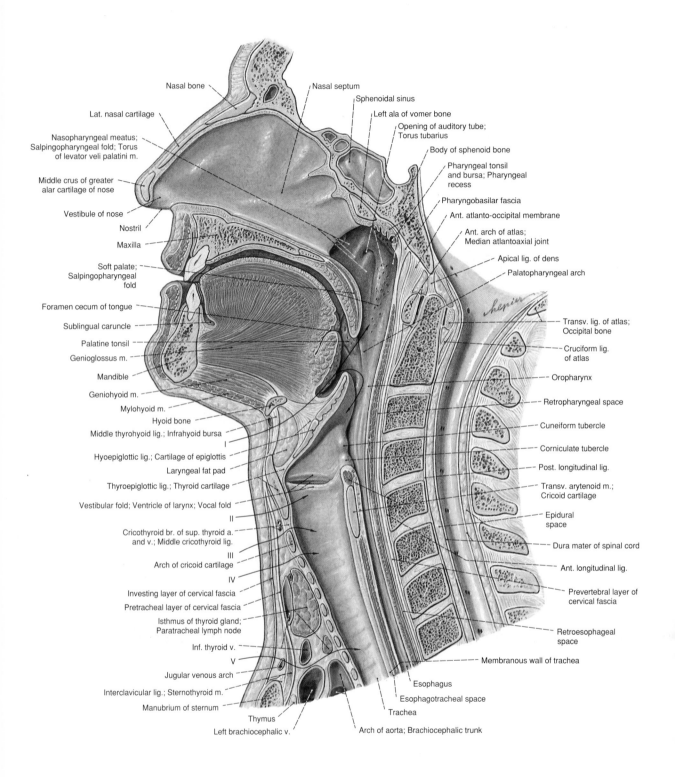

Nasal bone

Lat. nasal cartilage

Nasopharyngeal meatus;
Salpingopharyngeal fold; Torus
of levator veli palatini m.

Middle crus of greater
alar cartilage of nose

Vestibule of nose

Nostril

Maxilla

Soft palate;
Salpingopharyngeal
fold

Foramen cecum of tongue

Sublingual caruncle

Palatine tonsil

Genioglossus m.

Mandible

Geniohyoid m.

Mylohyoid m.

Hyoid bone

Middle thyrohyoid lig.; Infrahyoid bursa

Hyoepiglottic lig.; Cartilage of epiglottis

Laryngeal fat pad

Thyroepiglottic lig.; Thyroid cartilage

Vestibular fold; Ventricle of larynx; Vocal fold

Cricothyroid br. of sup. thyroid a.
and v.; Middle cricothyroid lig.

Arch of cricoid cartilage

Investing layer of cervical fascia

Pretracheal layer of cervical fascia

Isthmus of thyroid gland;
Paratracheal lymph node

Inf. thyroid v.

Jugular venous arch

Interclavicular lig.; Sternothyroid m.

Manubrium of sternum

Thymus

Left brachiocephalic v.

Nasal septum

Sphenoidal sinus

Left ala of vomer bone

Opening of auditory tube;
Torus tubarius

Body of sphenoid bone

Pharyngeal tonsil
and bursa; Pharyngeal
recess

Pharyngobasilar fascia

Ant. atlanto-occipital membrane

Ant. arch of atlas;
Median atlantoaxial joint

Apical lig. of dens

Palatopharyngeal arch

Transv. lig. of atlas;
Occipital bone

Cruciform lig.
of atlas

Oropharynx

Retropharyngeal space

Cuneiform tubercle

Corniculate tubercle

Post. longitudinal lig.

Transv. arytenoid m.;
Cricoid cartilage

Epidural
space

Dura mater of spinal cord

Ant. longitudinal lig.

Prevertebral layer of
cervical fascia

Retroesophageal
space

Membranous wall of trachea

Esophagus

Esophagotracheal space

Trachea

Arch of aorta; Brachiocephalic trunk

Fig. 286. Paramedian sagittal section through the face, head and neck. The arrows I-V indicate surgical approaches to the pharynx, larynx and trachea: I = subhyoid pharyngotomy; II = laryngotomy; III = coniotomy (cricothyrotomy) between the thyroid and cricoid cartilages through the middle cricothyroid ligament and the conus elasticus; IV = superior tracheotomy below the arch of the cricoid cartilage and above the isthmus of the thyroid gland; V = inferior tracheotomy below the isthmus of the thyroid gland.

Fig. 287. The pharyngeal and laryngeal muscles and portions ▶ of the facial muscles, lateral view. 1, 2, 3, 4 are the parts of the superior pharyngeal constrictor muscle: pterygopharyngeal, buccopharyngeal, mylopharyngeal and glossopharyngeal, respectively.

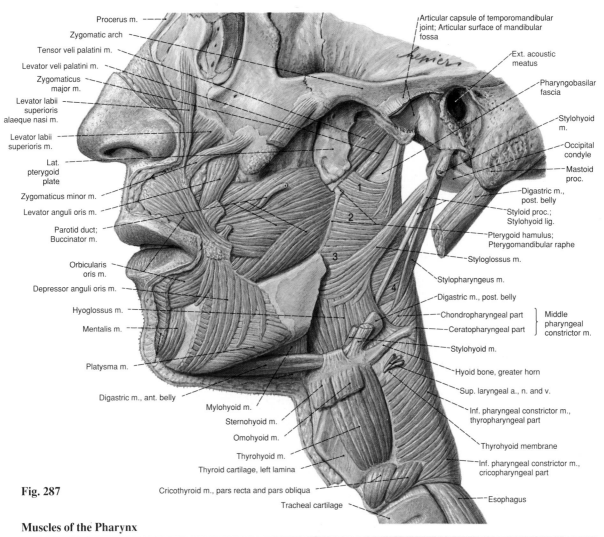

Procerus m.
Zygomatic arch
Tensor veli palatini m.
Levator veli palatini m.
Zygomaticus major m.
Levator labii superioris alaeque nasi m.
Levator labii superioris m.
Lat. pterygoid plate
Zygomaticus minor m.
Levator anguli oris m.
Parotid duct; Buccinator m.
Orbicularis oris m.
Depressor anguli oris m.
Hyoglossus m.
Mentalis m.
Platysma m.
Digastric m., ant. belly
Mylohyoid m.
Sternohyoid m.
Omohyoid m.
Thyrohyoid m.
Thyroid cartilage, left lamina
Cricothyroid m., pars recta and pars obliqua
Tracheal cartilage

Articular capsule of temporomandibular joint; Articular surface of mandibular fossa
Ext. acoustic meatus
Pharyngobasilar fascia
Stylohyoid m.
Occipital condyle
Mastoid proc.
Digastric m., post. belly
Styloid proc.; Stylohyoid lig.
Pterygoid hamulus; Pterygomandibular raphe
Styloglossus m.
Stylopharyngeus m.
Digastric m., post. belly
Chondropharyngeal part ⎫ Middle
Ceratopharyngeal part ⎭ pharyngeal constrictor m.
Stylohyoid m.
Hyoid bone, greater horn
Sup. laryngeal a., n. and v.
Inf. pharyngeal constrictor m., thyropharyngeal part
Thyrohyoid membrane
Inf. pharyngeal constrictor m., cricopharyngeal part
Esophagus

Fig. 287

Muscles of the Pharynx

Name	Origin	Insertion	Innervation	Function
1. Superior constrictor muscle		Pharyngobasilar fascia, median raphe	1,2,3,5 and 6 from pharyngeal plexus of the glossopharyngeal (IX), vagus (X) and accessory (XI) nerves; 4 from only glosso-pharyngeal nerve (IX)	1-3 constrict the pharynx; in swallowing, 1 arches the mucosa forward (closing the nasopharynx with the soft palate); 4 elevates the pharynx toward the base of the skull
Pterygopharyngeal part	Pterygoid hamulus of sphenoid bone			
Buccopharyngeal part	Pterygomandibular raphe, buccinator muscle			
Mylopharyngeal part	Mylohyoid line of mandible			
Glossopharyngeal part	Transverse muscle of the tongue			
2. Middle constrictor muscle				
Chondropharyngeal part	Lesser horn of hyoid bone			
Ceratopharyngeal part	Greater horn of hyoid bone			
3. Inferior constrictor muscle			Additionally, from external laryngeal and recurrent nerves	
Thyropharyngeal part	Oblique line of thyroid cartilage			
Cricopharyngeal part	Lateral margin of cricoid cartilage			
4. Stylopharyngeus muscle	Styloid process of temporal bone	Lateral pharyngeal wall		Opens auditory tube and closes it during swallow-ing and yawning
5. Salpingopharyngeus muscle	Cartilage of auditory tube	Lateral pharyngeal wall		
6. Palatopharyngeus muscle (described with muscles of the palate, p. 79)				

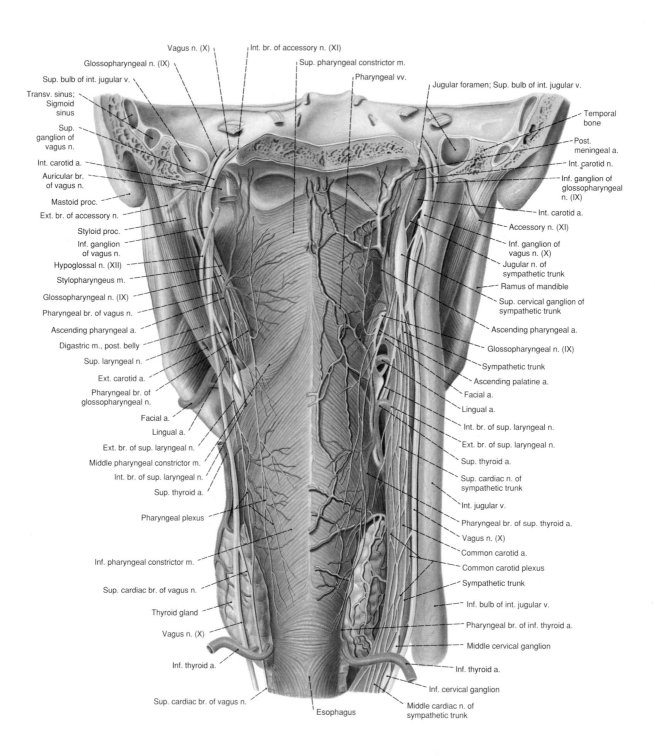

Fig. 288. Nerves and blood vessels on the dorsal and lateral walls of the pharynx.

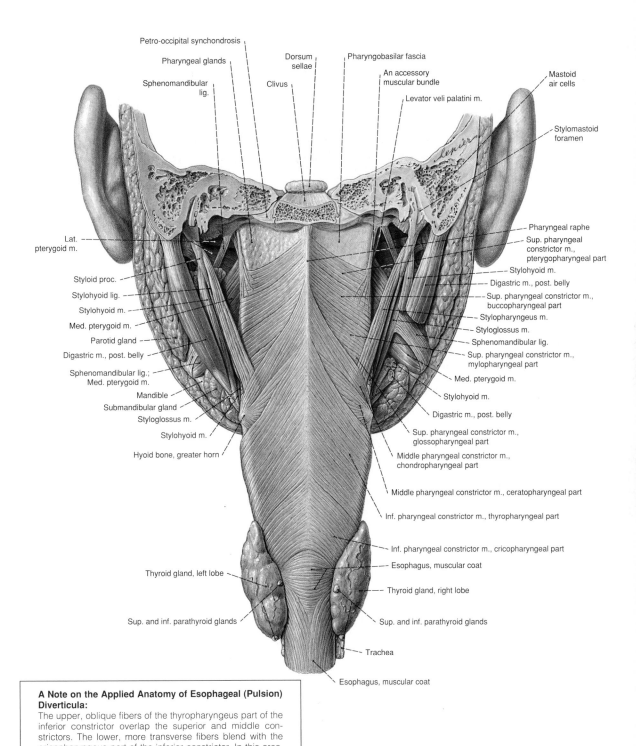

Petro-occipital synchondrosis

Pharyngeal glands

Sphenomandibular lig.

Dorsum sellae

Clivus

Pharyngobasilar fascia

An accessory muscular bundle

Levator veli palatini m.

Mastoid air cells

Stylomastoid foramen

Lat. pterygoid m.

Styloid proc.

Stylohyoid lig.

Stylohyoid m.

Med. pterygoid m.

Parotid gland

Digastric m., post. belly

Sphenomandibular lig.; Med. pterygoid m.

Mandible

Submandibular gland

Styloglossus m.

Stylohyoid m.

Hyoid bone, greater horn

Thyroid gland, left lobe

Sup. and inf. parathyroid glands

Pharyngeal raphe

Sup. pharyngeal constrictor m., pterygopharyngeal part

Stylohyoid m.

Digastric m., post. belly

Sup. pharyngeal constrictor m., buccopharyngeal part

Stylopharyngeus m.

Styloglossus m.

Sphenomandibular lig.

Sup. pharyngeal constrictor m., mylopharyngeal part

Med. pterygoid m.

Stylohyoid m.

Digastric m., post. belly

Sup. pharyngeal constrictor m., glossopharyngeal part

Middle pharyngeal constrictor m., chondropharyngeal part

Middle pharyngeal constrictor m., ceratopharyngeal part

Inf. pharyngeal constrictor m., thyropharyngeal part

Inf. pharyngeal constrictor m., cricopharyngeal part

Esophagus, muscular coat

Thyroid gland, right lobe

Sup. and inf. parathyroid glands

Trachea

Esophagus, muscular coat

A Note on the Applied Anatomy of Esophageal (Pulsion) Diverticula:

The upper, oblique fibers of the thyropharyngeus part of the inferior constrictor overlap the superior and middle constrictors. The lower, more transverse fibers blend with the cricopharyngeus part of the inferior constrictor. In this area, where the posterior wall of the pharynx is only a single sheet of inferior constrictor muscle, is the dehiscence of KILLIAN. The lowest cricopharyngeal fibers blend with the circular fibers of the esophagus in an area which is also deficient of muscle fibers (area of LAIMER). KILLIAN observed that a pharyngeal diverticulum (the "esophageal" diverticulum) herniates through the dehiscence, always above and never below the cricopharyngeal part of the inferior constrictor (From: R. J. Last, Anatomy, Regional and Applied, 7th edition, Churchill Livingstone, 1984).

Fig. 289. Dorsal view of the muscular wall of the pharynx after removal of the fascia. Muscles of the upper parapharyngeal region. Note the areas of sparse muscle fibers at the beginning of the esophagus (dehiscence of KILLIAN and area of LAIMER).

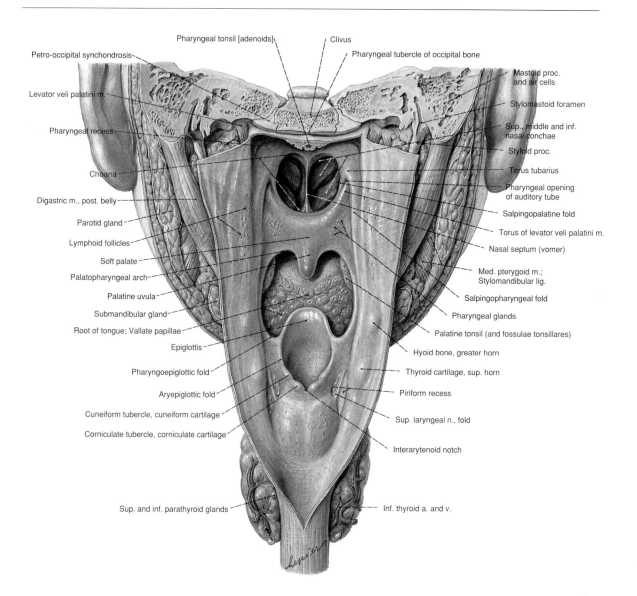

Pharyngeal tonsil [adenoids]

Clivus

Petro-occipital synchondrosis

Pharyngeal tubercle of occipital bone

Levator veli palatini m.

Mastoid proc. and air cells

Stylomastoid foramen

Pharyngeal recess

Sup., middle and inf. nasal conchae

Styloid proc.

Choana

Torus tubarius

Digastric m., post. belly

Pharyngeal opening of auditory tube

Parotid gland

Salpingopalatine fold

Lymphoid follicles

Torus of levator veli palatini m.

Soft palate

Nasal septum (vomer)

Palatopharyngeal arch

Med. pterygoid m.; Stylomandibular lig.

Palatine uvula

Salpingopharyngeal fold

Submandibular gland

Pharyngeal glands

Root of tongue; Vallate papillae

Palatine tonsil (and fossulae tonsillares)

Epiglottis

Hyoid bone, greater horn

Pharyngoepiglottic fold

Thyroid cartilage, sup. horn

Aryepiglottic fold

Piriform recess

Cuneiform tubercle, cuneiform cartilage

Sup. laryngeal n., fold

Corniculate tubercle, corniculate cartilage

Interarytenoid notch

Sup. and inf. parathyroid glands

Inf. thyroid a. and v.

Fig. 290. Internal view of the pharynx. The posterior wall has been opened by a longitudinal midline incision and by transverse cut near its origin at the base of the skull. The skull has been sectioned in the plane of the mastoid process and the clivus, and the occipital portion has been removed.

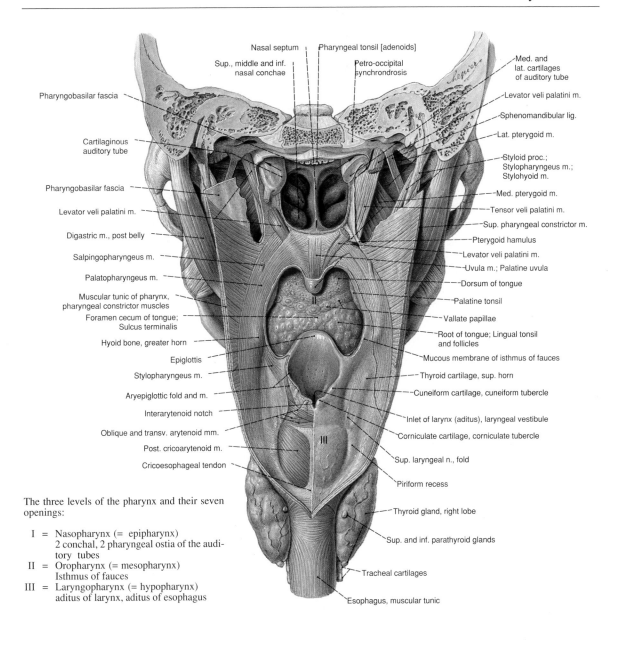

Nasal septum
Sup., middle and inf. nasal conchae
Pharyngeal tonsil [adenoids]
Petro-occipital synchrondrosis
Med. and lat. cartilages of auditory tube
Pharyngobasilar fascia
Levator veli palatini m.
Sphenomandibular lig.
Cartilaginous auditory tube
Lat. pterygoid m.
Styloid proc.; Stylopharyngeus m.; Stylohyoid m.
Pharyngobasilar fascia
Med. pterygoid m.
Levator veli palatini m.
Tensor veli palatini m.
Digastric m., post belly
Sup. pharyngeal constrictor m.
Salpingopharyngeus m.
Pterygoid hamulus
Levator veli palatini m.
Palatopharyngeus m.
Uvula m.; Palatine uvula
Dorsum of tongue
Muscular tunic of pharynx, pharyngeal constrictor muscles
Palatine tonsil
Foramen cecum of tongue; Sulcus terminalis
Vallate papillae
Root of tongue; Lingual tonsil and follicles
Hyoid bone, greater horn
Mucous membrane of isthmus of fauces
Epiglottis
Thyroid cartilage, sup. horn
Stylopharyngeus m.
Cuneiform cartilage, cuneiform tubercle
Aryepiglottic fold and m.
Inlet of larynx (aditus), laryngeal vestibule
Interarytenoid notch
Corniculate cartilage, corniculate tubercle
Oblique and transv. arytenoid mm.
Sup. laryngeal n., fold
Post. cricoarytenoid m.
Piriform recess
Cricoesophageal tendon
Thyroid gland, right lobe
Sup. and inf. parathyroid glands
Tracheal cartilages
Esophagus, muscular tunic

The three levels of the pharynx and their seven openings:

I = Nasopharynx (= epipharynx)
   2 conchal, 2 pharyngeal ostia of the auditory tubes
II = Oropharynx (= mesopharynx)
   Isthmus of fauces
III = Laryngopharynx (= hypopharynx)
   aditus of larynx, aditus of esophagus

Fig. 291. The same dissection as in Fig. 290, after extensive removal of the pharyngeal mucosa. By opening the posterior wall of the pharynx, the three parts of the pharynx with their boundaries and openings have been exposed. The muscles of the soft palate, pharynx and the dorsal aspect of the larynx are displayed. The auditory tube has been cut on the right side to expose the lateral and medial laminae of the tubal cartilage as well as the slit-like opening of the auditory tube. The right and left arrows on the roof of the pharynx indicate the paired pharyngeal recesses.

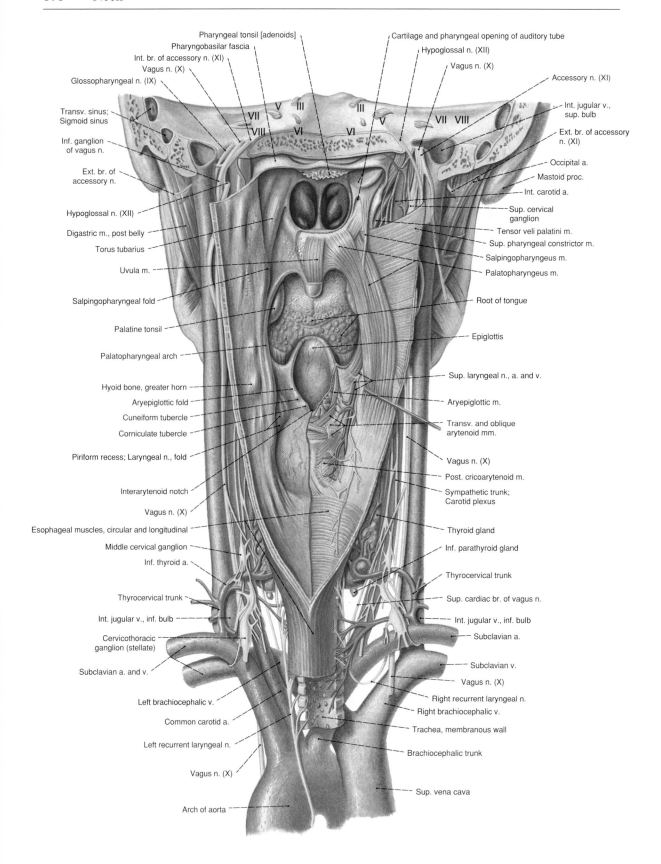

Pharyngeal tonsil [adenoids]
Pharyngobasilar fascia
Int. br. of accessory n. (XI)
Vagus n. (X)
Glossopharyngeal n. (IX)
Transv. sinus; Sigmoid sinus
Inf. ganglion of vagus n.
Ext. br. of accessory n.
Hypoglossal n. (XII)
Digastric m., post belly
Torus tubarius
Uvula m.
Salpingopharyngeal fold
Palatine tonsil
Palatopharyngeal arch
Hyoid bone, greater horn
Aryepiglottic fold
Cuneiform tubercle
Corniculate tubercle
Piriform recess; Laryngeal n., fold
Interarytenoid notch
Vagus n. (X)
Esophageal muscles, circular and longitudinal
Middle cervical ganglion
Inf. thyroid a.
Thyrocervical trunk
Int. jugular v., inf. bulb
Cervicothoracic ganglion (stellate)
Subclavian a. and v.
Left brachiocephalic v.
Common carotid a.
Left recurrent laryngeal n.
Vagus n. (X)
Arch of aorta

Cartilage and pharyngeal opening of auditory tube
Hypoglossal n. (XII)
Vagus n. (X)
Accessory n. (XI)
Int. jugular v., sup. bulb
Ext. br. of accessory n. (XI)
Occipital a.
Mastoid proc.
Int. carotid a.
Sup. cervical ganglion
Tensor veli palatini m.
Sup. pharyngeal constrictor m.
Salpingopharyngeus m.
Palatopharyngeus m.
Root of tongue
Epiglottis
Sup. laryngeal n., a. and v.
Aryepiglottic m.
Transv. and oblique arytenoid mm.
Vagus n. (X)
Post. cricoarytenoid m.
Sympathetic trunk; Carotid plexus
Thyroid gland
Inf. parathyroid gland
Thyrocervical trunk
Sup. cardiac br. of vagus n.
Int. jugular v., inf. bulb
Subclavian a.
Subclavian v.
Vagus n. (X)
Right recurrent laryngeal n.
Right brachiocephalic v.
Trachea, membranous wall
Brachiocephalic trunk
Sup. vena cava

V   III   III
VII   V
VIII VI   VI   VII VIII

Fig. 292. Muscles, blood vessels and nerves of the larynx and the parapharyngeal spaces. Dorsal view of the pharynx after opening its posterior wall. The mucous membrane of the right side of the pharynx has been removed. The posterior portion of the skull has been sectioned at the level of the jugular foramina. Roman numerals III–VIII = cranial nerves III–VIII.

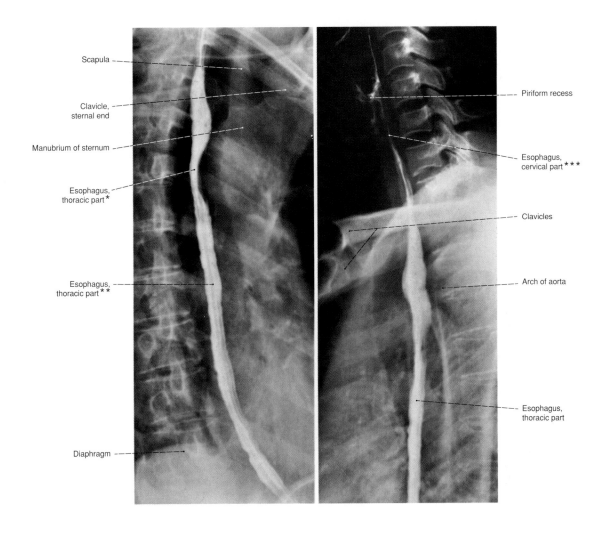

Scapula

Clavicle,
sternal end

Manubrium of sternum

Esophagus,
thoracic part *

Esophagus,
thoracic part **

Diaphragm

Piriform recess

Esophagus,
cervical part ***

Clavicles

Arch of aorta

Esophagus,
thoracic part

Figs. 293a, b. Radiographs of the esophagus, left and right anterior oblique projections using contrast media. (From: WICKE, Atlas of Radiologic Anatomy, 3rd edition, Urban & Schwarzenberg, Munich-Vienna-Baltimore, 1985).

  * 2nd esophageal constriction, caused by the arch of the aorta
 ** Retrocardial portion of the esophagus
*** 1st esophageal constriction = inlet of esophagus

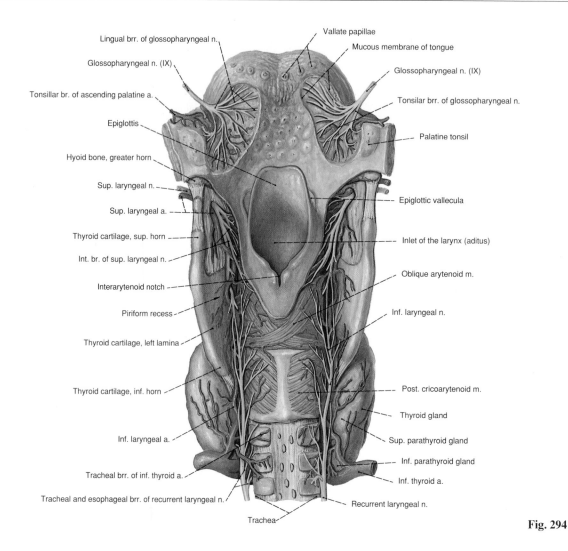

Lingual brr. of glossopharyngeal n.
Glossopharyngeal n. (IX)
Tonsillar br. of ascending palatine a.
Epiglottis
Hyoid bone, greater horn
Sup. laryngeal n.
Sup. laryngeal a.
Thyroid cartilage, sup. horn
Int. br. of sup. laryngeal n.
Interarytenoid notch
Piriform recess
Thyroid cartilage, left lamina
Thyroid cartilage, inf. horn
Inf. laryngeal a.
Tracheal brr. of inf. thyroid a.
Tracheal and esophageal brr. of recurrent laryngeal n.
Trachea

Vallate papillae
Mucous membrane of tongue
Glossopharyngeal n. (IX)
Tonsilar brr. of glossopharyngeal n.
Palatine tonsil
Epiglottic vallecula
Inlet of the larynx (aditus)
Oblique arytenoid m.
Inf. laryngeal n.
Post. cricoarytenoid m.
Thyroid gland
Sup. parathyroid gland
Inf. parathyroid gland
Inf. thyroid a.
Recurrent laryngeal n.

Fig. 294

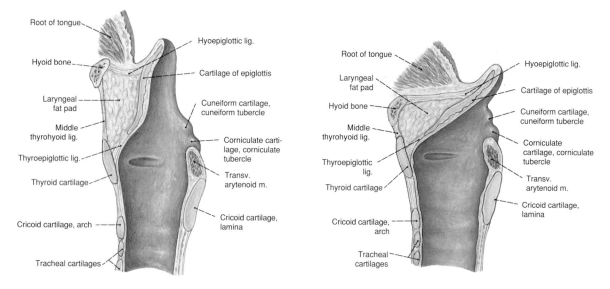

Root of tongue
Hyoid bone
Laryngeal fat pad
Middle thyrohyoid lig.
Thyroepiglottic lig.
Thyroid cartilage
Cricoid cartilage, arch
Tracheal cartilages

Hyoepiglottic lig.
Cartilage of epiglottis
Cuneiform cartilage, cuneiform tubercle
Corniculate cartilage, corniculate tubercle
Transv. arytenoid m.
Cricoid cartilage, lamina

Root of tongue
Laryngeal fat pad
Hyoid bone
Middle thyrohyoid lig.
Thyroepiglottic lig.
Thyroid cartilage
Cricoid cartilage, arch
Tracheal cartilages

Hyoepiglottic lig.
Cartilage of epiglottis
Cuneiform cartilage, cuneiform tubercle
Corniculate cartilage, corniculate tubercle
Transv. arytenoid m.
Cricoid cartilage, lamina

Fig. 295. Position of the epiglottis during respiration.

Fig. 296. Position of the epiglottis during swallowing. In swallowing, the distance between the hyoid bone and the thyroid cartilage is shortened, and the laryngeal fat pad pushes the epiglottis downwards.

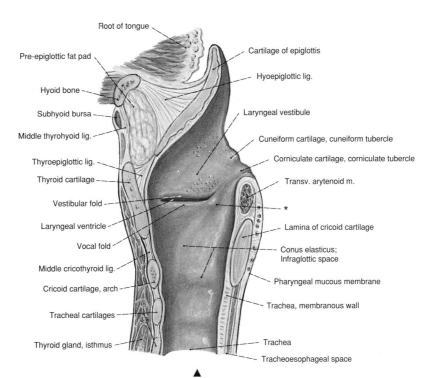

Root of tongue
Pre-epiglottic fat pad
Hyoid bone
Subhyoid bursa
Middle thyrohyoid lig.
Thyroepiglottic lig.
Thyroid cartilage
Vestibular fold
Laryngeal ventricle
Vocal fold
Middle cricothyroid lig.
Cricoid cartilage, arch
Tracheal cartilages
Thyroid gland, isthmus

Cartilage of epiglottis
Hyoepiglottic lig.
Laryngeal vestibule
Cuneiform cartilage, cuneiform tubercle
Corniculate cartilage, corniculate tubercle
Transv. arytenoid m.
*
Lamina of cricoid cartilage
Conus elasticus;
Infraglottic space
Pharyngeal mucous membrane
Trachea, membranous wall
Trachea
Tracheoesophageal space

◀ **Fig. 294.** Nerves and arteries of the larynx, root of the tongue, and the palatine tonsils, dorsal view.

Fig. 297. Midsagittal section of the larynx. View of the inside of the right half.

* Traditionally known as the macula flava where the elastic tissue appears yellow

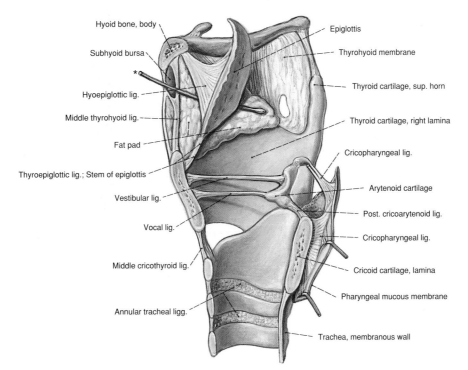

Hyoid bone, body
Subhyoid bursa
*
Hyoepiglottic lig.
Middle thyrohyoid lig.
Fat pad
Thyroepiglottic lig.; Stem of epiglottis
Vestibular lig.
Vocal lig.
Middle cricothyroid lig.
Annular tracheal ligg.

Epiglottis
Thyrohyoid membrane
Thyroid cartilage, sup. horn
Thyroid cartilage, right lamina
Cricopharyngeal lig.
Arytenoid cartilage
Post. cricoarytenoid lig.
Cricopharyngeal lig.
Cricoid cartilage, lamina
Pharyngeal mucous membrane
Trachea, membranous wall

* The probe is between the epiglottis and the laryngeal fat pad.

Fig. 298. The right half of the larynx in midsagittal section, showing the cartilages and the vestibular and vocal ligaments.

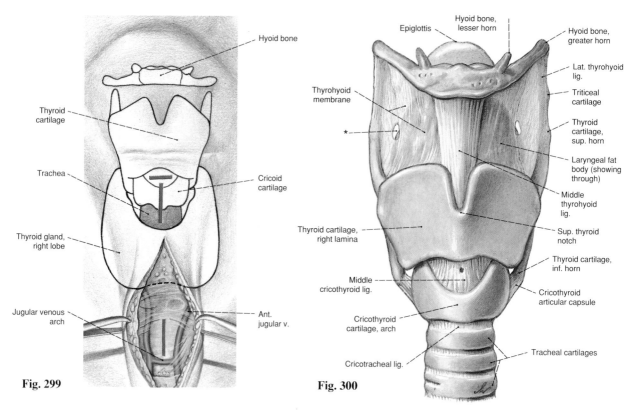

Fig. 299

Fig. 300

Fig. 299. Projection of the larynx, hyoid bone and thyroid gland onto the anterior surface of the neck. The skin over the jugular notch has been opened by a longitudinal incision to gain access to the suprasternal space with the jugular venous arch and the anterior jugular vein on each side (first layer of an inferior tracheotomy). The trachea lies 4-5 cm beneath the jugular notch. Shown also is the transverse incision for a laryngotomy in the region of the cricothyroid ligament (conus elasticus, thus also coniotomy), as well as the longitudinal incision through the arch of the cricoid cartilage (cricotomy). In coniotomy, the trachea is entered about 1 cm below the vocal cords. All three approaches to the trachea are marked in red (cf. Fig. 286).

Fig. 300. The cartilages, ligaments and articulations of the larynx, including the hyoid bone and the upper part of the trachea. Ventral view. Note the laryngeal fat body between the layers of the thyrohyoid membrane.

* Foramen for the superior laryngeal artery and the internal branch of the superior laryngeal nerve.

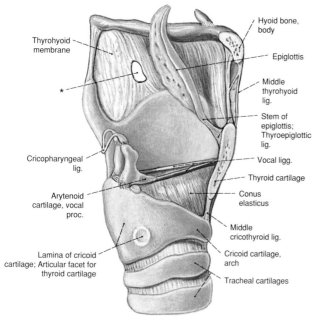

Fig. 301. Median section through the upper left part of the larynx. The right halves of the hyoid bone, epiglottis and thyroid cartilage have been removed. Both arytenoid cartilages, the cricoid cartilage and the two upper tracheal cartilages remain. The lateral portions of the conus elasticus, an elastic membrane which lies beneath the mucous membrane of the larynx, extend from the superior border of the cricoid cartilage to the inferior margin of the vocal ligaments with which it is continuous.

* Foramen for the superior laryngeal artery and the internal branch of the superior laryngeal nerve.

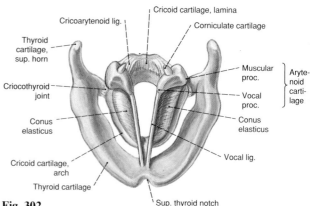

Fig. 302

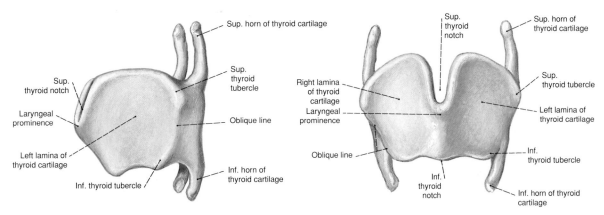

Fig. 303. Thyroid cartilage, lateral view.

Fig. 304. Thyroid cartilage, anterior view.

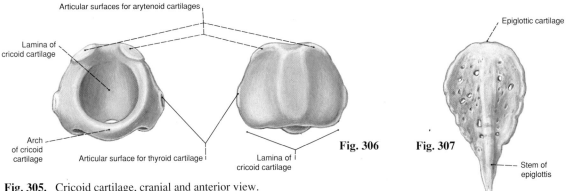

Fig. 305. Cricoid cartilage, cranial and anterior view.

Fig. 306. Lamina of cricoid cartilage, posterior view.

Fig. 307. Epiglottis, posterior view.

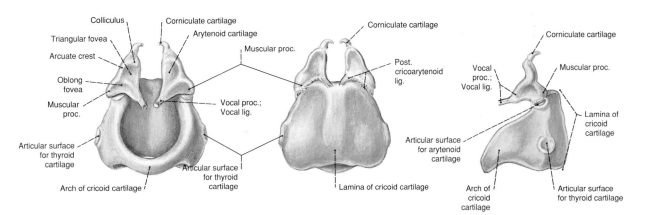

Fig. 308. Cranial and anterior view.

Fig. 309. Posterior view.

Fig. 310. Left lateral view.

Figs. 308–310. Articulated cricoid, arytenoid and corniculate cartilages.

◀ **Fig. 302.** View into the cartilaginous frame of the larynx from above onto the vocal ligaments and conus elasticus.

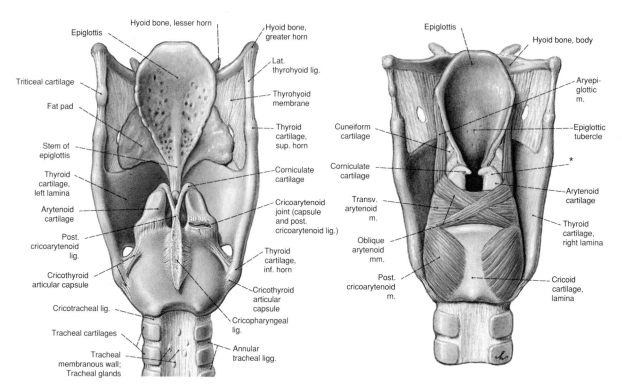

Fig. 311. The cartilages, ligaments and articulations of the larynx, as well as the hyoid bone and the upper part of the trachea. Posterior view. The larynx illustrated in Figs. 311 and 312 is from a juvenile. During puberty, calcification and ossification of the laryngeal cartilages proceeds from the thyroid, to the cricoid and then to the arytenoid cartilages (earlier in males than in females).

Fig. 312. Muscles of the larynx, posterior view.

* Traditionally: arycorniculate syndesmosis = connective tissue between the arytenoid and the corniculate cartilages, occasionally cartilaginous or with an articular space

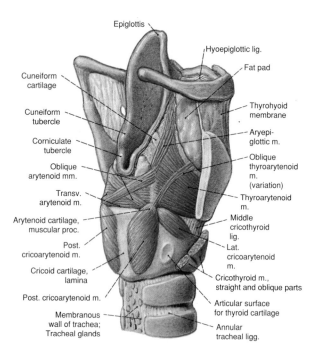

Fig. 313. Laryngeal musculature, posterolateral view.

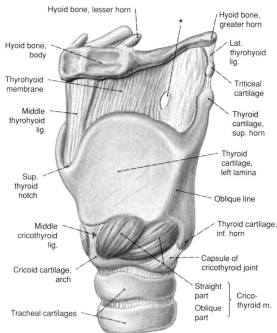

Fig. 314. Larynx and hyoid bone, left side viewed anterolaterally.

* Foramen for superior laryngeal artery and internal branch of the superior laryngeal nerve.

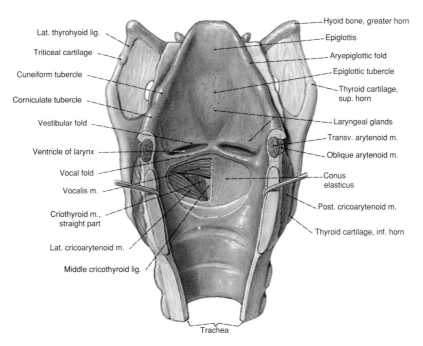

Lat. thyrohyoid lig.
Triticeal cartilage
Cuneiform tubercle
Corniculate tubercle
Vestibular fold
Ventricle of larynx
Vocal fold
Vocalis m.
Criothyroid m., straight part
Lat. cricoarytenoid m.
Middle cricothyroid lig.

Hyoid bone, greater horn
Epiglottis
Aryepiglottic fold
Epiglottic tubercle
Thyroid cartilage, sup. horn
Laryngeal glands
Transv. arytenoid m.
Oblique arytenoid m.
Conus elasticus
Post. cricoarytenoid m.
Thyroid cartilage, inf. horn

Trachea

Fig. 315. Larynx opened posteriorly and its walls retracted laterally. Beneath the vocal fold the mucous membrane and conus elasticus were removed to the superior border of the arch of the cricoid cartilage in order to display the muscles on the left from inside the larynx.

Intrinsic Muscles of the Larynx

Name	Origin	Insertion	Innervation	Function
Cricothyroid muscle superficial: pars recta deep: pars obliqua	Outer surface of cricoid cartilage	Caudal margin and inferior horn of thyroid cartilage	Cricothyroid muscle: external branch of the superior laryngeal nerve; the remaining laryngeal muscles: from the recurrent laryngeal nerve	Cricothyroid muscle tenses the vocal cords. Posterior cricoarytenoid muscle dilates, and the lateral cricoarytenoid muscle narrows the rima glottidis. The remaining are essentially sphincter muscles for the entrance of the larynx and the rima glottidis or the regulation of the tension in the vocalis muscle.
Posterior cricoarytenoid muscle	Dorsal surface of cricoid cartilage	Muscular process on the posterior surface of arytenoid cartilage		
Lateral cricoarytenoid muscle	Cranial margin of lateral part of cricoid cartilage			
Transverse arytenoid muscle	Lateral edge and dorsal surface of arytenoid cartilage	To the same part on the opposite side (crosses transversely between the 2 cartilages)		
Oblique arytenoid muscle	Muscular process on the posterior surface of one arytenoid cartilage	Apex of muscular process of opposite arytenoid cartilage		
Aryepiglottic muscle	Variable fibers of the oblique arytenoid muscle in the aryepiglottic folds	Lateral border of epiglottis		
Thyroarytenoid muscle	Inner surface of angle of the thyroid cartilage	Muscular process and lateral surface of arytenoid cartilage		
Vocalis muscle	From medial fibers of thyroarytenoid muscle adherent to the vocal ligament	Along vocal ligament extending to the vocal process of arytenoid cartilage		
Thyroepiglottic muscle	Continuation of thyroarytenoid muscle in the aryepiglottic fold	Margin of epiglottis		

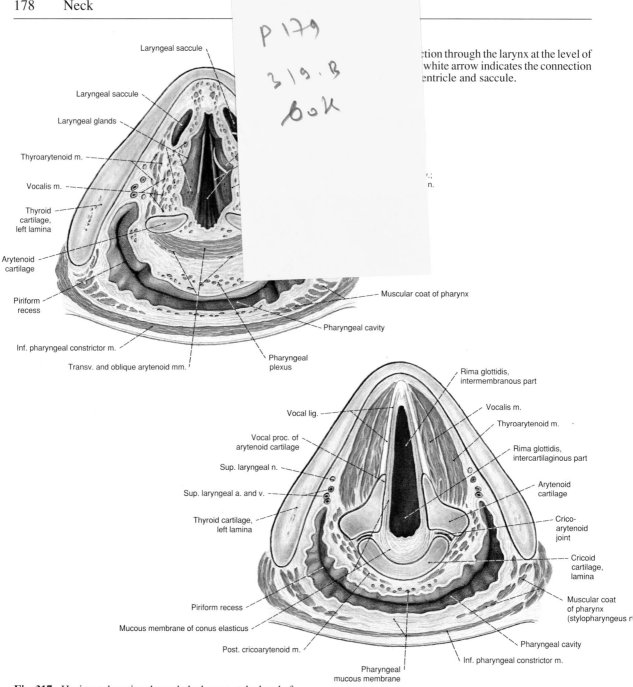

Laryngeal saccule

Laryngeal saccule

Laryngeal glands

Thyroarytenoid m.

Vocalis m.

Thyroid
cartilage,
left lamina

Arytenoid
cartilage

Piriform
recess

Inf. pharyngeal constrictor m.

Transv. and oblique arytenoid mm.

P 179
319.B
Book

...ction through the larynx at the level of
...white arrow indicates the connection
...entricle and saccule.

Muscular coat of pharynx

Pharyngeal cavity

Pharyngeal
plexus

Vocal lig.

Vocal proc. of
arytenoid cartilage

Sup. laryngeal n.

Sup. laryngeal a. and v.

Thyroid cartilage,
left lamina

Piriform recess

Mucous membrane of conus elasticus

Post. cricoarytenoid m.

Pharyngeal
mucous membrane

Rima glottidis,
intermembranous part

Vocalis m.

Thyroarytenoid m.

Rima glottidis,
intercartilaginous part

Arytenoid
cartilage

Crico-
arytenoid
joint

Cricoid
cartilage,
lamina

Muscular coat
of pharynx
(stylopharyngeus m)

Pharyngeal cavity

Inf. pharyngeal constrictor m.

Fig. 317. Horizontal section through the larynx at the level of
the rima glottidis and through the laryngeal part of the pharynx.

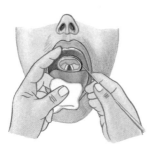

Fig. 318a. Laryngoscopy with tongue pulled forward. View
of the vocal folds (= indirect laryngoscopy).

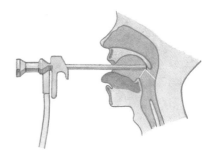

Fig. 318b. Laryngoscope in situ for an endoscopic examination of the interior of the larynx (= direct laryngoscopy).

Epiglottis   Vocal fold   Vestibular fold

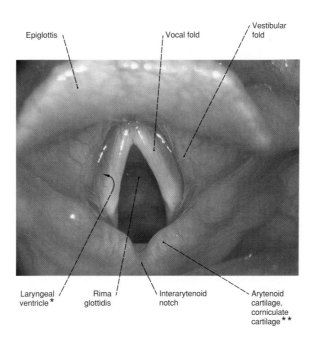

Laryngeal ventricle*   Rima glottidis   Interarytenoid notch   Arytenoid cartilage, corniculate cartilage**

Fig. 319a. Indirect laryngoscopy: Respiratory position with open rima glottidis, masculine larynx.

* Clinical eponym: Sinus of MORGAGNI
** Clinical eponym: Cartilage of SANTORINI

Epiglottis   Rima glottidis   Vocal fold

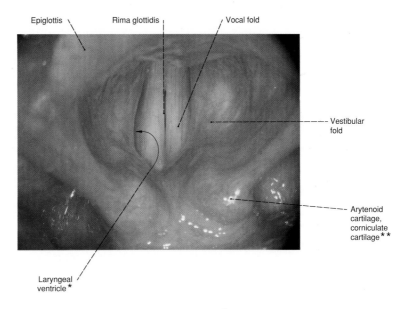

Vestibular fold

Arytenoid cartilage, corniculate cartilage**

Laryngeal ventricle*

Fig. 319b. Indirect laryngoscopy: Position during phonation with closed rima glottidis, masculine larynx. (Figs. 319a and b from: Dr. E. KRUSE, Center for Head-Neck-Ear Therapeutics, Philipps University, Marburg.)

* Clinical eponym: Sinus of MORGAGNI
** Clinical eponym: Cartilage of SANTORINI

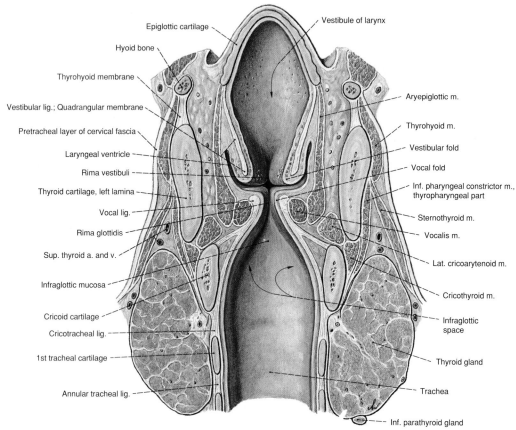

Fig. 320. Frontal section through the larynx and thyroid gland, posterior view.

- Epiglottic cartilage
- Vestibule of larynx
- Hyoid bone
- Thyrohyoid membrane
- Vestibular lig.; Quadrangular membrane
- Pretracheal layer of cervical fascia
- Laryngeal ventricle
- Rima vestibuli
- Thyroid cartilage, left lamina
- Vocal lig.
- Rima glottidis
- Sup. thyroid a. and v.
- Infraglottic mucosa
- Cricoid cartilage
- Cricotracheal lig.
- 1st tracheal cartilage
- Annular tracheal lig.
- Aryepiglottic m.
- Thyrohyoid m.
- Vestibular fold
- Vocal fold
- Inf. pharyngeal constrictor m., thyropharyngeal part
- Sternothyroid m.
- Vocalis m.
- Lat. cricoarytenoid m.
- Cricothyroid m.
- Infraglottic space
- Thyroid gland
- Trachea
- Inf. parathyroid gland

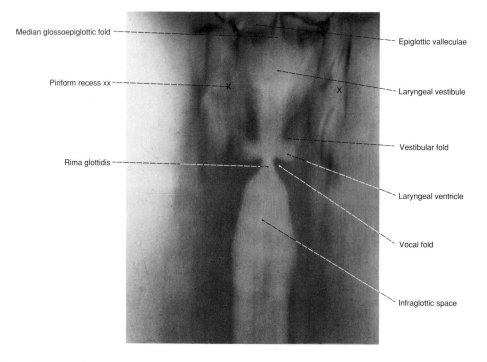

Fig. 321. Radiograph of the larynx, sagittal radiation direction.

- Median glossoepiglottic fold
- Piriform recess xx
- Rima glottidis
- Epiglottic valleculae
- Laryngeal vestibule
- Vestibular fold
- Laryngeal ventricle
- Vocal fold
- Infraglottic space

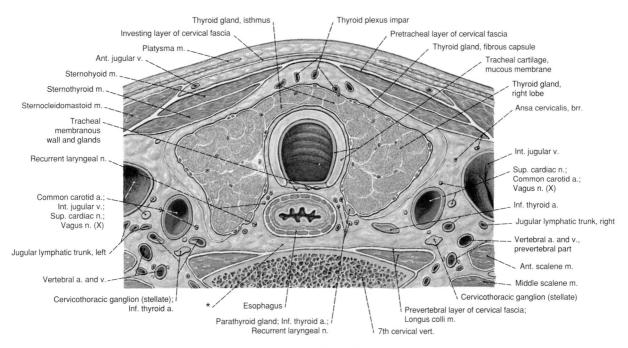

Fig. 322. Horizontal section through the cervical viscera at the level of the second tracheal cartilage.

* Clinically: retroesophageal space

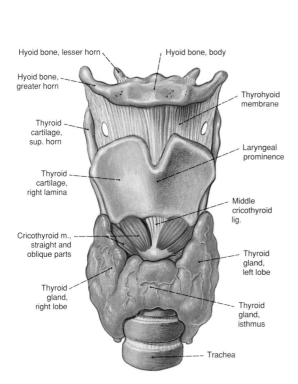

Fig. 323. Hyoid bone, larynx, thyroid gland and upper part of the trachea, anterior view.

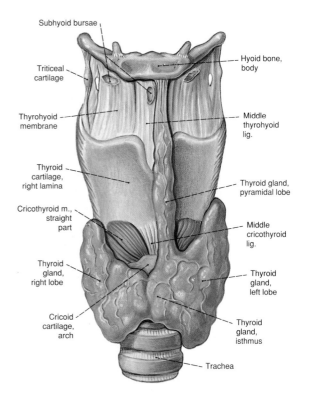

Fig. 324. Shape and position of the thyroid gland in relation to the trachea and larynx. The thyroid gland in this specimen contains a middle (pyramidal) lobe which extends to the median thyrohyoid ligament and continues as connective tissue to the hyoid bone. Anterior view.

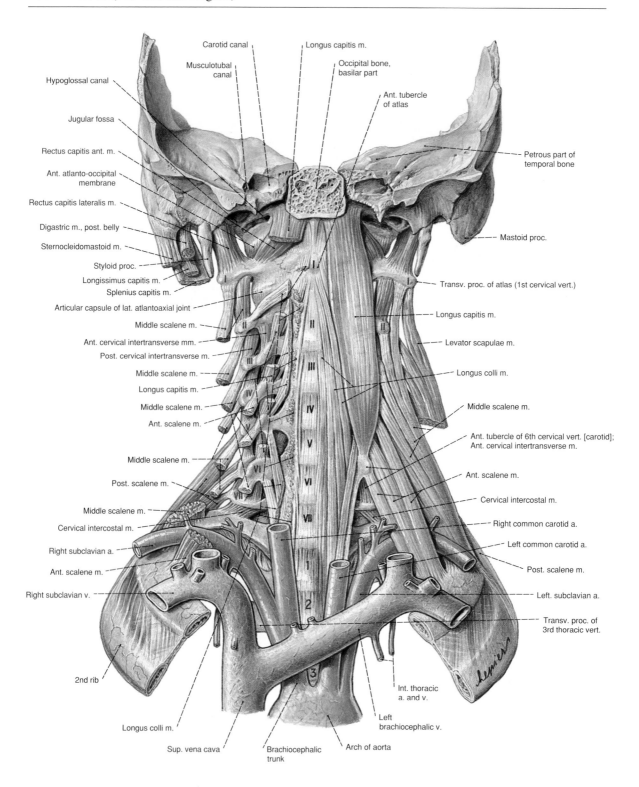

Fig. 325. The deep cervical muscles after removal of the cervical viscera. Anterior view. I-VII = first to seventh cervical vertebrae, 1-3 = first to third thoracic vertebrae. The base of the skull has been sectioned at the level of the basilar part of the occipital bone. The first and second ribs were sectioned near the bony-cartilaginous junction. The superior vena cava and arch of the aorta are shown with their large trunks: the brachiocephalic trunk, the common carotid artery, the left subclavian artery as it passes between the anterior and middle scalene muscles.

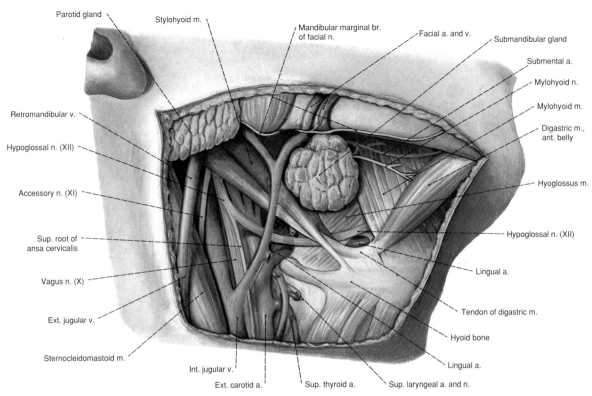

Fig. 326. Topography of the submandibular triangle. The small area bounded by the hypoglossal nerve above, the intermediate tendon of the digastric muscle below and the margin of the mylohyoid muscle medially is also called the "lingual triangle".

Anterior Vertebral Muscles

Name	Origin	Insertion	Innervation	Function
1. Longus colli muscle Vertical portion Superior oblique Inferior oblique	Bodies of first 3 thoracic and last 3 cervical vertebrae; Anterior tubercles of transverse processes of cervical vertebrae C3–C5; Anterior surface of bodies of first 2 or 3 thoracic vertebrae	Bodies of cervical vertebrae C2–C4; Tubercle on anterior arch of the atlas and body of next cervical vertebra; Anterior tubercles of the transverse processes of 5th and 6th cervical vertebrae;	Cervical plexus	Flex the cervical vertebral column or the head forward; unilaterally incline and rotate the head toward the same side; 3 and 4 assist in stabilizing atlanto-occipital joint
2. Longus capitis muscle	Anterior tubercles of transverse processes of 3rd–6th cervical vertebrae	Inferior border of basilar part of occipital bone		
3. Rectus capitis anterior muscle	Lateral mass of atlas and root of its transverse process	Inferior border of basilar part of occipital bone		
4. Rectus capitis lateralis muscle	Superior surface of transverse process of atlas	Inferior surface of jugular process of occipital bone		

Lateral Vertebral Muscles

Name	Origin	Insertion	Innervation	Function
1. Anterior scalene muscle	Anterior tubercles of transverse processes of 3rd to 6th cervical vertebrae	Scalene tubercle on 1st rib	Cervical and brachial plexuses	Elevate 1st and 2nd ribs (muscles of inspiration); flex and laterally rotate vertebral column
2. Middle scalene muscle	Posterior tubercles of transverse processes of last six cervical vertebrae	Cranial surface of 1st rib, behind subclavian groove		
3. Posterior scalene muscle	Posterior tubercles of transverse processes of 4th to 6th cervical vertebrae	Outer surface of 2nd rib		

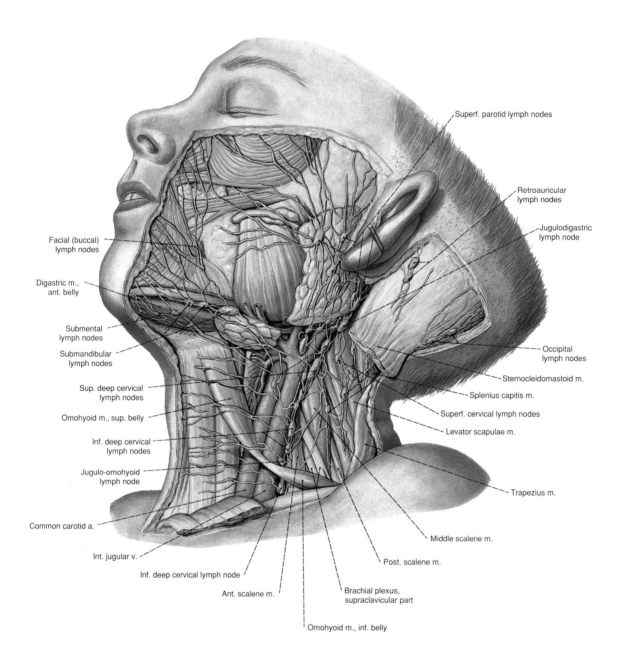

Superf. parotid lymph nodes

Retroauricular
lymph nodes

Jugulodigastric
lymph node

Facial (buccal)
lymph nodes

Digastric m.,
ant. belly

Submental
lymph nodes

Submandibular
lymph nodes

Sup. deep cervical
lymph nodes

Omohyoid m., sup. belly

Inf. deep cervical
lymph nodes

Jugulo-omohyoid
lymph node

Common carotid a.

Int. jugular v.

Inf. deep cervical lymph node

Ant. scalene m.

Omohyoid m., inf. belly

Brachial plexus,
supraclavicular part

Post. scalene m.

Middle scalene m.

Trapezius m.

Levator scapulae m.

Superf. cervical lymph nodes

Splenius capitis m.

Sternocleidomastoid m.

Occipital
lymph nodes

Fig. 327. The superficial lymphatic vessels and lymph nodes
of the face and the deep cervical lymph nodes of an 8-year-old
boy. The platysma has been removed and the sternocleidomastoid muscle sectioned. Along the carotid sheath the
deep cervical lymph nodes form a chain of "cervical glands".

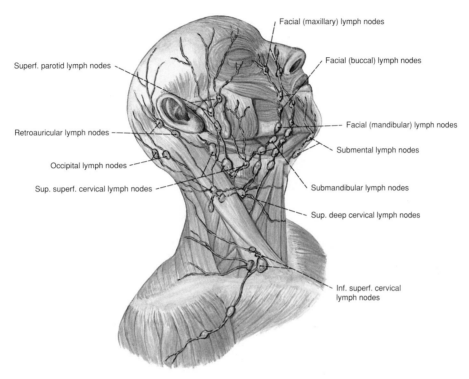

Fig. 328. Superficial lymph nodes and course of the lymphatic vessels in the head and neck. (From: LEIBER: The Human Lymph Nodes, Urban & Schwarzenberg, Munich-Berlin, 1961.)

Lymphatic Vessels and Lymph Nodes of the Head and Neck Region

The lymph nodes (formerly also known as lymph glands) are filtration organs situated at varying intervals along the lymphatic vessels, thereby interrupting their course. At specific points the lymph nodes may be grouped together (e.g., in the head and neck region, in the axilla, at the hilus of the lung, in the mesenteries, in the inguinal region), while other regions (e.g., the limbs) may be devoid of lymph nodes or may have only a few. In the limbs, as in the head, superficial nodes outnumber deep nodes. In the neck, the numbers of superficial and deep nodes are about equal, while in the trunk, the deep nodes are more numerous than the superficial. Lymph nodes are designated according to the region of the body in which they are located. Nodes vary in size and shape, the smallest being microscopic in size, while others may be the size of a pea, hazelnut or even larger. Most lymph nodes have an elongated, ellipsoidal, frequently bean-shaped appearance. As a rule the hilus can be recognized. Depending upon their size, the nodes contain a variable number of lymphatic vessels (afferent vessels). Generally, not more than one or two efferent vessels exit from the hilus of the node. Lymph is filtered in the nodes. Nodes therefore function as sieves for both animate (e.g., bacteria, tumor cells) and inanimate agents (e.g., cellular debris, metabolites, exogenous pigments). Additionally, they are the source and storage area of the B- and T-lymphocytes and other immunologically active protective agents. They participate in all inflammatory processes: in the head region, for example, in diseases of the tonsils, nasal cavity and sinuses, ear, teeth and salivary glands.

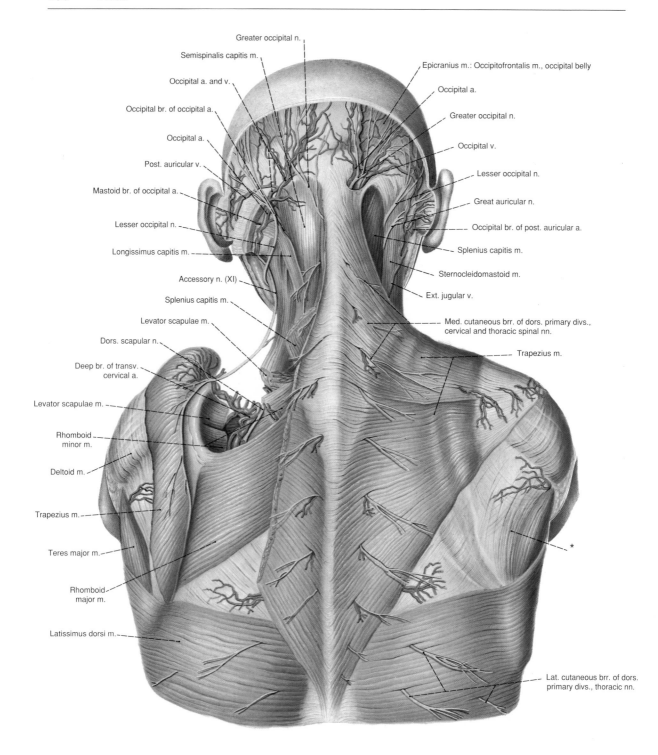

Greater occipital n.

Semispinalis capitis m.

Occipital a. and v.

Occipital br. of occipital a.

Occipital a.

Post. auricular v.

Mastoid br. of occipital a.

Lesser occipital n.

Longissimus capitis m.

Accessory n. (XI)

Splenius capitis m.

Levator scapulae m.

Dors. scapular n.

Deep br. of transv. cervical a.

Levator scapulae m.

Rhomboid minor m.

Deltoid m.

Trapezius m.

Teres major m.

Rhomboid major m.

Latissimus dorsi m.

Epicranius m.: Occipitofrontalis m., occipital belly

Occipital a.

Greater occipital n.

Occipital v.

Lesser occipital n.

Great auricular n.

Occipital br. of post. auricular a.

Splenius capitis m.

Sternocleidomastoid m.

Ext. jugular v.

Med. cutaneous brr. of dors. primary divs., cervical and thoracic spinal nn.

Trapezius m.

*

Lat. cutaneous brr. of dors. primary divs., thoracic nn.

Fig. 329. Nerves and blood vessels of the posterior neck and the upper back. On the left side, the trapezius, sternocleidomastoid, splenius capitis and levator scapulae muscles have been resected.

* Traditionally: Infraspinatus fascia

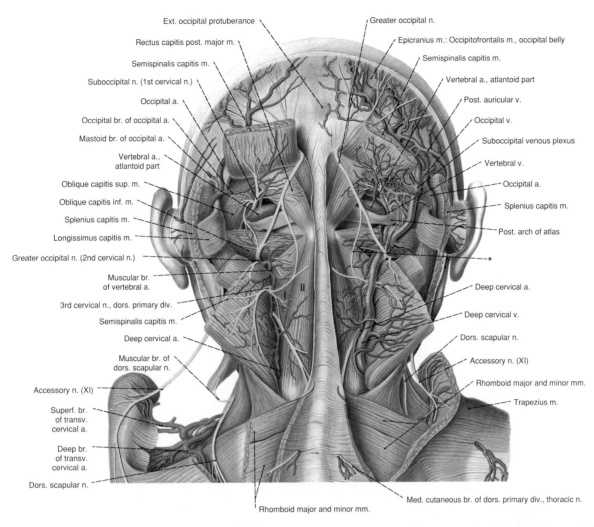

Ext. occipital protuberance
Rectus capitis post. major m.
Semispinalis capitis m.
Suboccipital n. (1st cervical n.)
Occipital a.
Occipital br. of occipital a.
Mastoid br. of occipital a.
Vertebral a., atlantoid part
Oblique capitis sup. m.
Oblique capitis inf. m.
Splenius capitis m.
Longissimus capitis m.
Greater occipital n. (2nd cervical n.)
Muscular br. of vertebral a.
3rd cervical n., dors. primary div.
Semispinalis capitis m.
Deep cervical a.
Muscular br. of dors. scapular n.
Accessory n. (XI)
Superf. br. of transv. cervical a.
Deep br. of transv. cervical a.
Dors. scapular n.

Greater occipital n.
Epicranius m.: Occipitofrontalis m., occipital belly
Semispinalis capitis m.
Vertebral a., atlantoid part
Post. auricular v.
Occipital v.
Suboccipital venous plexus
Vertebral v.
Occipital a.
Splenius capitis m.
Post. arch of atlas
*
Deep cervical a.
Deep cervical v.
Dors. scapular n.
Accessory n. (XI)
Rhomboid major and minor mm.
Trapezius m.
Med. cutaneous br. of dors. primary div., thoracic n.
Rhomboid major and minor mm.

Fig. 330. Nerves and blood vessels of the occipital and nuchal regions. Deep layer. I = Multifidus muscles, II = Semispinalis cervicis muscle. In Figs. 329 and 330, the occipital artery lies between the splenius capitis and longissimus capitis muscles, while in many cases it courses under the longissimus capitis muscle.

* Unnamed communicating nerve between dorsal primary divisions of the second and third cervical nerves

Splenius Capitis and Cervicis Muscles

The two muscles are close together at their origins from the spinous processes. Gradually as they approach their individual insertions, they become separated into two specific muscles.

Name	Origin	Insertion	Innervation	Function
1. **Splenius capitis muscle** The splenius muscles are deep back muscles and are innervated accordingly	Nuchal ligament, spinous processes of 3rd (4th) to 7th cervical and first 3 thoracic vertebrae	Lateral third of the superior nuchal line to the mastoid process of the temporal bone	Dorsal primary divisions of nerves C1–4 (5), lateral branches	Muscles of both sides acting together, extend the neck and head. Muscles of one side rotate the upper cervical vertebrae and, by its action on the atlas, rotate the head to the same side; seldom act alone
2. **Splenius cervicis muscle**	Spines of 3rd (or 4th) to 6th thoracic vertebrae	Posterior tubercles of the transverse processes of the first 3 cervical vertebrae		

Note: The atlantiod part of the vertebral artery is deep within the suboccipital muscular triangle bounded by the oblique capitis superior and inferior muscles and the rectus capitis posterior major muscle. After passing through the transverse foramen of the atlas, the vertebral artery runs medially in its groove in the atlas, passing through the posterior atlanto-occipital membrane and the dura mater to enter the cranium through the foramen magnum of the occipital bone and contribute to the brain's blood supply (Figs. 269, 562).

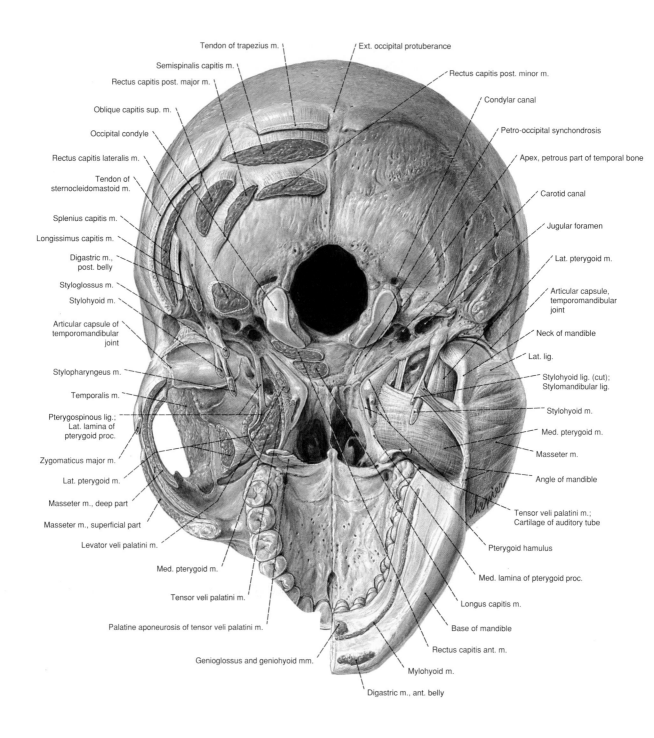

Tendon of trapezius m.

Semispinalis capitis m.

Rectus capitis post. major m.

Oblique capitis sup. m.

Occipital condyle

Rectus capitis lateralis m.

Tendon of
sternocleidomastoid m.

Splenius capitis m.

Longissimus capitis m.

Digastric m.,
post. belly

Styloglossus m.

Stylohyoid m.

Articular capsule of
temporomandibular
joint

Stylopharyngeus m.

Temporalis m.

Pterygospinous lig.;
Lat. lamina of
pterygoid proc.

Zygomaticus major m.

Lat. pterygoid m.

Masseter m., deep part

Masseter m., superficial part

Levator veli palatini m.

Med. pterygoid m.

Tensor veli palatini m.

Palatine aponeurosis of tensor veli palatini m.

Genioglossus and geniohyoid mm.

Digastric m., ant. belly

Ext. occipital protuberance

Rectus capitis post. minor m.

Condylar canal

Petro-occipital synchondrosis

Apex, petrous part of temporal bone

Carotid canal

Jugular foramen

Lat. pterygoid m.

Articular capsule,
temporomandibular
joint

Neck of mandible

Lat. lig.

Stylohyoid lig. (cut);
Stylomandibular lig.

Stylohyoid m.

Med. pterygoid m.

Masseter m.

Angle of mandible

Tensor veli palatini m.;
Cartilage of auditory tube

Pterygoid hamulus

Med. lamina of pterygoid proc.

Longus capitis m.

Base of mandible

Rectus capitis ant. m.

Mylohyoid m.

Fig. 331. Muscle origins on the base of the skull. The left side
of the mandible has been removed. On the right side, the
muscles of mastication are retained.

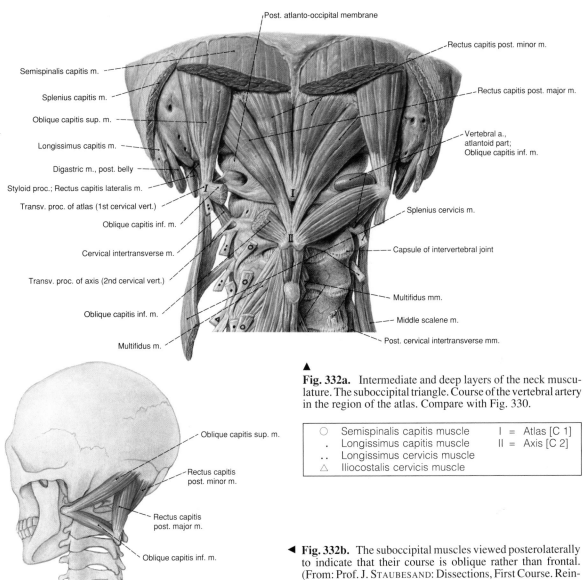

Fig. 332a. Intermediate and deep layers of the neck musculature. The suboccipital triangle. Course of the vertebral artery in the region of the atlas. Compare with Fig. 330.

○	Semispinalis capitis muscle	I = Atlas [C 1]
.	Longissimus capitis muscle	II = Axis [C 2]
..	Longissimus cervicis muscle	
△	Iliocostalis cervicis muscle	

◄ Fig. 332b. The suboccipital muscles viewed posterolaterally to indicate that their course is oblique rather than frontal. (From: Prof. J. STAUBESAND: Dissections, First Course. Reinhardt, Munich-Basel, 1965.)

Suboccipital Muscles

Name	Origin	Insertion	Innervation	Function
1. **Rectus capitis posterior major muscle**	Spinous process of the axis	Inferior nuchal line (of occipital bone)	Suboccipital nerve (dorsal primary division of the first cervical nerve)	Extend and rotate head. The rectus capitis posterior major and oblique capitis inferior muscles turn the head to the same side
2. **Rectus capitis posterior minor muscle**	Posterior arch tubercle of atlas	Medial part of the inferior nuchal line of the occipital bone		
3. **Oblique capitis superior muscle**	Transverse process of the atlas	Occipital bone, above the inferior nuchal line		
4. **Oblique capitis inferior muscle**	Spinous process of the axis	Transverse process of the atlas		

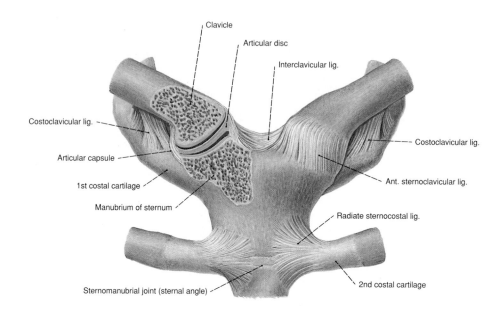

Fig. 333.   The sternoclavicular joints and the first two sternocostal joints. Anterior view. The right sternoclavicular joint has been opened by a frontal cut.

Articulations of the Upper Limb

Joint	Type of Joint	Possible Movements
Sternoclavicular joint	Double gliding; special feature: articular disc	Limited caudal movement of clavicle in this joint, about 10 cm ventrally, 3 cm dorsally and 10 cm cranially; rotation of clavicle about 30° around its longitudinal axis
Acromioclavicular joint	Plane or gliding	Translation ventrally and dorsally, cranially and caudally; rotation of the clavicle in the transverse plane

Articulations of the Upper Limb (Continued)

Joint	Type of Joint	Possible Movements
Shoulder joint	Ball-and-socket joint	Flexion Extension Abduction Adduction Inward rotation Outward rotation
Elbow joint a) Humeroulnar articulation	Hinge joint	Flexion Extension
b) Humeroradial articulation	Ball-and-socket joint (functionally restricted)	Flexion Extension Rotation
c) Proximal radioulnar articulation	Pivot joint	Pronation Supination
Distal radioulnar articulation	Pivot joint	
Carpal joints a) Radiocarpal (wrist) articulation	Ellipsoid joint	Flexion, Extension Abduction, Adduction Circumduction
b) Midcarpal articulation	Compound joint: saddle, gliding	Flexion, Extension Abduction Rotation
Carpometacarpal joint articulation of thumb	Saddle joint	Flexion Extension Abduction Adduction Opposition Rotation Circumduction
Metacarpophalangeal joints articulations of fingers	Ball-and-socket joints (functionally restricted)	Flexion Extension Abduction Adduction Circumduction
Finger joints interphalangeal articulations	Hinge joints	Flexion Extension

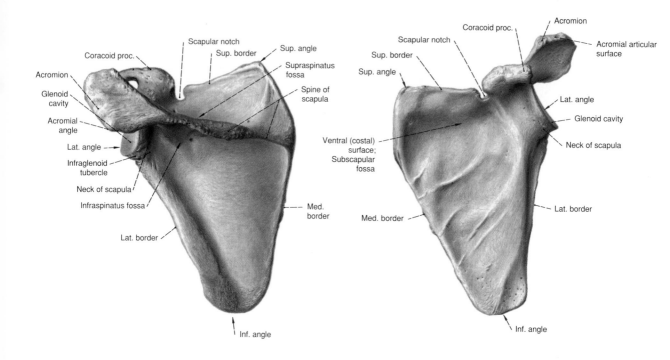

Fig. 334.   The left scapula. View of dorsal surface.

Fig. 335.   The left scapula. View of ventral surface.

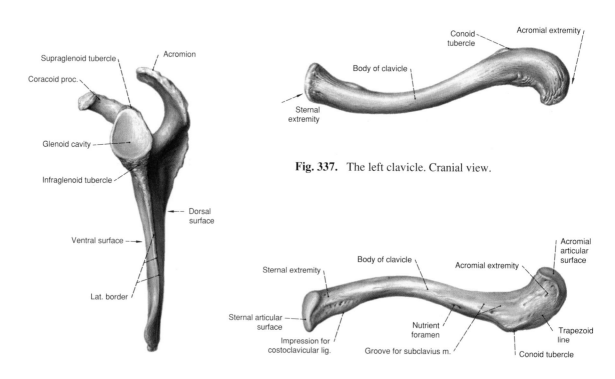

Fig. 336.   The left scapula. Lateral view.

Fig. 337.   The left clavicle. Cranial view.

Fig. 338.   The left clavicle. Caudal view.

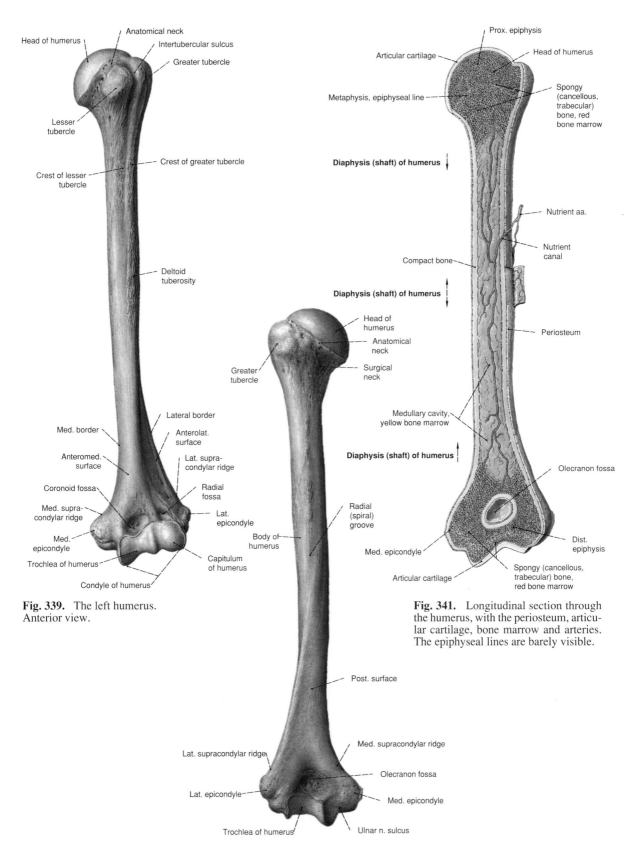

Fig. 339. The left humerus.
Anterior view.

Fig. 340. The left humerus.
Posterior view.

Fig. 341. Longitudinal section through the humerus, with the periosteum, articular cartilage, bone marrow and arteries. The epiphyseal lines are barely visible.

Fig. 339 labels:
Head of humerus
Anatomical neck
Intertubercular sulcus
Greater tubercle
Lesser tubercle
Crest of greater tubercle
Crest of lesser tubercle
Deltoid tuberosity
Lateral border
Med. border
Anterolat. surface
Anteromed. surface
Lat. supra-condylar ridge
Coronoid fossa
Radial fossa
Med. supra-condylar ridge
Lat. epicondyle
Med. epicondyle
Trochlea of humerus
Capitulum of humerus
Condyle of humerus

Fig. 340 labels:
Head of humerus
Anatomical neck
Surgical neck
Greater tubercle
Radial (spiral) groove
Body of humerus
Post. surface
Lat. supracondylar ridge
Med. supracondylar ridge
Lat. epicondyle
Olecranon fossa
Med. epicondyle
Trochlea of humerus
Ulnar n. sulcus

Fig. 341 labels:
Prox. epiphysis
Articular cartilage
Head of humerus
Metaphysis, epiphyseal line
Spongy (cancellous, trabecular) bone, red bone marrow
Diaphysis (shaft) of humerus
Nutrient aa.
Nutrient canal
Compact bone
Periosteum
Medullary cavity, yellow bone marrow
Olecranon fossa
Med. epicondyle
Dist. epiphysis
Articular cartilage
Spongy (cancellous, trabecular) bone, red bone marrow

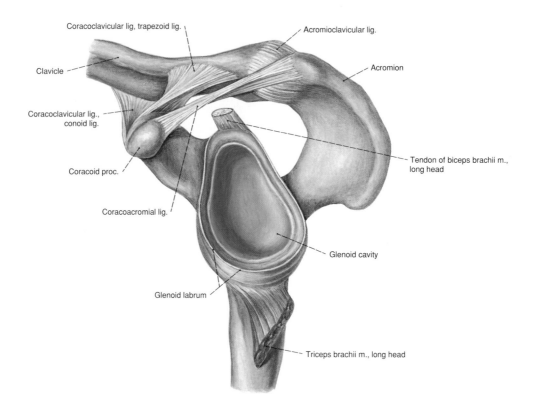

Coracoclavicular lig, trapezoid lig.
Acromioclavicular lig.
Clavicle
Acromion
Coracoclavicular lig., conoid lig.
Coracoid proc.
Tendon of biceps brachii m., long head
Coracoacromial lig.
Glenoid cavity
Glenoid labrum
Triceps brachii m., long head

Fig. 342. The glenoid cavity of the left shoulder. The articular capsule was removed at the glenoid labrum; the origin of the long head of the triceps brachii muscle and the tendon of the long head of the biceps brachii muscle remain.

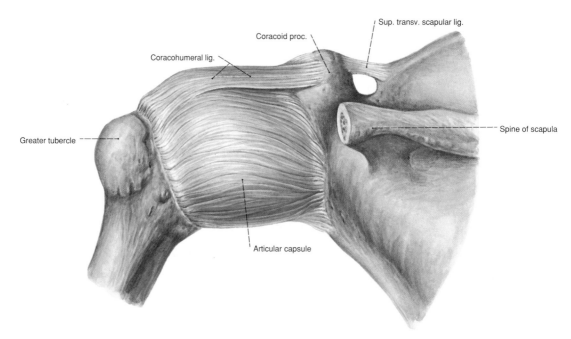

Coracoid proc.
Sup. transv. scapular lig.
Coracohumeral lig.
Greater tubercle
Spine of scapula
Articular capsule

Fig. 343. Articular capsule of the left shoulder joint. The acromion has been removed. Posterior cranial view.

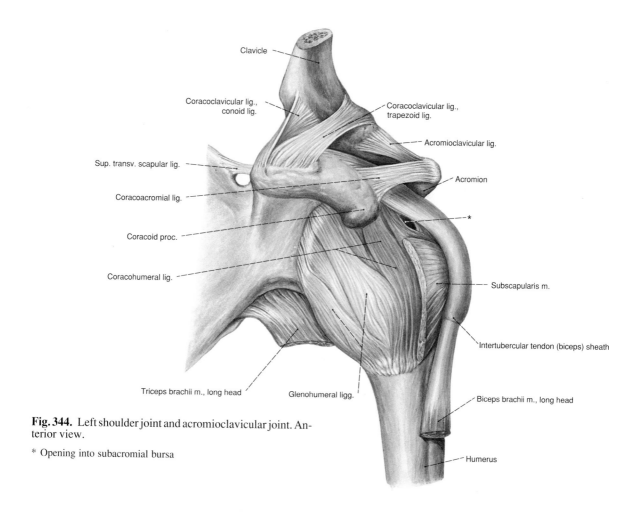

Clavicle

Coracoclavicular lig.,
conoid lig.

Coracoclavicular lig.,
trapezoid lig.

Acromioclavicular lig.

Sup. transv. scapular lig.

Acromion

Coracoacromii lig.

*

Coracoid proc.

Coracohumeral lig.

Subscapularis m.

Intertubercular tendon (biceps) sheath

Triceps brachii m., long head

Glenohumeral ligg.

Biceps brachii m., long head

Humerus

Fig. 344. Left shoulder joint and acromioclavicular joint. Anterior view.

* Opening into subacromial bursa

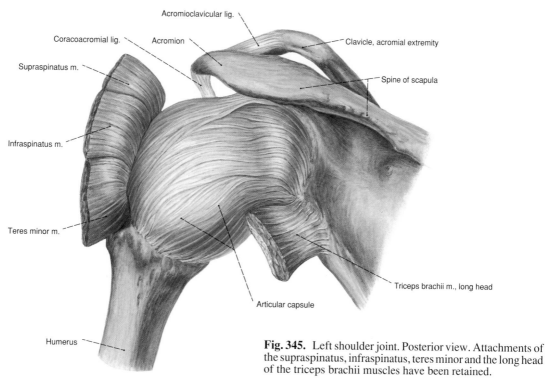

Acromioclavicular lig.

Coracoacromial lig.   Acromion

Clavicle, acromial extremity

Supraspinatus m.

Spine of scapula

Infraspinatus m.

Teres minor m.

Triceps brachii m., long head

Humerus

Articular capsule

Fig. 345. Left shoulder joint. Posterior view. Attachments of the supraspinatus, infraspinatus, teres minor and the long head of the triceps brachii muscles have been retained.

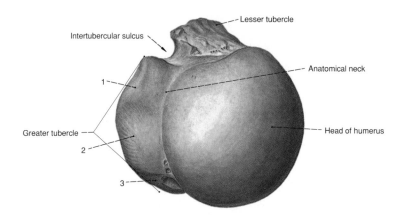

Fig. 346. The head of the left humerus, viewed from above.

1 = Greater tubercle facets for the supraspinatus muscle
2 = for the infraspinatus muscle
3 = for the teres minor muscle

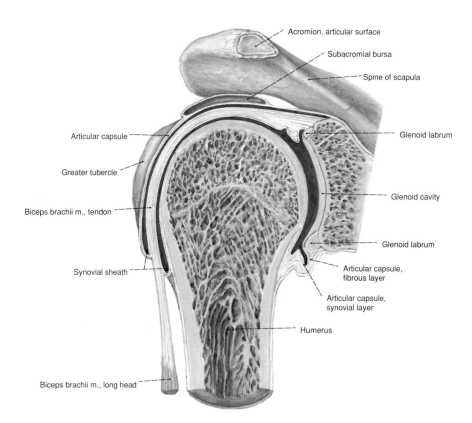

Fig. 347. Frontal section through the right shoulder joint; posterior half.

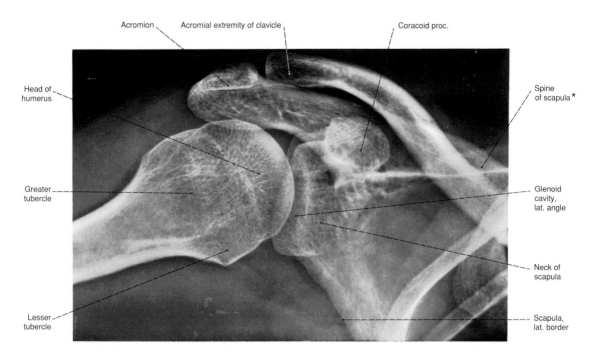

Acromion   Acromial extremity of clavicle   Coracoid proc.

Head of
humerus

Spine
of scapula *

Greater
tubercle

Glenoid
cavity,
lat. angle

Neck of
scapula

Lesser
tubercle

Scapula,
lat. border

Fig. 348a. Radiograph of the right shoulder joint, radiation in anteroposterior direction. Supine position, arm abducted and rotated medially. Central ray perpendicular to the shoulder joint. In this position, unlike Fig. 348b, the lesser tubercle projects into the axilla and the greater tubercle projects onto the anatomical neck of the humerus. The apex of the coracoid process is projected onto the upper part of the shoulder joint. (From R. BIRKNER: Normal Radiologic Patterns and Variances of the Human Skeleton, Urban & Schwarzenberg, Munich-Vienna-Baltimore, 1977.)

* Base = line of attachment

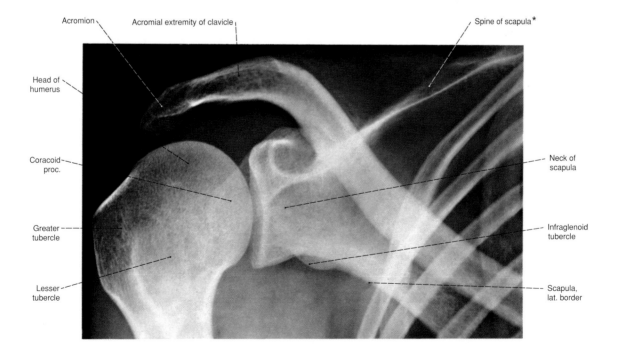

Acromion   Acromial extremity of clavicle   Spine of scapula *

Head of
humerus

Coracoid
proc.

Neck of
scapula

Greater
tubercle

Infraglenoid
tubercle

Lesser
tubercle

Scapula,
lat. border

Fig. 348b. Radiograph of the right shoulder joint, radiation in the cranial anteroposterior direction. Supine position, arm rotated medially. (From R. BIRKNER: Normal Radiologic Patterns and Variances of the Human Skeleton, Urban & Schwarzenberg, Munich-Vienna-Baltimore, 1977.)

* Base = line of attachment

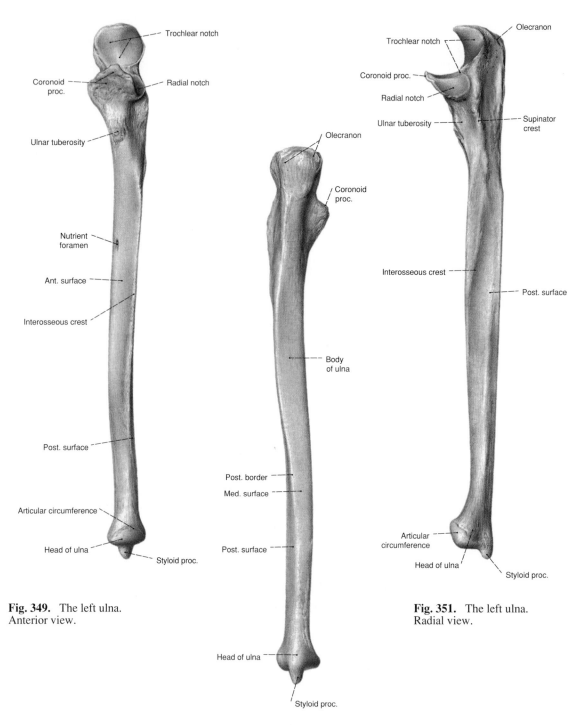

Trochlear notch
Coronoid proc.
Radial notch
Ulnar tuberosity
Nutrient foramen
Ant. surface
Interosseous crest
Post. surface
Articular circumference
Head of ulna
Styloid proc.

Fig. 349. The left ulna. Anterior view.

Olecranon
Coronoid proc.
Body of ulna
Post. border
Med. surface
Post. surface
Head of ulna
Styloid proc.

Fig. 350. The left ulna. Posterior view.

Olecranon
Trochlear notch
Coronoid proc.
Radial notch
Ulnar tuberosity
Supinator crest
Interosseous crest
Post. surface
Articular circumference
Head of ulna
Styloid proc.

Fig. 351. The left ulna. Radial view.

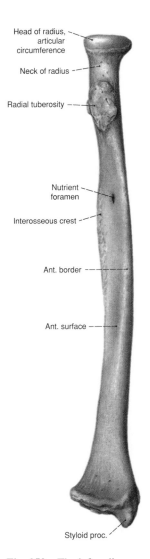

Head of radius, articular circumference

Neck of radius

Radial tuberosity

Nutrient foramen

Interosseous crest

Ant. border

Ant. surface

Styloid proc.

Fig. 352. The left radius.
Anterior view.

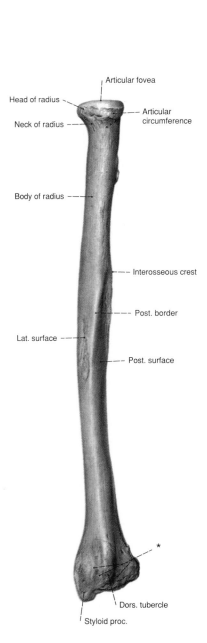

Articular fovea

Head of radius

Neck of radius

Articular circumference

Body of radius

Interosseous crest

Post. border

Lat. surface

Post. surface

*

Dors. tubercle

Styloid proc.

Fig. 353. The left radius.
Posterior view.

* Grooves and crests for the extensor tendons

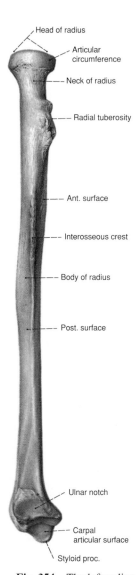

Head of radius

Articular circumference

Neck of radius

Radial tuberosity

Ant. surface

Interosseous crest

Body of radius

Post. surface

Ulnar notch

Carpal articular surface

Styloid proc.

Fig. 354. The left radius.
Ulnar, medial view.

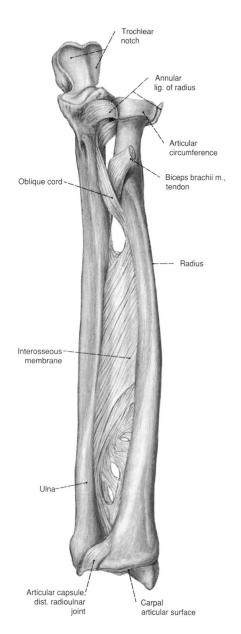

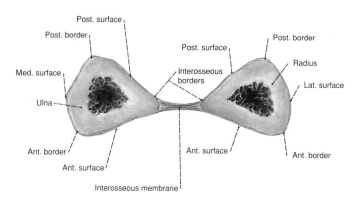

Fig. 356. Surfaces and borders of the bones of the left forearm, sectioned near the middle of its long axis.

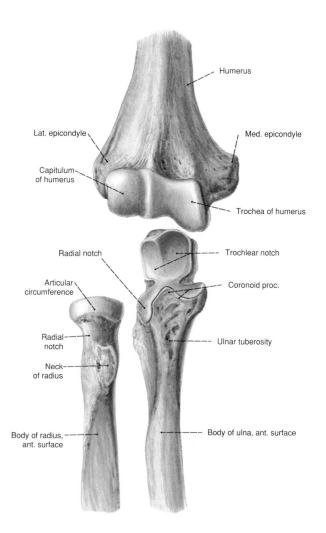

Fig. 355. The ulna, radius, and interosseous membrane of the left forearm. Anterior view with forearm supinated. The elbow joint has been opened, the capsule removed; the annular ligament encircling the head of the radius has been cut.

Fig. 357. The disarticulated joints of the three bones which form the right elbow joint. Anterior view.

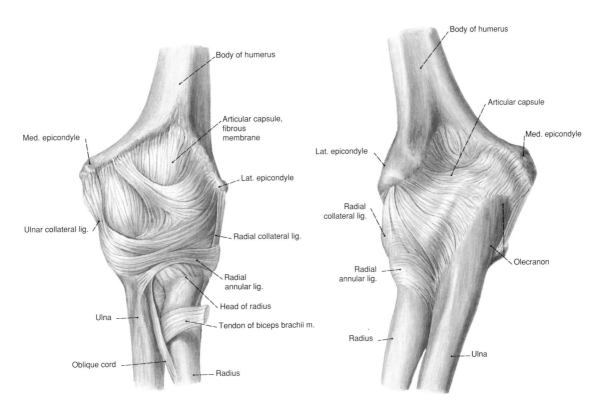

Fig. 358. The left elbow joint. Anterior view.

Fig. 359. The left elbow joint. Posterolateral view.

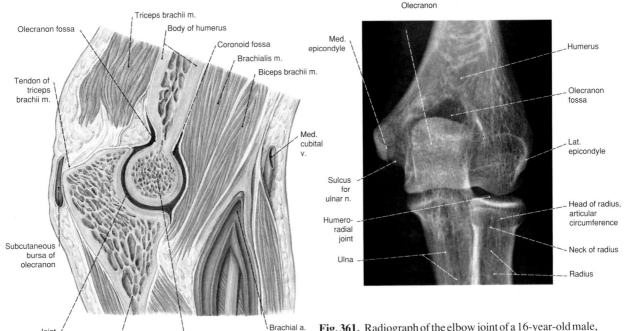

Fig. 360. Sagittal section through the left elbow joint.

Fig. 361. Radiograph of the elbow joint of a 16-year-old male, radiation in sagittal direction. The epiphyseal plates remain as visible lines. Fusion of the epiphyses of the head of the radius and of the olecranon occurs one year before epiphyseal closure at the distal end of the humerus. (From R. Birkner: Normal Radiologic Patterns and Variances of the Human Skeleton, Urban & Schwarzenberg, Munich-Vienna-Baltimore, 1977.)

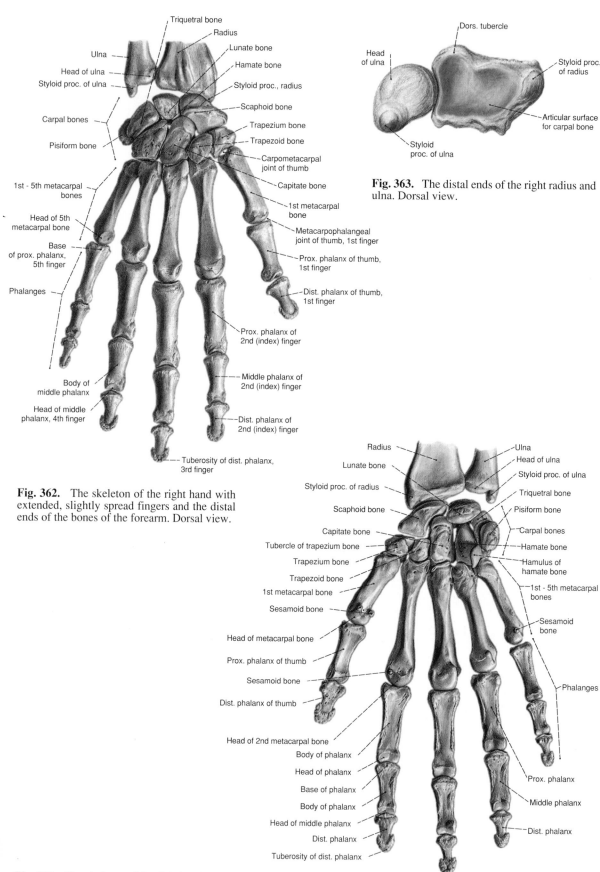

Triquetral bone
Radius
Lunate bone
Ulna
Hamate bone
Head of ulna
Styloid proc. of ulna
Styloid proc., radius
Carpal bones
Scaphoid bone
Pisiform bone
Trapezium bone
Trapezoid bone
Carpometacarpal joint of thumb
1st - 5th metacarpal bones
Capitate bone
Head of 5th metacarpal bone
1st metacarpal bone
Metacarpophalangeal joint of thumb, 1st finger
Base of prox. phalanx, 5th finger
Prox. phalanx of thumb, 1st finger
Phalanges
Dist. phalanx of thumb, 1st finger
Prox. phalanx of 2nd (index) finger
Middle phalanx of 2nd (index) finger
Body of middle phalanx
Head of middle phalanx, 4th finger
Dist. phalanx of 2nd (index) finger
Tuberosity of dist. phalanx, 3rd finger

Fig. 362. The skeleton of the right hand with extended, slightly spread fingers and the distal ends of the bones of the forearm. Dorsal view.

Dors. tubercle
Head of ulna
Styloid proc. of radius
Styloid proc. of ulna
Articular surface for carpal bone

Fig. 363. The distal ends of the right radius and ulna. Dorsal view.

Radius
Ulna
Lunate bone
Head of ulna
Styloid proc. of radius
Styloid proc. of ulna
Scaphoid bone
Triquetral bone
Pisiform bone
Capitate bone
Carpal bones
Tubercle of trapezium bone
Hamate bone
Trapezium bone
Hamulus of hamate bone
Trapezoid bone
1st - 5th metacarpal bones
1st metacarpal bone
Sesamoid bone
Sesamoid bone
Head of metacarpal bone
Prox. phalanx of thumb
Sesamoid bone
Phalanges
Dist. phalanx of thumb
Head of 2nd metacarpal bone
Body of phalanx
Head of phalanx
Prox. phalanx
Base of phalanx
Middle phalanx
Body of phalanx
Head of middle phalanx
Dist. phalanx
Dist. phalanx
Tuberosity of dist. phalanx

Fig. 364. The skeleton of the right hand (same specimen as in Fig. 362). Palmar view.

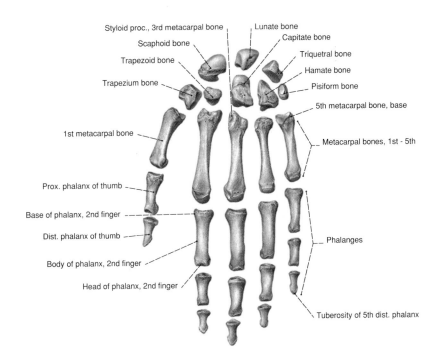

Styloid proc., 3rd metacarpal bone
Scaphoid bone
Trapezoid bone
Trapezium bone
Lunate bone
Capitate bone
Triquetral bone
Hamate bone
Pisiform bone
5th metacarpal bone, base
1st metacarpal bone
Metacarpal bones, 1st – 5th
Prox. phalanx of thumb
Base of phalanx, 2nd finger
Dist. phalanx of thumb
Body of phalanx, 2nd finger
Head of phalanx, 2nd finger
Phalanges
Tuberosity of 5th dist. phalanx

Fig. 365. The bones of the left hand, dorsal view. The carpal and metacarpal bones and the phalanges have been disarticulated for didactic purposes (cf. Figs. 371 and 372).

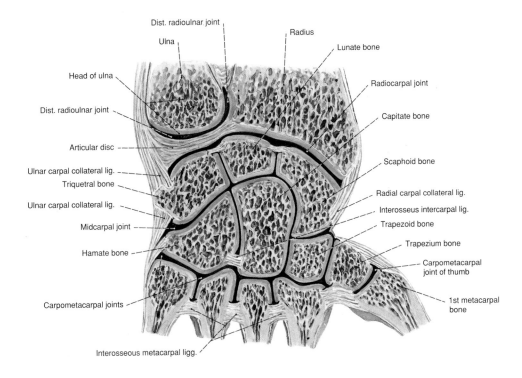

Dist. radioulnar joint
Ulna
Radius
Lunate bone
Head of ulna
Radiocarpal joint
Dist. radioulnar joint
Capitate bone
Articular disc
Ulnar carpal collateral lig.
Triquetral bone
Scaphoid bone
Ulnar carpal collateral lig.
Radial carpal collateral lig.
Interosseus intercarpal lig.
Trapezoid bone
Midcarpal joint
Trapezium bone
Hamate bone
Carpometacarpal joint of thumb
1st metacarpal bone
Carpometacarpal joints
Interosseous metacarpal ligg.

Fig. 366. Longitudinal section through the left wrist joints, cut parallel to the back of the hand.

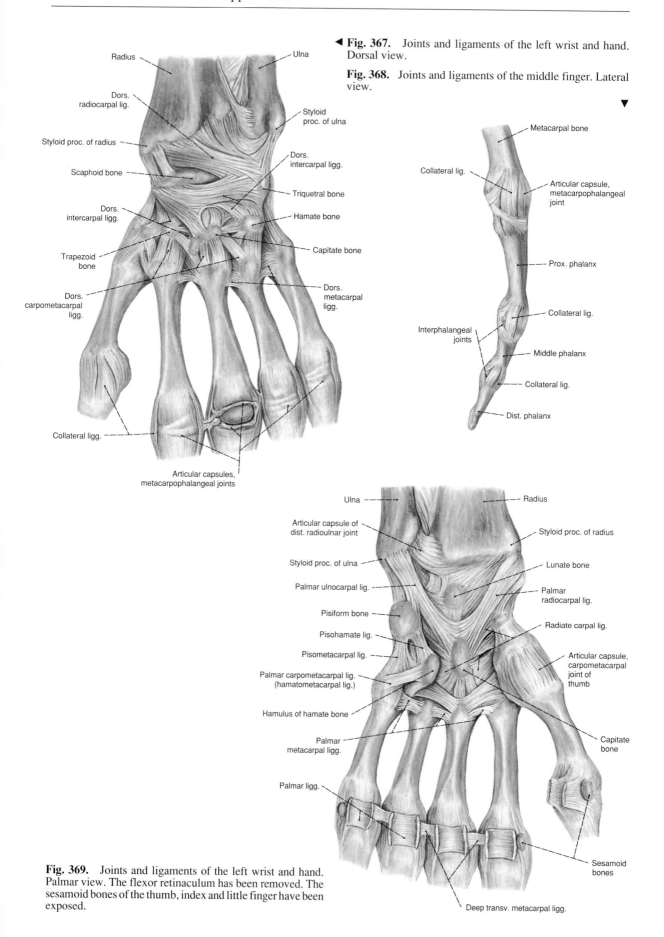

◀ **Fig. 367.**　Joints and ligaments of the left wrist and hand. Dorsal view.

Fig. 368.　Joints and ligaments of the middle finger. Lateral view.

▼

Radius

Ulna

Dors. radiocarpal lig.

Styloid proc. of ulna

Styloid proc. of radius

Dors. intercarpal ligg.

Scaphoid bone

Triquetral bone

Dors. intercarpal ligg.

Hamate bone

Capitate bone

Trapezoid bone

Dors. carpometacarpal ligg.

Dors. metacarpal ligg.

Collateral ligg.

Articular capsules, metacarpophalangeal joints

Metacarpal bone

Collateral lig.

Articular capsule, metacarpophalangeal joint

Prox. phalanx

Collateral lig.

Interphalangeal joints

Middle phalanx

Collateral lig.

Dist. phalanx

Ulna

Radius

Articular capsule of dist. radioulnar joint

Styloid proc. of radius

Styloid proc. of ulna

Lunate bone

Palmar ulnocarpal lig.

Palmar radiocarpal lig.

Pisiform bone

Radiate carpal lig.

Pisohamate lig.

Pisometacarpal lig.

Articular capsule, carpometacarpal joint of thumb

Palmar carpometacarpal lig. (hamatometacarpal lig.)

Hamulus of hamate bone

Capitate bone

Palmar metacarpal ligg.

Palmar ligg.

Sesamoid bones

Deep transv. metacarpal ligg.

Fig. 369.　Joints and ligaments of the left wrist and hand. Palmar view. The flexor retinaculum has been removed. The sesamoid bones of the thumb, index and little finger have been exposed.

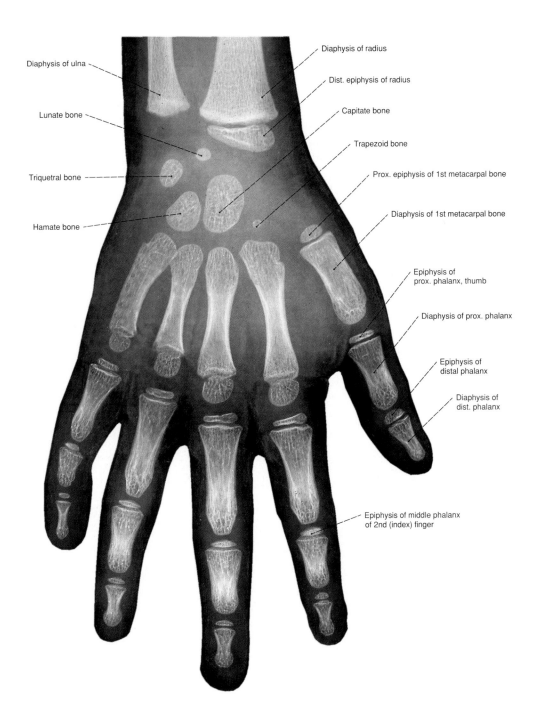

Diaphysis of ulna

Lunate bone

Triquetral bone

Hamate bone

Diaphysis of radius

Dist. epiphysis of radius

Capitate bone

Trapezoid bone

Prox. epiphysis of 1st metacarpal bone

Diaphysis of 1st metacarpal bone

Epiphysis of
prox. phalanx, thumb

Diaphysis of prox. phalanx

Epiphysis of
distal phalanx

Diaphysis of
dist. phalanx

Epiphysis of middle phalanx
of 2nd (index) finger

Fig. 370. Radiograph of the hand of a 5¹/₂-year-old boy. The phalanges of the fingers have only one proximal epiphysis; the second to fifth metacarpals have only one distal epiphysis. The first metacarpal bone appears here as the proximal phalanx of the thumb. The centers of ossification of the trapezoid and lunate bones are still very small; that of the scaphoid bone is absent. It is, therefore, possible to determine "bone age" by examining the extent of ossification.

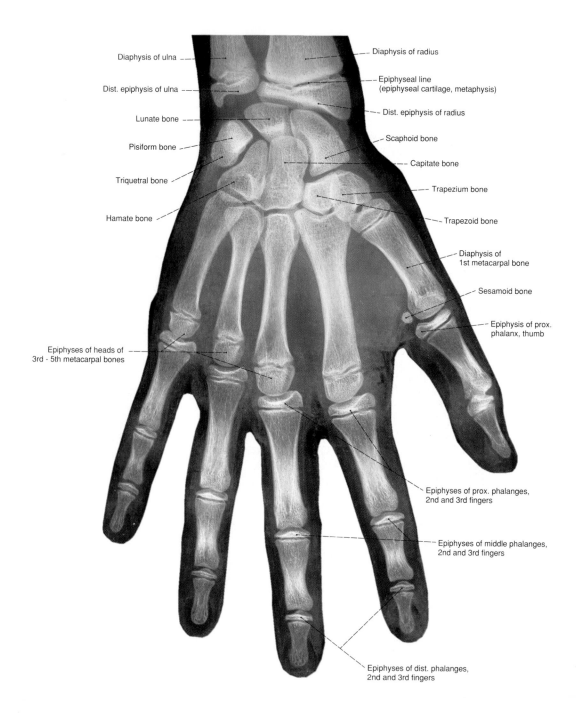

Diaphysis of ulna

Dist. epiphysis of ulna

Lunate bone

Pisiform bone

Triquetral bone

Hamate bone

Epiphyses of heads of
3rd - 5th metacarpal bones

Diaphysis of radius

Epiphyseal line
(epiphyseal cartilage, metaphysis)

Dist. epiphysis of radius

Scaphoid bone

Capitate bone

Trapezium bone

Trapezoid bone

Diaphysis of
1st metacarpal bone

Sesamoid bone

Epiphysis of prox.
phalanx, thumb

Epiphyses of prox. phalanges,
2nd and 3rd fingers

Epiphyses of middle phalanges,
2nd and 3rd fingers

Epiphyses of dist. phalanges,
2nd and 3rd fingers

Fig. 371. Radiograph of the left hand of a 15$^1/_2$-year-old boy.
The distal epiphyses of the radius and ulna, as well as the distal
and proximal epiphyses of the metacarpals and phalanges, are
not yet united with their respective diaphyses (distinct epiphyseal lines).

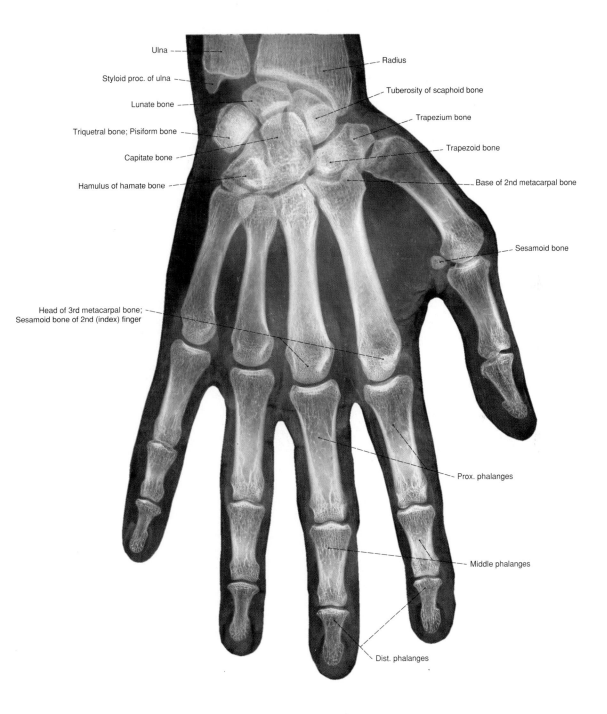

Ulna

Styloid proc. of ulna

Lunate bone

Triquetral bone; Pisiform bone

Capitate bone

Hamulus of hamate bone

Head of 3rd metacarpal bone;
Sesamoid bone of 2nd (index) finger

Radius

Tuberosity of scaphoid bone

Trapezium bone

Trapezoid bone

Base of 2nd metacarpal bone

Sesamoid bone

Prox. phalanges

Middle phalanges

Dist. phalanges

Fig. 372. Radiograph of the left hand of an adult. The shadow of the pisiform bone overlies that of the triquetral bone. The soft parts are seen only as faint shadows. The thicker calcium-containing bones appear darker than the thin or calcium-deficient bones and articular cartilages.

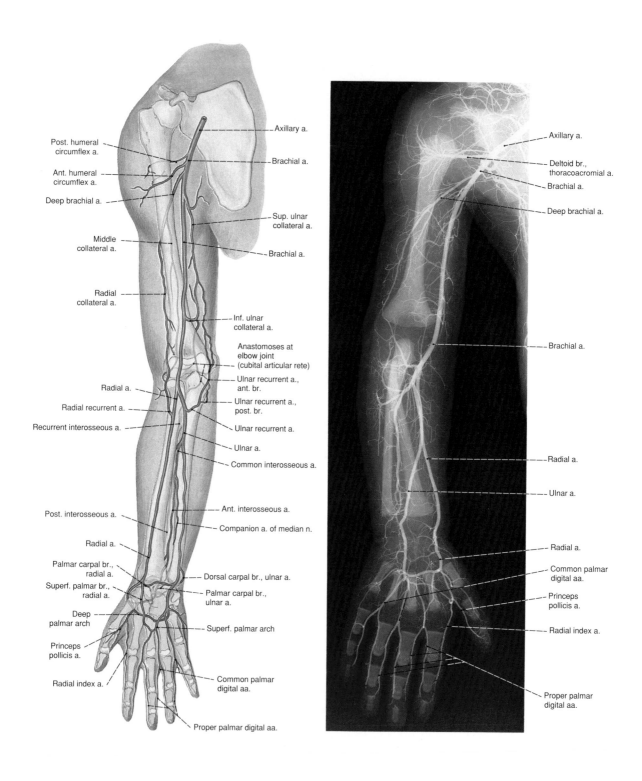

Fig. 373. Arteries of the upper limb. Schematic overview, anterior view.

Fig. 374. Angiographic demonstration of the arteries of the upper limb of an infant (stillborn). (From: Dr. G.W. KAUFFMANN, Radiology Center of the Clinic of Freiburg i. Br.)

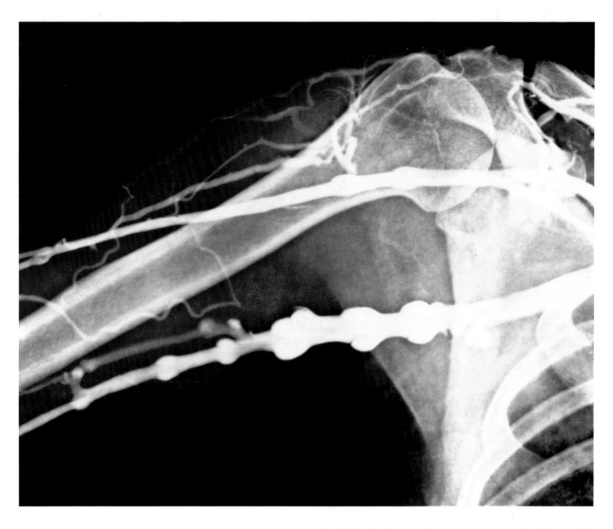

Fig. 375. Radiograph (veno- or phlebogram) of the brachial vein, the axillary vein, as well as the cephalic vein, and some of their tributaries. Note the succession of valvular segments clearly visible in the axillary vein.

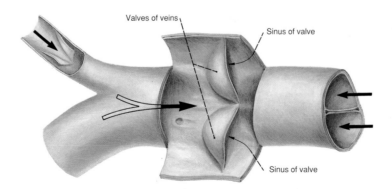

Valves of veins

Sinus of valve

Sinus of valve

Fig. 376. Schematic venous valves. The arrows pointing to the right indicate the normal direction of blood flow. Regurgitation (arrows pointing to the left) causes the valves to close.

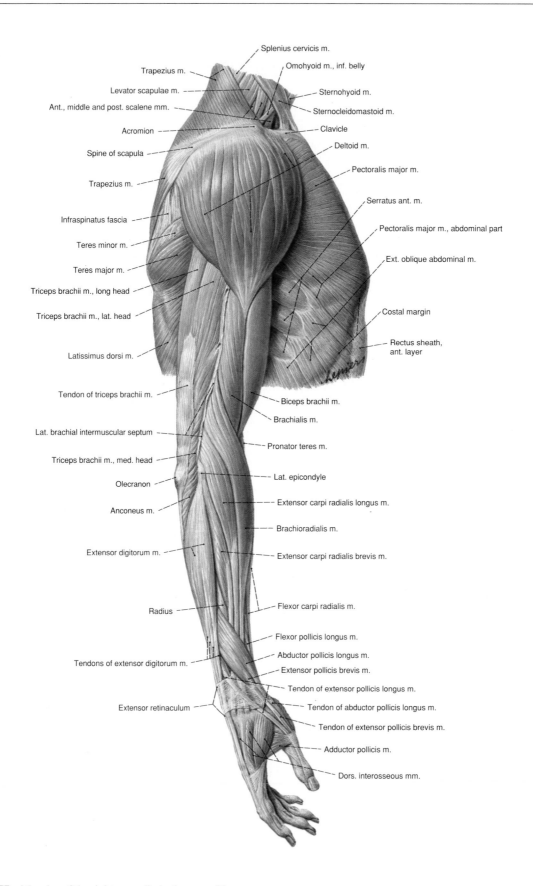

Fig. 377. Muscles of the right upper limb, thorax and lower neck region. Lateral view.

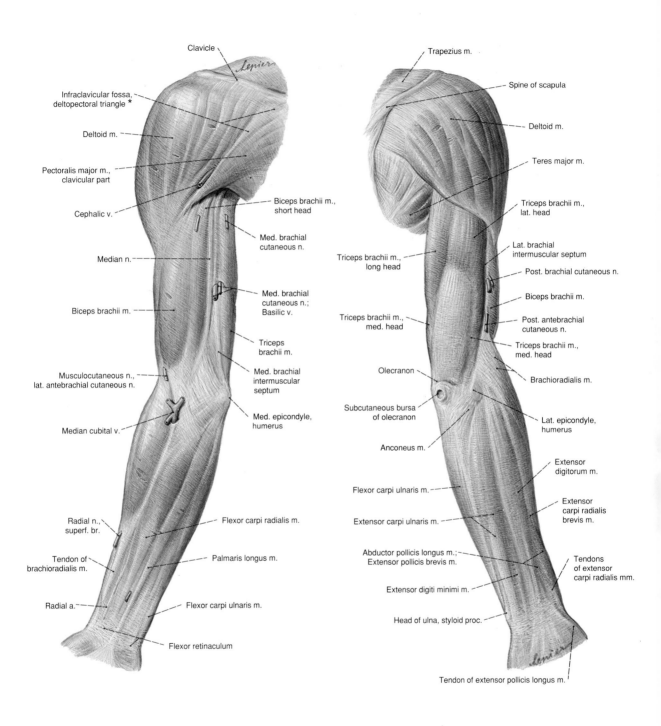

Clavicle

Infraclavicular fossa, deltopectoral triangle *

Deltoid m.

Pectoralis major m., clavicular part

Cephalic v.

Median n.

Biceps brachii m.

Musculocutaneous n., lat. antebrachial cutaneous n.

Median cubital v.

Radial n., superf. br.

Tendon of brachioradialis m.

Radial a.

Biceps brachii m., short head

Med. brachial cutaneous n.

Med. brachial cutaneous n.; Basilic v.

Triceps brachii m.

Med. brachial intermuscular septum

Med. epicondyle, humerus

Flexor carpi radialis m.

Palmaris longus m.

Flexor carpi ulnaris m.

Flexor retinaculum

Trapezius m.

Spine of scapula

Deltoid m.

Teres major m.

Triceps brachii m., lat. head

Lat. brachial intermuscular septum

Post. brachial cutaneous n.

Biceps brachii m.

Post. antebrachial cutaneous n.

Triceps brachii m., med. head

Brachioradialis m.

Lat. epicondyle, humerus

Extensor digitorum m.

Extensor carpi radialis brevis m.

Tendons of extensor carpi radialis mm.

Triceps brachii m., long head

Triceps brachii m., med. head

Olecranon

Subcutaneous bursa of olecranon

Anconeus m.

Flexor carpi ulnaris m.

Extensor carpi ulnaris m.

Abductor pollicis longus m.; Extensor pollicis brevis m.

Extensor digiti minimi m.

Head of ulna, styloid proc.

Tendon of extensor pollicis longus m.

Fig. 378. The fascia overlying the muscles of the right upper limb, anterior aspect.

Fig. 379. The fascia overlying the muscles of the right upper limb, posterior aspect.

* Clinical eponym: Mohrenheim's fossa

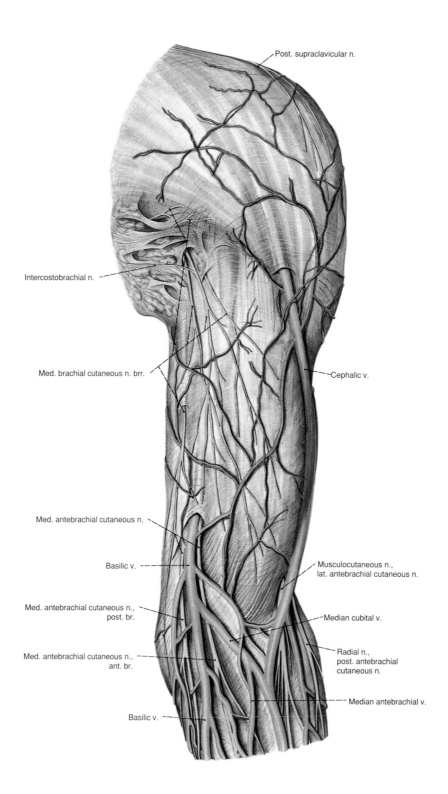

Post. supraclavicular n.

Intercostobrachial n.

Med. brachial cutaneous n. brr.

Cephalic v.

Med. antebrachial cutaneous n.

Basilic v.

Musculocutaneous n.,
lat. antebrachial cutaneous n.

Med. antebrachial cutaneous n.,
post. br.

Median cubital v.

Radial n.,
post. antebrachial
cutaneous n.

Med. antebrachial cutaneous n.,
ant. br.

Median antebrachial v.

Basilic v.

Fig. 380. Cutaneous nerves and superficial veins of the anterior aspect of the left arm. Skin and subcutaneous fat have been removed; the fascia has been retained.

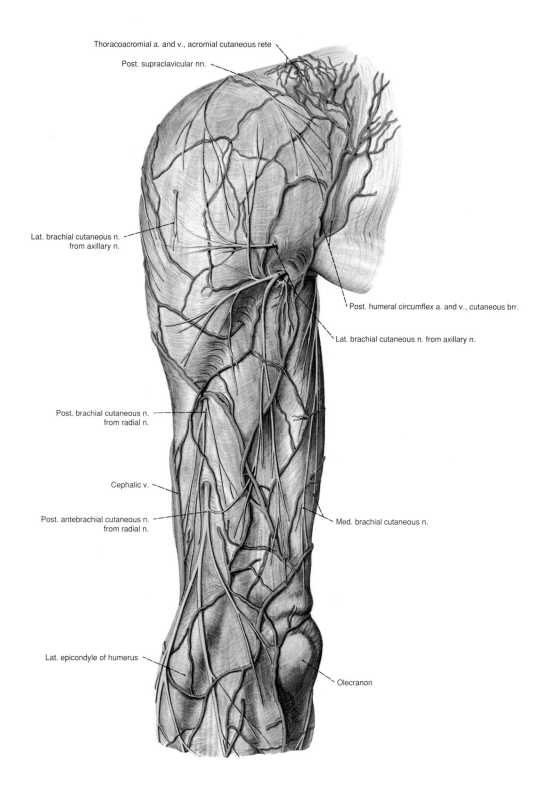

Thoracoacromial a. and v., acromial cutaneous rete

Post. supraclavicular nn.

Lat. brachial cutaneous n.
from axillary n.

Post. humeral circumflex a. and v., cutaneous brr.

Lat. brachial cutaneous n. from axillary n.

Post. brachial cutaneous n.
from radial n.

Cephalic v.

Post. antebrachial cutaneous n.
from radial n.

Med. brachial cutaneous n.

Lat. epicondyle of humerus

Olecranon

Fig. 381. Cutaneous nerves and superficial veins of the posterior aspect of the left arm.

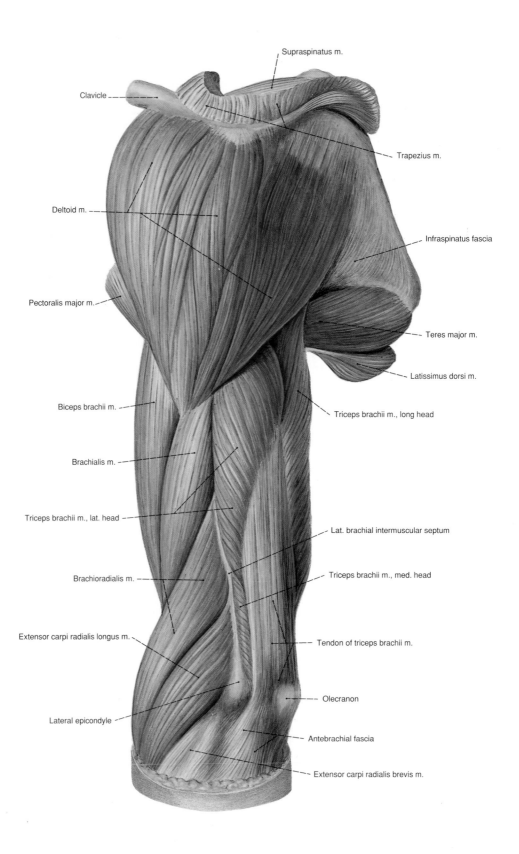

Supraspinatus m.

Clavicle

Trapezius m.

Deltoid m.

Infraspinatus fascia

Pectoralis major m.

Teres major m.

Latissimus dorsi m.

Biceps brachii m.

Triceps brachii m., long head

Brachialis m.

Triceps brachii m., lat. head

Lat. brachial intermuscular septum

Brachioradialis m.

Triceps brachii m., med. head

Extensor carpi radialis longus m.

Tendon of triceps brachii m.

Olecranon

Lateral epicondyle

Antebrachial fascia

Extensor carpi radialis brevis m.

Fig. 382.   The deltoid muscle and muscles of the left arm, superficial layer. Posterolateral view.

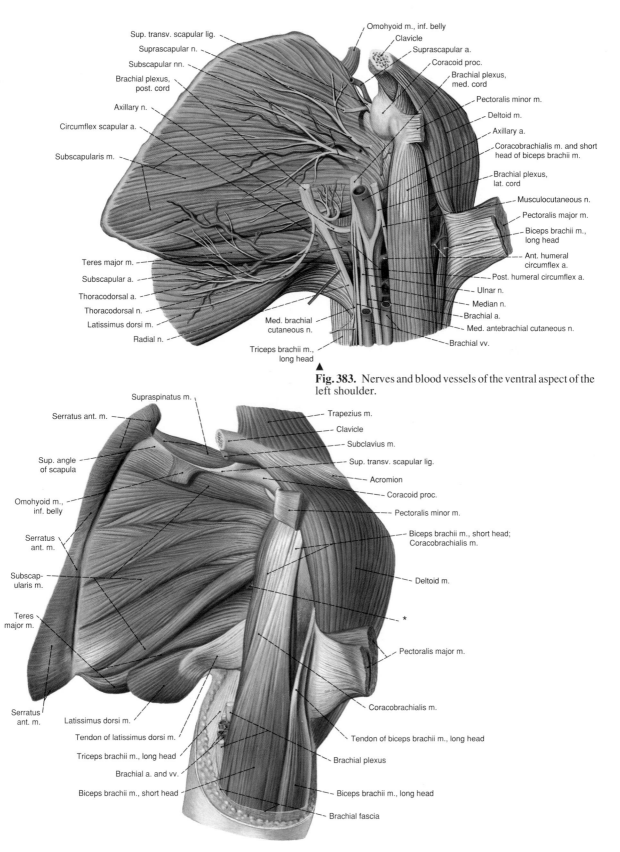

Sup. transv. scapular lig.
Suprascapular n.
Subscapular nn.
Brachial plexus, post. cord
Axillary n.
Circumflex scapular a.
Subscapularis m.
Teres major m.
Subscapular a.
Thoracodorsal a.
Thoracodorsal n.
Latissimus dorsi m.
Radial n.
Med. brachial cutaneous n.
Triceps brachii m., long head

Omohyoid m., inf. belly
Clavicle
Suprascapular a.
Coracoid proc.
Brachial plexus, med. cord
Pectoralis minor m.
Deltoid m.
Axillary a.
Coracobrachialis m. and short head of biceps brachii m.
Brachial plexus, lat. cord
Musculocutaneous n.
Pectoralis major m.
Biceps brachii m., long head
Ant. humeral circumflex a.
Post. humeral circumflex a.
Ulnar n.
Median n.
Brachial a.
Med. antebrachial cutaneous n.
Brachial vv.

Fig. 383. Nerves and blood vessels of the ventral aspect of the left shoulder.

Supraspinatus m.
Serratus ant. m.
Sup. angle of scapula
Omohyoid m., inf. belly
Serratus ant. m.
Subscapularis m.
Teres major m.
Serratus ant. m.
Latissimus dorsi m.
Tendon of latissimus dorsi m.
Triceps brachii m., long head
Brachial a. and vv.
Biceps brachii m., short head

Trapezius m.
Clavicle
Subclavius m.
Sup. transv. scapular lig.
Acromion
Coracoid proc.
Pectoralis minor m.
Biceps brachii m., short head; Coracobrachialis m.
Deltoid m.
*
Pectoralis major m.
Coracobrachialis m.
Tendon of biceps brachii m., long head
Brachial plexus
Biceps brachii m., long head
Brachial fascia

Fig. 384. Muscles of the costal surface of the left scapula. The pectoral, cervical and back muscles have been cut close to their attachments.

* Triangular space

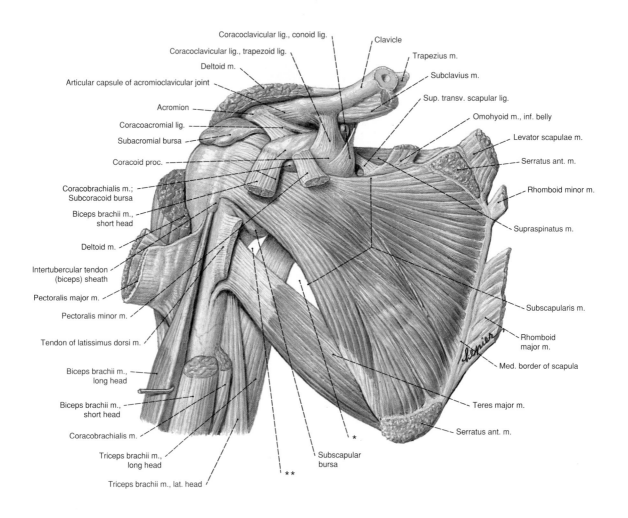

Coracoclavicular lig., conoid lig.
Clavicle
Coracoclavicular lig., trapezoid lig.
Trapezius m.
Deltoid m.
Subclavius m.
Articular capsule of acromioclavicular joint
Sup. transv. scapular lig.
Acromion
Omohyoid m., inf. belly
Coracoacromial lig.
Levator scapulae m.
Subacromial bursa
Serratus ant. m.
Coracoid proc.
Rhomboid minor m.
Coracobrachialis m.;
Subcoracoid bursa
Supraspinatus m.
Biceps brachii m.,
short head
Deltoid m.
Intertubercular tendon
(biceps) sheath
Subscapularis m.
Pectoralis major m.
Rhomboid
major m.
Pectoralis minor m.
Med. border of scapula
Tendon of latissimus dorsi m.
Biceps brachii m.,
long head
Teres major m.
Biceps brachii m.,
short head
Serratus ant. m.
Coracobrachialis m.
Triceps brachii m.,
long head
Subscapular
bursa
Triceps brachii m., lat. head
**

Fig. 385. Muscles of the ventral aspect of the right shoulder joint, deep layer. Various muscles have been cut near their attachments.

* Triangular space
** Quadrangular space

Triangular and Quadrangular Spaces (Figs. 385, 386)
1. Quadrangular space is bounded laterally by the humerus, medially by the long head of the triceps, inferiorly by the teres major and superiorly by the teres minor and capsule of the shoulder joint. It transmits the posterior humeral circumflex vessels and the axillary nerve.
2. Triangular space is bounded inferiorly by the teres major, superiorly by the teres minor, and laterally by the long head of the triceps. It transmits the scapular circumflex artery.

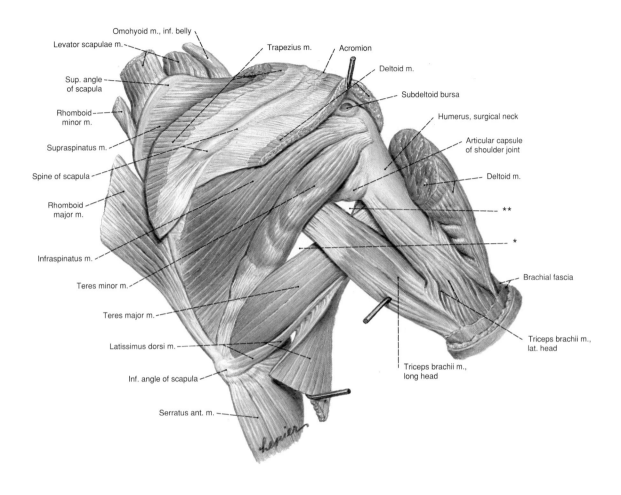

Fig. 386. Muscles of the dorsal aspect of the right shoulder joint.

* Triangular space
** Quadrangular space

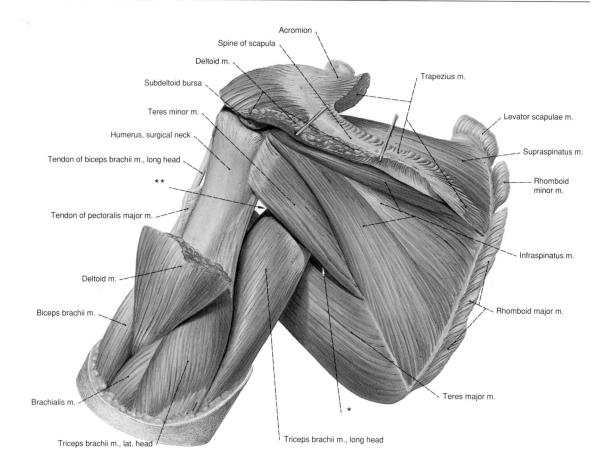

Fig. 387. Muscles of the dorsal surface of the left scapula.

* Triangular space
** Quadrangular space

Muscles of the Shoulder (Fig. 387)

Name	Origin	Insertion	Innervation	Function
Deltoid muscle	Acromial third of clavicle, the acromion, and spine of scapula	Deltoid tuberosity of humerus (subdeltoid bursa between muscle and greater tubercle)	Axillary nerve (5th, 6th cervical nerves)	Ventral (clavicular) fibers assist in flexion and medial rotation of humerus. Dorsal (scapular) fibers assist in extension and lateral rotation of humerus. Middle or central part is a powerful abductor of humerus, lifting it to the horizontal. Anterior and posterior fibers brace limb to steady abducted arm.
Supraspinatus muscle	Supraspinatous fossa of the scapula	Highest facet of greater tubercle of humerus (tendinous)	Suprascapular nerve from brachial plexus (4th, 5th, 6th cervical nerves)	Supraspinatus muscle abducts humerus; starts abduction and assists the deltoid muscle; lateral rotator of humerus.
Infraspinatus muscle	Caudal margin of the spine of the scapula, infraspinatous fossa	Middle facet of greater tubercle of humerus (tendinous)		Infraspinatus muscle rotates humerus laterally; assists in holding head of humerus in glenoid cavity.

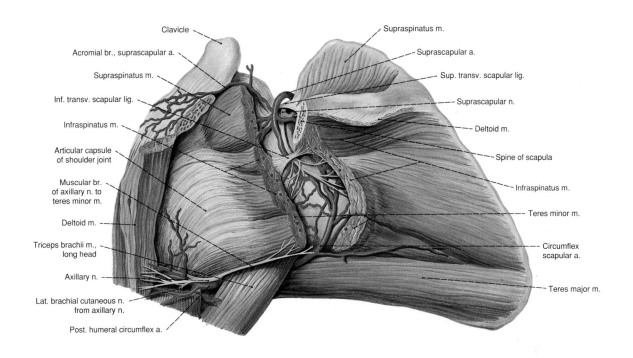

Fig. 388. Nerves and blood vessels of the dorsal aspect of the left shoulder. The deltoid muscle has been partially removed and reflected. A portion of the acromion has been removed. The supraspinatus, infraspinatus and teres minor muscles have been sectioned and somewhat pulled apart.

Muscles of the Shoulder (continued, Fig. 387)

Name	Origin	Insertion	Innervation	Function
Teres minor muscle	Caudal section of infraspinatous fossa and lateral margin (medial third) of scapula	Lower facet of greater tubercle of humerus (tendinous)	Axillary nerve from brachial plexus (4th, 5th cervical nerves)	Lateral rotator and adductor of humerus
Teres major muscle	Lateral border of scapula near inferior angle (medial third)	Tendon on crest of lesser tubercle of humerus, dorsal to that of the latissimus dorsi muscle (the 2 tendons separated by a bursa)	Lower subscapular nerve lower (5th, 6th cervical) of the brachial plexus, infraclavicular part	Adductor, medial rotator and extensor of arm; assists latissimus dorsi muscle
Subscapularis muscle (under its insertion: the subscapular bursa)	Costal surface of scapula; subscapular fossa	Short, broad tendon into the lesser tubercle and its adjacent crest; blends with capsule of shoulder joint	Upper and lower subscapular nerves (5th, 6th cervical) from brachial plexus, infraclavicular part	Adductor and medial rotator of arm; assists in flexion and extension

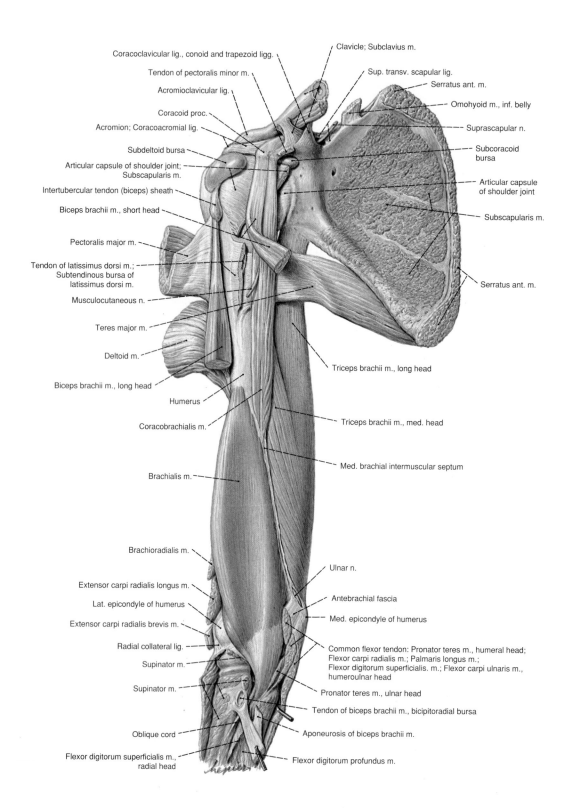

Coracoclavicular lig., conoid and trapezoid ligg.

Tendon of pectoralis minor m.

Acromioclavicular lig.

Coracoid proc.

Acromion; Coracoacromial lig.

Subdeltoid bursa

Articular capsule of shoulder joint; Subscapularis m.

Intertubercular tendon (biceps) sheath

Biceps brachii m., short head

Pectoralis major m.

Tendon of latissimus dorsi m.; Subtendinous bursa of latissimus dorsi m.

Musculocutaneous n.

Teres major m.

Deltoid m.

Biceps brachii m., long head

Humerus

Coracobrachialis m.

Brachialis m.

Brachioradialis m.

Extensor carpi radialis longus m.

Lat. epicondyle of humerus

Extensor carpi radialis brevis m.

Radial collateral lig.

Supinator m.

Supinator m.

Oblique cord

Flexor digitorum superficialis m., radial head

Clavicle; Subclavius m.

Sup. transv. scapular lig.

Serratus ant. m.

Omohyoid m., inf. belly

Suprascapular n.

Subcoracoid bursa

Articular capsule of shoulder joint

Subscapularis m.

Serratus ant. m.

Triceps brachii m., long head

Triceps brachii m., med. head

Med. brachial intermuscular septum

Ulnar n.

Antebrachial fascia

Med. epicondyle of humerus

Common flexor tendon: Pronator teres m., humeral head; Flexor carpi radialis m.; Palmaris longus m.; Flexor digitorum superficialis. m.; Flexor carpi ulnaris m., humeroulnar head

Pronator teres m., ulnar head

Tendon of biceps brachii m., bicipitoradial bursa

Aponeurosis of biceps brachii m.

Flexor digitorum profundus m.

Fig. 389. Muscles of the ventral aspect of the right shoulder joint (deep layer, the subscapularis muscle essentially removed) and the flexor compartment of the arm after sectioning of the heads of the biceps brachii muscle and its removal.

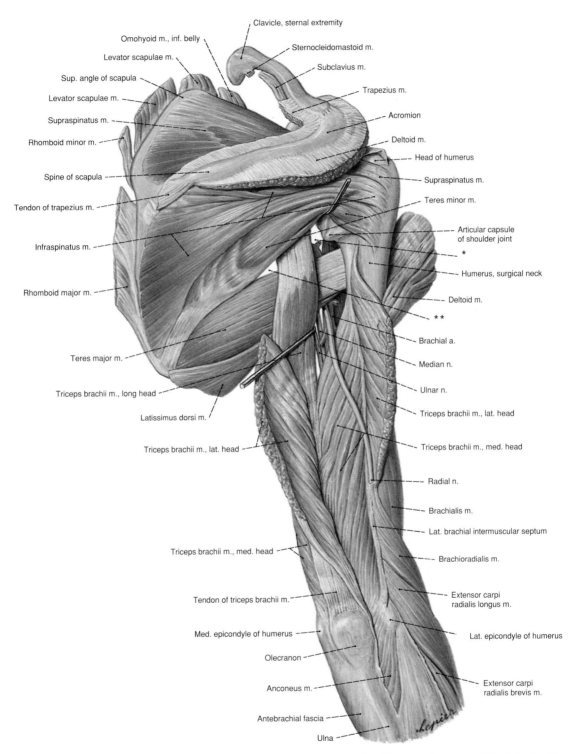

Clavicle, sternal extremity
Omohyoid m., inf. belly
Sternocleidomastoid m.
Levator scapulae m.
Subclavius m.
Sup. angle of scapula
Trapezius m.
Levator scapulae m.
Acromion
Supraspinatus m.
Deltoid m.
Rhomboid minor m.
Head of humerus
Spine of scapula
Supraspinatus m.
Teres minor m.
Tendon of trapezius m.
Articular capsule
of shoulder joint
*
Infraspinatus m.
Humerus, surgical neck
Deltoid m.
Rhomboid major m.
**
Brachial a.
Median n.
Teres major m.
Ulnar n.
Triceps brachii m., long head
Triceps brachii m., lat. head
Latissimus dorsi m.
Triceps brachii m., med. head
Triceps brachii m., lat. head
Radial n.
Brachialis m.
Lat. brachial intermuscular septum
Triceps brachii m., med. head
Brachioradialis m.
Extensor carpi
radialis longus m.
Tendon of triceps brachii m.
Med. epicondyle of humerus
Lat. epicondyle of humerus
Olecranon
Anconeus m.
Extensor carpi
radialis brevis m.
Antebrachial fascia
Ulna

Fig. 390. Muscles of the posterior aspect of the right shoulder
joint after removal of a large part of the trapezius and deltoid
muscles and sectioning of the lateral head of the triceps brachii
muscle in the extensor compartment of the arm to expose the
radial nerve.

  * Quadrangular space
 ** Triangular space

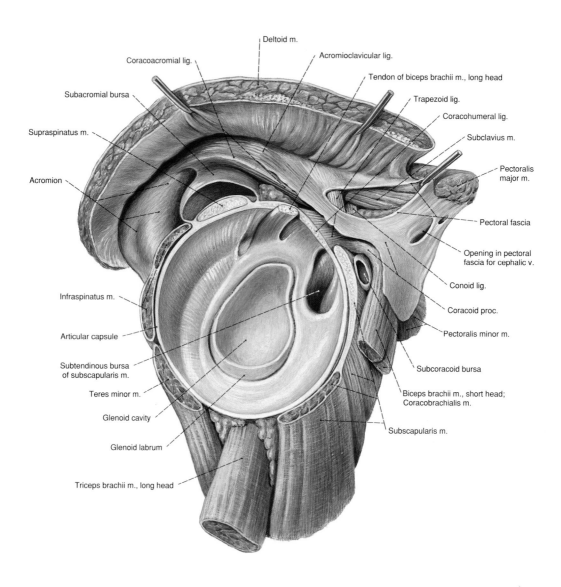

Fig. 391.   Lateral view of the opened socket of the right
shoulder joint showing the rotator cuff.

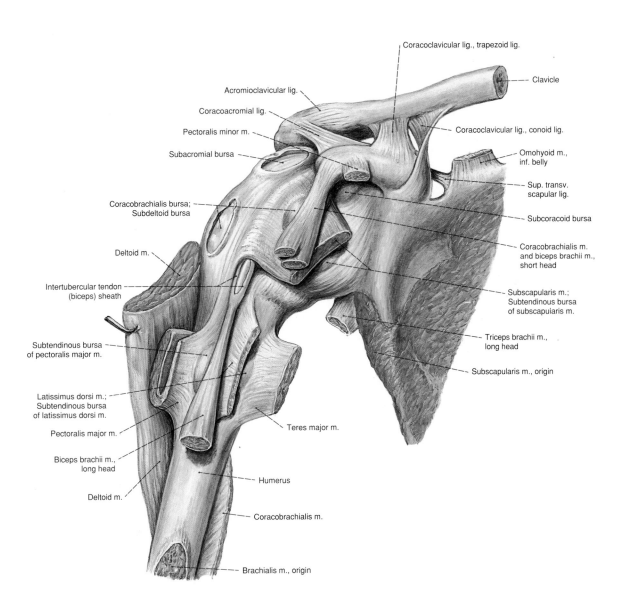

Coracoclavicular lig., trapezoid lig.

Clavicle

Acromioclavicular lig.

Coracoacromial lig.

Pectoralis minor m.

Subacromial bursa

Coracoclavicular lig., conoid lig.

Omohyoid m.,
inf. belly

Sup. transv.
scapular lig.

Coracobrachialis bursa;
Subdeltoid bursa

Subcoracoid bursa

Coracobrachialis m.
and biceps brachii m.,
short head

Deltoid m.

Intertubercular tendon
(biceps) sheath

Subscapularis m.;
Subtendinous bursa
of subscapularis m.

Triceps brachii m.,
long head

Subtendinous bursa
of pectoralis major m.

Subscapularis m., origin

Latissimus dorsi m.;
Subtendinous bursa
of latissimus dorsi m.

Teres major m.

Pectoralis major m.

Biceps brachii m.,
long head

Humerus

Deltoid m.

Coracobrachialis m.

Brachialis m., origin

Fig. 392. Muscle origins and insertions on the ventral aspect
of the right shoulder joint (cf. Fig. 391).

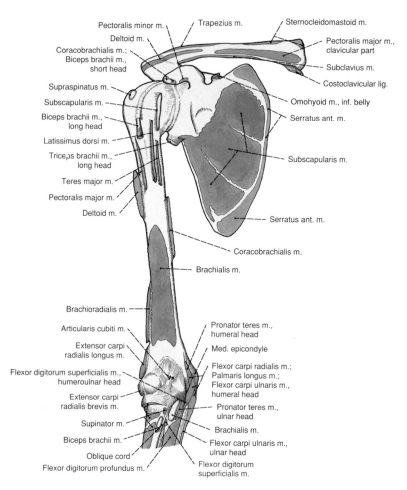

Pectoralis minor m.
Deltoid m.
Coracobrachialis m.;
Biceps brachii m.,
short head
Supraspinatus m.
Subscapularis m.
Biceps brachii m.,
long head
Latissimus dorsi m.
Triceps brachii m.,
long head
Teres major m.
Pectoralis major m.
Deltoid m.

Trapezius m.
Sternocleidomastoid m.
Pectoralis major m.,
clavicular part
Subclavius m.
Costoclavicular lig.
Omohyoid m., inf. belly
Serratus ant. m.

Subscapularis m.

Serratus ant. m.

Coracobrachialis m.

Brachialis m.

Brachioradialis m.
Articularis cubiti m.
Extensor carpi
radialis longus m.
Flexor digitorum superficialis m.,
humeroulnar head
Extensor carpi
radialis brevis m.
Supinator m.
Biceps brachii m.
Oblique cord
Flexor digitorum profundus m.

Pronator teres m.,
humeral head
Med. epicondyle
Flexor carpi radialis m.;
Palmaris longus m.;
Flexor carpi ulnaris m.,
humeral head
Pronator teres m.,
ulnar head
Brachialis m.
Flexor carpi ulnaris m.,
ulnar head
Flexor digitorum
superficialis m.

Fig. 393. Diagram of muscle origins and insertions on the right clavicle, scapula, humerus and proximal end of the ulna and radius. Anterior view.

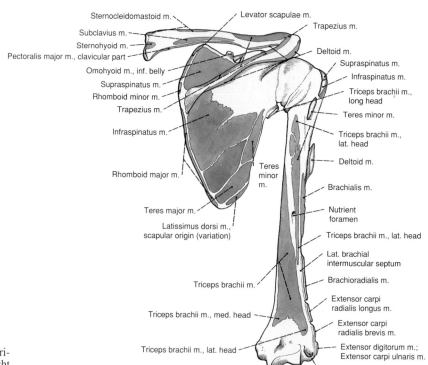

Sternocleidomastoid m.
Subclavius m.
Sternohyoid m.
Pectoralis major m., clavicular part
Omohyoid m., inf. belly
Supraspinatus m.
Rhomboid minor m.
Trapezius m.
Infraspinatus m.
Rhomboid major m.
Teres major m.
Latissimus dorsi m.,
scapular origin (variation)
Triceps brachii m.
Triceps brachii m., med. head
Triceps brachii m., lat. head

Levator scapulae m.
Trapezius m.
Deltoid m.
Supraspinatus m.
Infraspinatus m.
Triceps brachii m.,
long head
Teres minor m.
Triceps brachii m.,
lat. head
Deltoid m.
Teres minor m.
Brachialis m.
Nutrient foramen
Triceps brachii m., lat. head
Lat. brachial
intermuscular septum
Brachioradialis m.
Extensor carpi
radialis longus m.
Extensor carpi
radialis brevis m.
Extensor digitorum m.;
Extensor carpi ulnaris m.
Anconeus m.

Fig. 394. Diagram of muscle origins and insertions on the right clavicle, scapula and humerus. Posterior view.

Flexor Muscles of the Arm

Name	Origin	Insertion	Innervation	Function
1. Biceps brachii muscle (operates 2 joints) *Long head*	Tendon through shoulder joint to supra-glenoid tubercle of scapula (long tendon)	Posterior surface of radial tuberosity. Bicipital aponeurosis to antebrachial fascia	Musculocutaneous nerve for all three flexor muscles (5th, 6th, 7th cervical nerves)	Flexes and supinates the forearm; tenses antebrachial fascia. Long head aids in abduction at the shoulder joint and holds head of humerus in place; short head aids in adduction at shoulder joint
Short head	Short tendon from tip of coracoid process of scapula			
2. Coracobrachialis muscle (sometimes pierced by musculo-cutaneous nerve)	Apex of the coracoid process (fused with short head of biceps)	Ventral and medial surface of humerus near its middle		Flexes and adducts the arm
3. Brachialis muscle (operates 1 joint)	Distal half of anterior aspect of humerus, distal to deltoid tuberosity and the medial and lateral intermuscular septa	Tuberosity of ulna (short tendon) and rough depression on anterior surface of coronoid process		Primary flexor of forearm

Extensor Muscles of the Arm

Name	Origin	Insertion	Innervation	Function
1. Triceps brachii muscle *Long head* (operates 2 joints)	Infraglenoidal tubercle of scapula; separates the two axillary spaces; forms a distinct tendon	Fibers pass distally joining common tendon of insertion on the olecranon	Radial nerve (7th, 8th cervical nerves)	Extends forearm at the elbow joint; long head extends humerus at shoulder joint; braces extended elbow joint (when pushing an object)
Lateral head (operates 1 joint)	Lateral and posterior surface of body of humerus, posterior 2/3 of lateral inter-muscular septum	Joins the common tendon of insertion on the olecranon		
Medial head (operates 1 joint)	Whole length of medial intermuscular septum; dorsal surface of body of humerus (distal to crest of greater tubercle and groove for radial nerve); distal third of lateral intermuscular septum (to lateral epicondyle)	Posterior aspect of olecranon and into deep fascia on both sides of the forearm		
2. Anconeus muscle (situated in the forearm)	Lateral epicondyle of humerus; continuation of the lateral side of the medial head of the triceps brachii muscle	Posterior surface of ulna, slightly distal to olecranon		

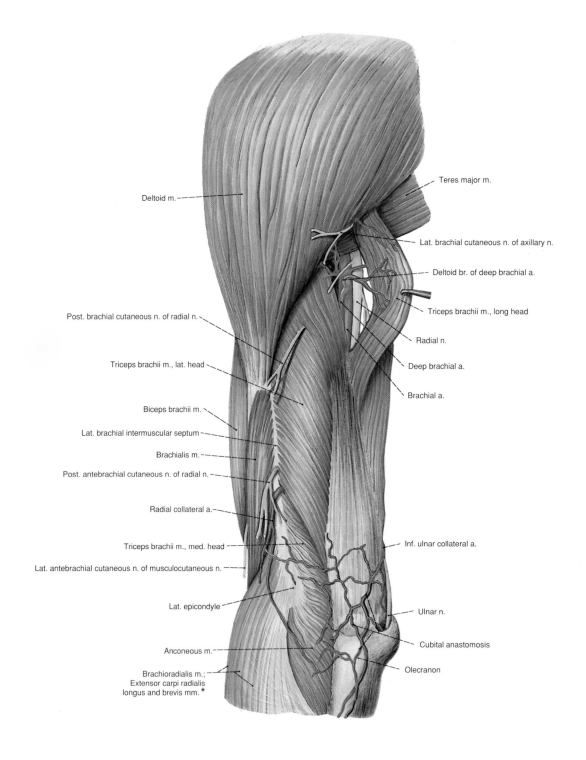

Teres major m.

Lat. brachial cutaneous n. of axillary n.

Deltoid m.

Deltoid br. of deep brachial a.

Triceps brachii m., long head

Post. brachial cutaneous n. of radial n.

Radial n.

Triceps brachii m., lat. head

Deep brachial a.

Brachial a.

Biceps brachii m.

Lat. brachial intermuscular septum

Brachialis m.

Post. antebrachial cutaneous n. of radial n.

Radial collateral a.

Triceps brachii m., med. head

Inf. ulnar collateral a.

Lat. antebrachial cutaneous n. of musculocutaneous n.

Lat. epicondyle

Ulnar n.

Cubital anastomosis

Anconeous m.

Olecranon

Brachioradialis m.;
Extensor carpi radialis
longus and brevis mm. *

Fig. 395. Nerves and arteries of the posterior aspect of the left arm (superficial layer).

* Radial group of forearm extensor muscles

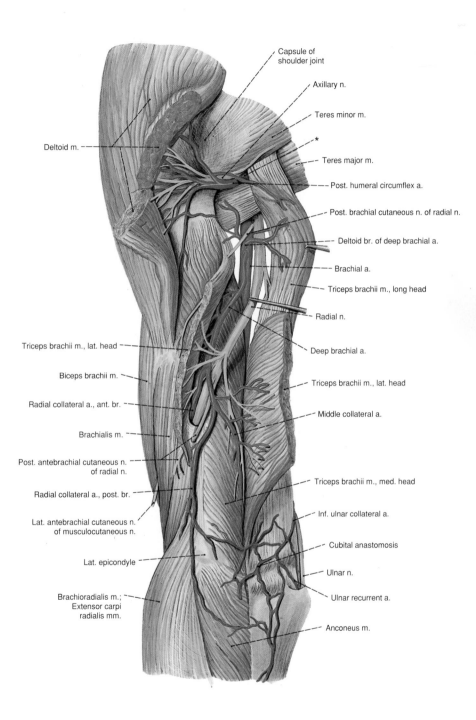

Capsule of
shoulder joint

Axillary n.

Teres minor m.

*

Teres major m.

Post. humeral circumflex a.

Post. brachial cutaneous n. of radial n.

Deltoid br. of deep brachial a.

Brachial a.

Triceps brachii m., long head

Radial n.

Deep brachial a.

Triceps brachii m., lat. head

Middle collateral a.

Triceps brachii m., med. head

Inf. ulnar collateral a.

Cubital anastomosis

Ulnar n.

Ulnar recurrent a.

Anconeus m.

Deltoid m.

Triceps brachii m., lat. head

Biceps brachii m.

Radial collateral a., ant. br.

Brachialis m.

Post. antebrachial cutaneous n.
of radial n.

Radial collateral a., post. br.

Lat. antebrachial cutaneous n.
of musculocutaneous n.

Lat. epicondyle

Brachioradialis m.;
Extensor carpi
radialis mm.

Fig. 396. The deep nerves and arteries of the posterior aspect
of the left arm. The lateral head of the triceps brachii muscle
has been sectioned and reflected. Note the course of the radial
nerve in the radial groove.

* Quadrangular space

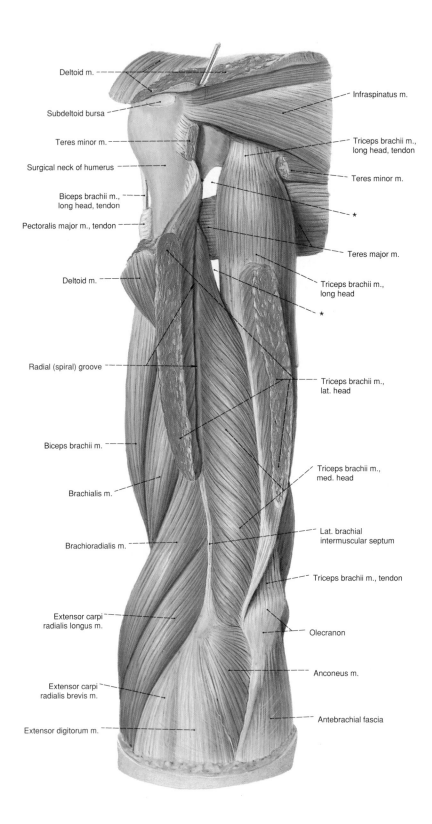

Deltoid m.

Subdeltoid bursa

Teres minor m.

Surgical neck of humerus

Biceps brachii m., long head, tendon

Pectoralis major m., tendon

Deltoid m.

Radial (spiral) groove

Biceps brachii m.

Brachialis m.

Brachioradialis m.

Extensor carpi radialis longus m.

Extensor carpi radialis brevis m.

Extensor digitorum m.

Infraspinatus m.

Triceps brachii m., long head, tendon

Teres minor m.

*

Teres major m.

Triceps brachii m., long head

*

Triceps brachii m., lat. head

Triceps brachii m., med. head

Lat. brachial intermuscular septum

Triceps brachii m., tendon

Olecranon

Anconeus m.

Antebrachial fascia

Fig. 397. Deep muscles of the left arm, posterolateral view. The deltoid muscle has been removed except for its attachments. The antebrachial fascia was removed where it covered the anconeus muscle. The teres minor muscle and lateral head of the triceps brachii muscle were cut and reflected.

* The quadrangular and triangular spaces

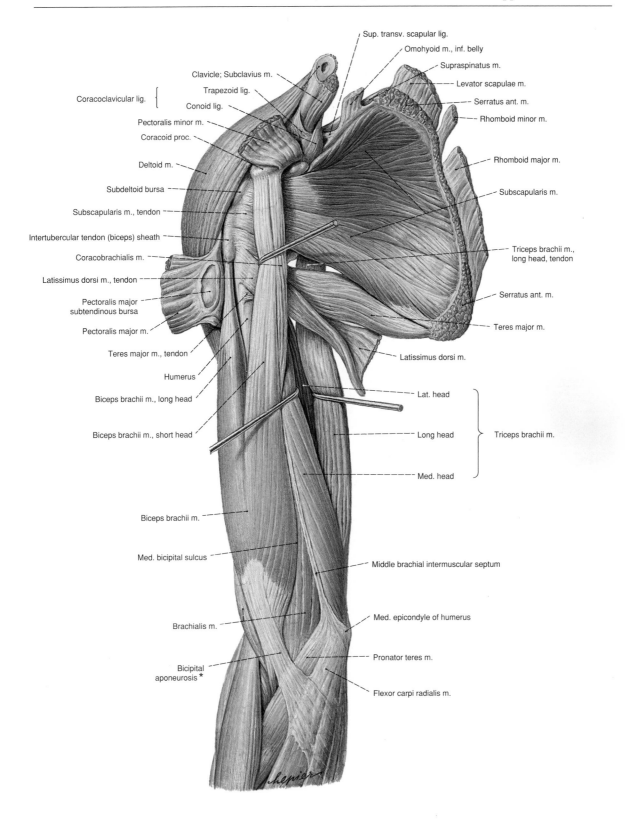

Sup. transv. scapular lig.
Omohyoid m., inf. belly
Supraspinatus m.
Levator scapulae m.
Serratus ant. m.
Rhomboid minor m.

Clavicle; Subclavius m.
Trapezoid lig.
Coracoclavicular lig.
Conoid lig.
Pectoralis minor m.
Coracoid proc.
Deltoid m.
Subdeltoid bursa
Subscapularis m., tendon
Intertubercular tendon (biceps) sheath
Coracobrachialis m.
Latissimus dorsi m., tendon
Pectoralis major subtendinous bursa
Pectoralis major m.
Teres major m., tendon
Humerus
Biceps brachii m., long head
Biceps brachii m., short head

Rhomboid major m.
Subscapularis m.
Triceps brachii m., long head, tendon
Serratus ant. m.
Teres major m.
Latissimus dorsi m.

Lat. head
Long head
Med. head
Triceps brachii m.

Biceps brachii m.
Med. bicipital sulcus
Brachialis m.
Bicipital aponeurosis *

Middle brachial intermuscular septum
Med. epicondyle of humerus
Pronator teres m.
Flexor carpi radialis m.

Fig. 398. The muscles of the costal surface of the right scapula and of the anterior surface of the arm. Superficial layer.

* Lacertus fibrosus

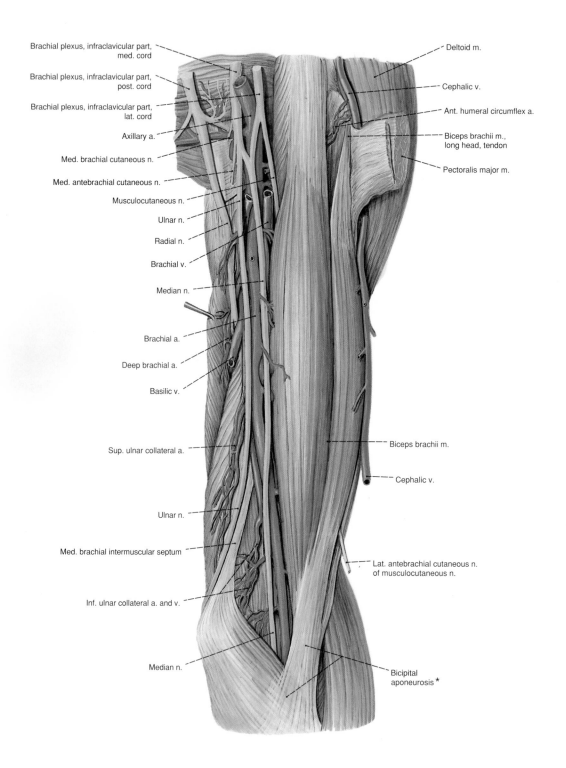

Brachial plexus, infraclavicular part, med. cord

Brachial plexus, infraclavicular part, post. cord

Brachial plexus, infraclavicular part, lat. cord

Axillary a.

Med. brachial cutaneous n.

Med. antebrachial cutaneous n.

Musculocutaneous n.

Ulnar n.

Radial n.

Brachial v.

Median n.

Brachial a.

Deep brachial a.

Basilic v.

Sup. ulnar collateral a.

Ulnar n.

Med. brachial intermuscular septum

Inf. ulnar collateral a. and v.

Median n.

Deltoid m.

Cephalic v.

Ant. humeral circumflex a.

Biceps brachii m., long head, tendon

Pectoralis major m.

Biceps brachii m.

Cephalic v.

Lat. antebrachial cutaneous n. of musculocutaneous n.

Bicipital aponeurosis *

Fig. 399. Nerves, blood vessels and muscles of the anterior arm (left).

* Lacertus fibrosus

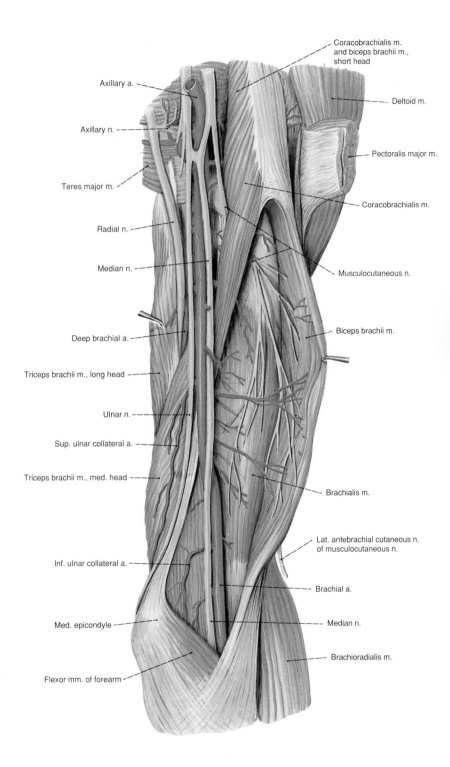

Coracobrachialis m.
and biceps brachii m.,
short head

Axillary a.

Axillary n.

Deltoid m.

Teres major m.

Pectoralis major m.

Radial n.

Coracobrachialis m.

Median n.

Musculocutaneous n.

Deep brachial a.

Biceps brachii m.

Triceps brachii m., long head

Ulnar n.

Sup. ulnar collateral a.

Triceps brachii m., med. head

Brachialis m.

Lat. antebrachial cutaneous n.
of musculocutaneous n.

Inf. ulnar collateral a.

Brachial a.

Med. epicondyle

Median n.

Brachioradialis m.

Flexor mm. of forearm

Fig. 400. Same dissection as in Fig. 399. The biceps brachii muscle has been retracted laterally and the veins removed.

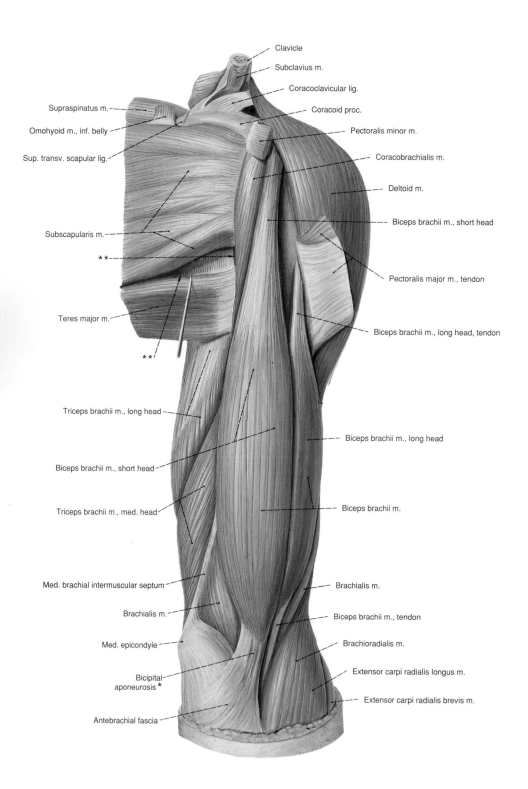

Clavicle

Subclavius m.

Coracoclavicular lig.

Coracoid proc.

Pectoralis minor m.

Coracobrachialis m.

Deltoid m.

Biceps brachii m., short head

Pectoralis major m., tendon

Biceps brachii m., long head, tendon

Biceps brachii m., long head

Biceps brachii m.

Brachialis m.

Biceps brachii m., tendon

Brachioradialis m.

Extensor carpi radialis longus m.

Extensor carpi radialis brevis m.

Supraspinatus m.

Omohyoid m., inf. belly

Sup. transv. scapular lig.

Subscapularis m.

**

Teres major m.

**

Triceps brachii m., long head

Biceps brachii m., short head

Triceps brachii m., med. head

Med. brachial intermuscular septum

Brachialis m.

Med. epicondyle

Bicipital aponeurosis *

Antebrachial fascia

Fig. 401. Muscles of the anterior aspect of the left arm, superficial layer.

* Lacertus fibrosus
** The triangular and quadrangular spaces

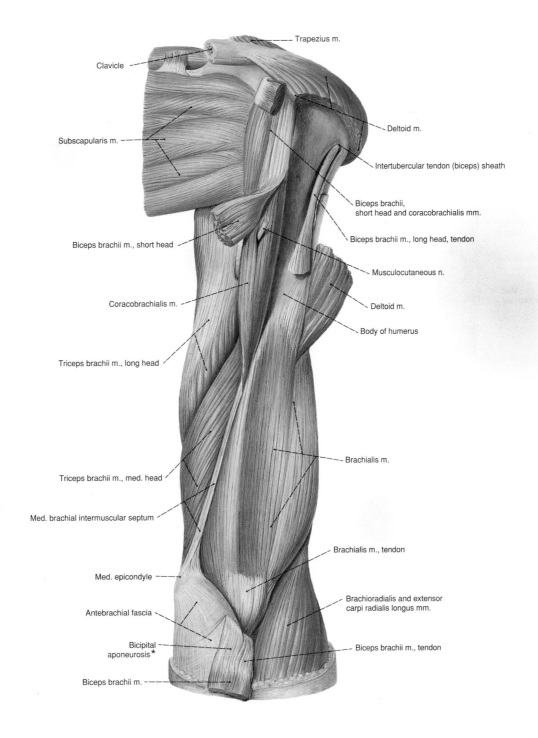

Trapezius m.

Clavicle

Subscapularis m.

Biceps brachii m., short head

Coracobrachialis m.

Triceps brachii m., long head

Triceps brachii m., med. head

Med. brachial intermuscular septum

Med. epicondyle

Antebrachial fascia

Bicipital aponeurosis*

Biceps brachii m.

Deltoid m.

Intertubercular tendon (biceps) sheath

Biceps brachii, short head and coracobrachialis mm.

Biceps brachii m., long head, tendon

Musculocutaneous n.

Deltoid m.

Body of humerus

Brachialis m.

Brachialis m., tendon

Brachioradialis and extensor carpi radialis longus mm.

Biceps brachii m., tendon

Fig. 402. Muscles of the anterior aspect of the left arm, deep layer. The deltoid and biceps brachii muscles have been sectioned (the middle segment of the biceps brachii muscle has been removed).

* Lacertus fibrosus

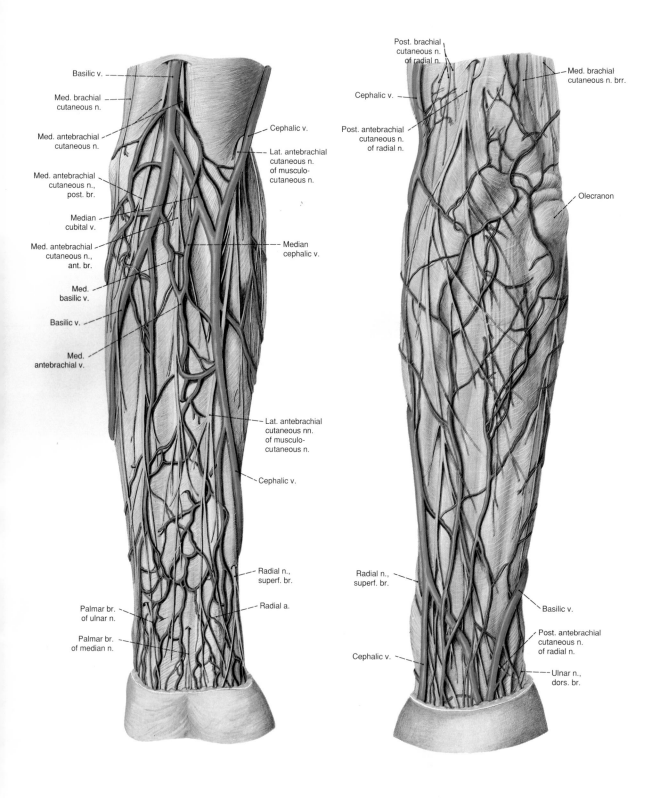

Basilic v.

Med. brachial cutaneous n.

Med. antebrachial cutaneous n.

Med. antebrachial cutaneous n., post. br.

Median cubital v.

Med. antebrachial cutaneous n., ant. br.

Med. basilic v.

Basilic v.

Med. antebrachial v.

Cephalic v.

Lat. antebrachial cutaneous n. of musculo-cutaneous n.

Median cephalic v.

Lat. antebrachial cutaneous nn. of musculo-cutaneous n.

Cephalic v.

Radial n., superf. br.

Palmar br. of ulnar n.

Radial a.

Palmar br. of median n.

Post. brachial cutaneous n. of radial n.

Cephalic v.

Post. antebrachial cutaneous n. of radial n.

Med. brachial cutaneous n. brr.

Olecranon

Radial n., superf. br.

Basilic v.

Post. antebrachial cutaneous n. of radial n.

Cephalic v.

Ulnar n., dors. br.

Fig. 403. Cutaneous nerves and superficial veins of the anterior aspect of the left forearm.

Fig. 404. Cutaneous nerves and superficial veins of the posterior aspect of the left forearm.

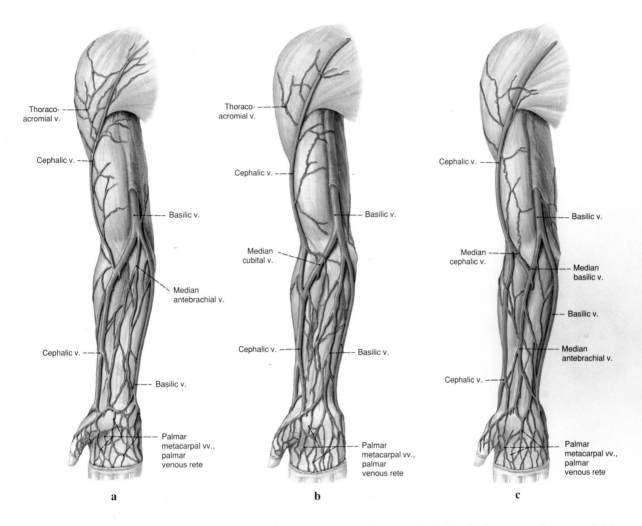

Thoraco-acromial v.

Cephalic v.

Basilic v.

Median antebrachial v.

Cephalic v.

Basilic v.

Palmar metacarpal vv., palmar venous rete

a

Thoraco-acromial v.

Cephalic v.

Basilic v.

Median cubital v.

Cephalic v.

Basilic v.

Palmar metacarpal vv., palmar venous rete

b

Cephalic v.

Median cephalic v.

Cephalic v.

Basilic v.

Median basilic v.

Basilic v.

Median antebrachial v.

Palmar metacarpal vv., palmar venous rete

c

Fig. 405a, b, c.   Variations in the pattern of the superficial veins of the upper limb.

Superficial Veins of the Upper Limb

Cephalic vein begins in the radial half of the **dorsal venous network** on the dorsum of the hand; receives tributaries from the palm of the hand via the intercapitular veins; courses proximally on the **radial** border of the forearm to the antecubital fossa where it anastomoses with the basilic vein. It continues in the arm, generally smaller than in the forearm, in the groove along the lateral border of the biceps brachii proximally to the deltopectoral triangle (clinical eponym: Mohrenheim's fossa) where it pierces the clavipectoral fascia and enters the axillary vein.

Basilic vein originates, like the cephalic vein, in the dorsal venous network but on the ulnar side of the dorsum of the hand; ascends on the posterior surface of the ulnar side of the forearm to the cubital fossa where it connects to the cephalic vein via the median cubital vein. Continuing proximally in the arm, generally larger than the cephalic vein, it runs in the medial bicipital groove to somewhat below the middle of the arm where it perforates the deep fascia and joins the brachial vein to form the axillary vein.

Median cubital vein is the oblique, highly variable anastomosis between the basilic and cephalic veins, and, in most instances, receives the median antebrachial vein which drains the venous plexus of the palmar surface of the hand and the forearm.

Median basilic vein and **median cephalic vein** form connections between the basilic, cephalic and median cubital veins. The superficial veins in the cubital fossa are important for withdrawal of blood and intravenous injections.

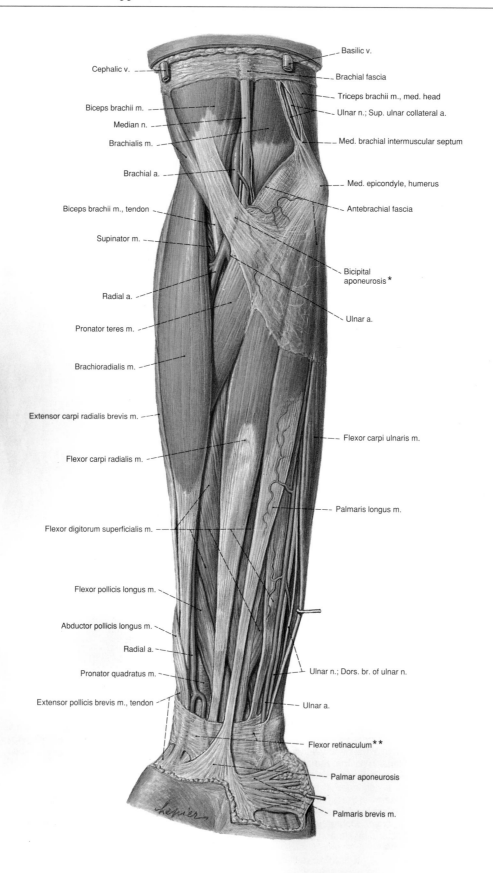

Cephalic v.

Biceps brachii m.

Median n.

Brachialis m.

Brachial a.

Biceps brachii m., tendon

Supinator m.

Radial a.

Pronator teres m.

Brachioradialis m.

Extensor carpi radialis brevis m.

Flexor carpi radialis m.

Flexor digitorum superficialis m.

Flexor pollicis longus m.

Abductor pollicis longus m.

Radial a.

Pronator quadratus m.

Extensor pollicis brevis m., tendon

Basilic v.

Brachial fascia

Triceps brachii m., med. head

Ulnar n.; Sup. ulnar collateral a.

Med. brachial intermuscular septum

Med. epicondyle, humerus

Antebrachial fascia

Bicipital aponeurosis *

Ulnar a.

Flexor carpi ulnaris m.

Palmaris longus m.

Ulnar n.; Dors. br. of ulnar n.

Ulnar a.

Flexor retinaculum **

Palmar aponeurosis

Palmaris brevis m.

Fig. 406. Superficial muscles of the right anterior forearm.
Dissection of the palmaris brevis muscle in the hypothenar
region of the palm of the hand.

  * Lacertus fibrosus
** Transverse carpal ligament

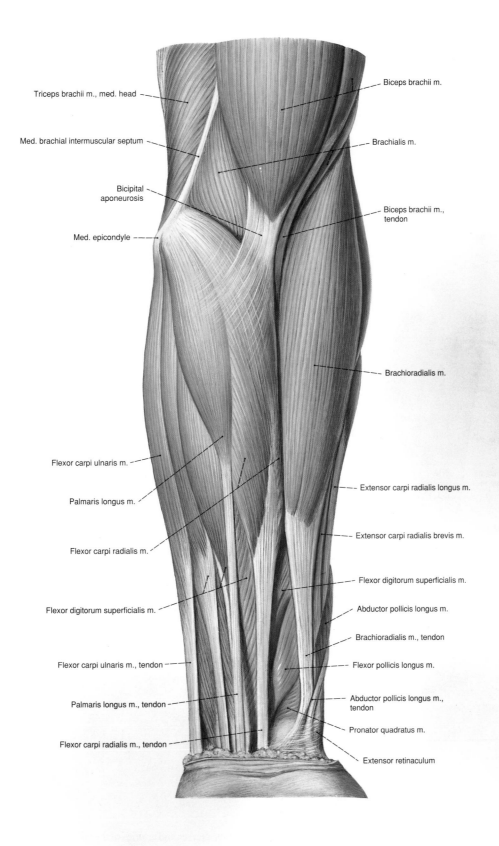

Triceps brachii m., med. head

Med. brachial intermuscular septum

Bicipital aponeurosis

Med. epicondyle

Flexor carpi ulnaris m.

Palmaris longus m.

Flexor carpi radialis m.

Flexor digitorum superficialis m.

Flexor carpi ulnaris m., tendon

Palmaris longus m., tendon

Flexor carpi radialis m., tendon

Biceps brachii m.

Brachialis m.

Biceps brachii m., tendon

Brachioradialis m.

Extensor carpi radialis longus m.

Extensor carpi radialis brevis m.

Flexor digitorum superficialis m.

Abductor pollicis longus m.

Brachioradialis m., tendon

Flexor pollicis longus m.

Abductor pollicis longus m., tendon

Pronator quadratus m.

Extensor retinaculum

Fig. 407. Superficial muscles of the left anterior forearm.

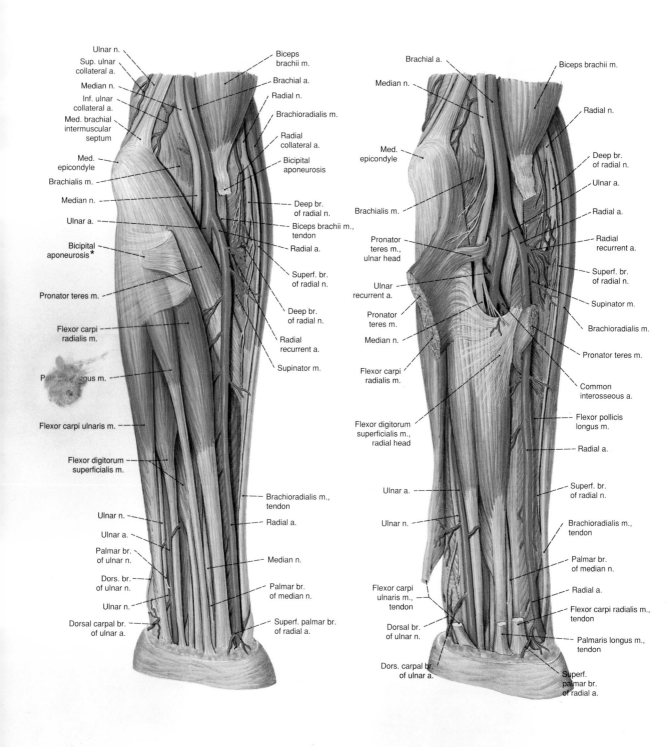

Fig. 408. Nerves and arteries of the left anterior forearm (superficial layer). The bicipital aponeurosis has been cut and reflected, and the brachioradialis muscle pulled laterally.

* Lacertus fibrosus

Fig. 409. Nerves and arteries of the left anterior forearm (deep layer). Pronator teres, palmaris longus, and flexor carpi radialis muscles have been partially removed.

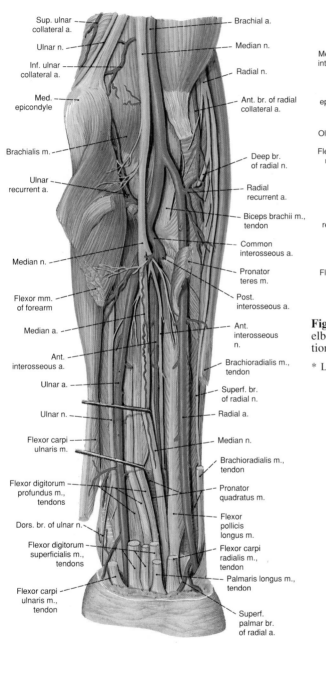

Fig. 410. Deep layer of nerves and arteries of the left anterior forearm. Flexors and pronators of the superficial layer have been sectioned.

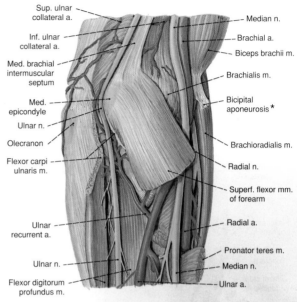

Fig. 411. Nerves and arteries on the ulnar side of the left elbow. Flexors and pronators of the forearm have been sectioned.

* Lacertus fibrosus

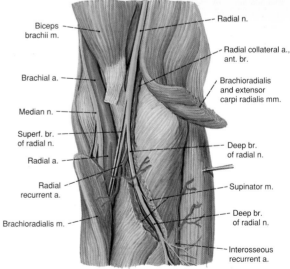

Fig. 412. Nerves and arteries on the radial side of the left elbow. The radial group of muscles of the forearm has been sectioned, the supinator muscle has been split along the deep branch of the radial nerve.

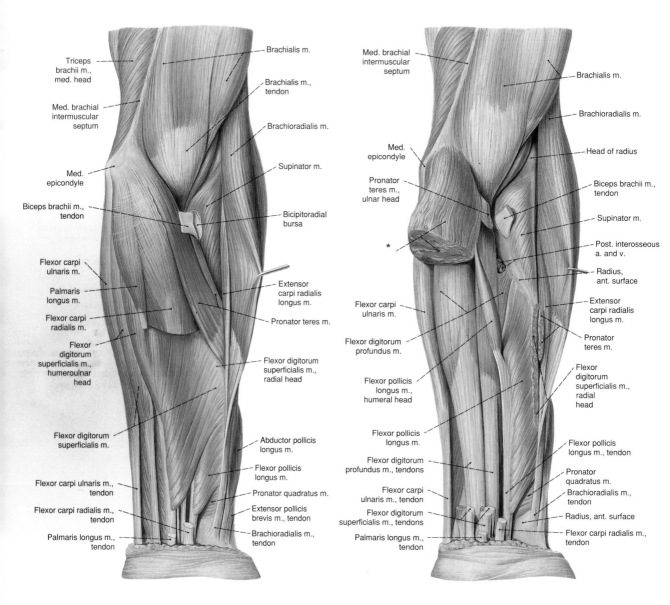

Triceps brachii m., med. head

Med. brachial intermuscular septum

Med. epicondyle

Biceps brachii m., tendon

Flexor carpi ulnaris m.

Palmaris longus m.

Flexor carpi radialis m.

Flexor digitorum superficialis m., humeroulnar head

Flexor digitorum superficialis m.

Flexor carpi ulnaris m., tendon

Flexor carpi radialis m., tendon

Palmaris longus m., tendon

Brachialis m.

Brachialis m., tendon

Brachioradialis m.

Supinator m.

Bicipitoradial bursa

Extensor carpi radialis longus m.

Pronator teres m.

Flexor digitorum superficialis m., radial head

Abductor pollicis longus m.

Flexor pollicis longus m.

Pronator quadratus m.

Extensor pollicis brevis m., tendon

Brachioradialis m., tendon

Med. brachial intermuscular septum

Med. epicondyle

Pronator teres m., ulnar head

*

Flexor carpi ulnaris m.

Flexor digitorum profundus m.

Flexor pollicis longus m., humeral head

Flexor pollicis longus m.

Flexor digitorum profundus m., tendons

Flexor carpi ulnaris m., tendon

Flexor digitorum superficialis m., tendons

Palmaris longus m., tendon

Brachialis m.

Brachioradialis m.

Head of radius

Biceps brachii m., tendon

Supinator m.

Post. interosseous a. and v.

Radius, ant. surface

Extensor carpi radialis longus m.

Pronator teres m.

Flexor digitorum superficialis m., radial head

Flexor pollicis longus m., tendon

Pronator quadratus m.

Brachioradialis m., tendon

Radius, ant. surface

Flexor carpi radialis m., tendon

Fig. 413. Flexor muscles of the left forearm after severing the palmaris longus and the flexor carpi radialis muscles.

Fig. 414. Deep layer of the flexor muscles of the left forearm. All superficial flexors, except the flexor carpi ulnaris muscle, were sectioned and removed.

* Common origin of the superficial flexor muscles

Flexor Muscles of the Forearm, Superficial Group (Figs. 406, 407)

Name	Origin	Insertion	Innervation	Function
Pronator teres muscle (Figs. 424, 425) *Humeral head*	Medial epicondyle of the humerus, antebrachial fascia	On the lateral and dorsal surface of radius (middle 1/3)	Median nerve (passes between the two heads) (6th, 7th cervical nerves)	Pronates and flexes the forearm
Ulnar head	Coronoid process of the ulna			
Flexor carpi radialis muscle	Medial epicondyle of the humerus, antebrachial fascia	Base of 2nd metacarpal bone, palmar surface	Median nerve (6th, 7th cervical nerves)	Flexes hand at wrist joint; abducts hand (radially flexes), pronates forearm (with elbow extended)
Palmaris longus muscle	Medial epicondyle of humerus, antebrachial fascia	Palmar aponeurosis	Median nerve (6th, 7th cervical nerves)	Tenses palmar aponeurosis; assists in flexion of elbow and hand at wrist joint
Flexor digitorum superficialis muscle (represents the deep part of the superficial flexor group) *Humeroulnar head*	Medial epicondyle of the humerus and coronoid process of ulna	4 long tendons to the middle phalanges of fingers 2 to 5	Median nerve (7th, 8th cervical, 1st thoracic nerves)	Flexes second phalanges of the 4 medial fingers; assists in flexion of first phalanx and hand (medially abducts)
Radial head	Upper half, anterior border of radius			
Flexor carpi ulnaris muscle *Humeral head*	Medial epicondyle of the humerus	Pisiform bone and, by ligaments (Fig. 424), to 5th metacarpal and hamate bones	Ulnar nerve (8th cervical, 1st thoracic nerves)	Flexes and adducts hand at wrist joint; flexor of elbow
Ulnar head	Olecranon and, via the antebrachial fascia, posterior border of ulna (upper 2/3)			

Flexor Muscles of the Forearm, Deep Group (Fig. 416)

Name	Origin	Insertion	Innervation	Function
Flexor digitorum profundus muscle	Anterior and medial surface of ulna; interosseous membrane	Distal phalanges of fingers 2 to 5	Ulnar nerve for ulnar side, median nerve for radial side (8th cervical, 1st thoracic nerves)	Flexes joints of fingers 2 to 5 (especially the distal phalanges but also the other phalanges) and hand at wrist joint
Flexor pollicis longus muscle *Radial head* (the major part)	Anterior surface of radius (distal to insertion of supinator muscle); interosseus membrane	Terminal phalanx of thumb	Anterior interosseous branch of median nerve (8th cervical, 1st thoracic nerves)	Flexes the hand and the distal phalanx of thumb
Humeral head	Medial epicondyle of humerus			
Pronator quadratus muscle (Fig. 424)	Anterior border of ulna, distal 1/4	Anterior border and surface of radius	Anterior interosseus branch of median nerve (8th cervical, 1st thoracic nerves)	Pronates the hand

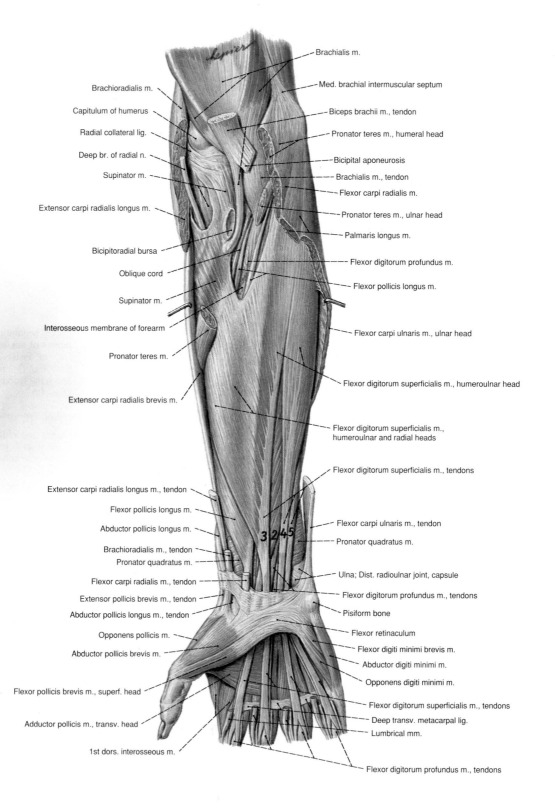

Brachialis m. — Brachialis m.

Brachioradialis m.

Capitulum of humerus

Radial collateral lig.

Deep br. of radial n.

Supinator m.

Extensor carpi radialis longus m.

Bicipitoradial bursa

Oblique cord

Supinator m.

Interosseous membrane of forearm

Pronator teres m.

Extensor carpi radialis brevis m.

Extensor carpi radialis longus m., tendon

Flexor pollicis longus m.

Abductor pollicis longus m.

Brachioradialis m., tendon

Pronator quadratus m.

Flexor carpi radialis m., tendon

Extensor pollicis brevis m., tendon

Abductor pollicis longus m., tendon

Opponens pollicis m.

Abductor pollicis brevis m.

Flexor pollicis brevis m., superf. head

Adductor pollicis m., transv. head

1st dors. interosseous m.

Med. brachial intermuscular septum

Biceps brachii m., tendon

Pronator teres m., humeral head

Bicipital aponeurosis

Brachialis m., tendon

Flexor carpi radialis m.

Pronator teres m., ulnar head

Palmaris longus m.

Flexor digitorum profundus m.

Flexor pollicis longus m.

Flexor carpi ulnaris m., ulnar head

Flexor digitorum superficialis m., humeroulnar head

Flexor digitorum superficialis m., humeroulnar and radial heads

Flexor digitorum superficialis m., tendons

Flexor carpi ulnaris m., tendon

Pronator quadratus m.

3 2 4 5

Ulna; Dist. radioulnar joint, capsule

Flexor digitorum profundus m., tendons

Pisiform bone

Flexor retinaculum

Flexor digiti minimi brevis m.

Abductor digiti minimi m.

Opponens digiti minimi m.

Flexor digitorum superficialis m., tendons

Deep transv. metacarpal lig.

Lumbrical mm.

Flexor digitorum profundus m., tendons

Fig. 415. Anterior aspect of the right forearm. Middle layer after exposure of the flexor digitorum superficialis muscle following removal of the superficial radial group of extensor muscles and the pronator teres, flexor carpi radialis and palmaris longus muscles.

Numbers 2-5 refer to the parts of the flexor digitorum superficialis muscle which are distributed to the corresponding fingers.

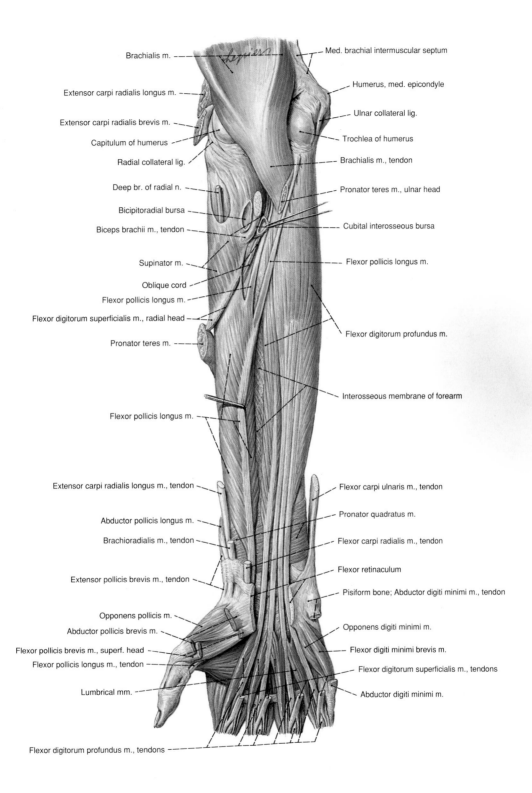

Brachialis m.

Extensor carpi radialis longus m.

Extensor carpi radialis brevis m.

Capitulum of humerus

Radial collateral lig.

Deep br. of radial n.

Bicipitoradial bursa

Biceps brachii m., tendon

Supinator m.

Oblique cord

Flexor pollicis longus m.

Flexor digitorum superficialis m., radial head

Pronator teres m.

Flexor pollicis longus m.

Extensor carpi radialis longus m., tendon

Abductor pollicis longus m.

Brachioradialis m., tendon

Extensor pollicis brevis m., tendon

Opponens pollicis m.

Abductor pollicis brevis m.

Flexor pollicis brevis m., superf. head

Flexor pollicis longus m., tendon

Lumbrical mm.

Flexor digitorum profundus m., tendons

Med. brachial intermuscular septum

Humerus, med. epicondyle

Ulnar collateral lig.

Trochlea of humerus

Brachialis m., tendon

Pronator teres m., ulnar head

Cubital interosseous bursa

Flexor pollicis longus m.

Flexor digitorum profundus m.

Interosseous membrane of forearm

Flexor carpi ulnaris m., tendon

Pronator quadratus m.

Flexor carpi radialis m., tendon

Flexor retinaculum

Pisiform bone; Abductor digiti minimi m., tendon

Opponens digiti minimi m.

Flexor digiti minimi brevis m.

Flexor digitorum superficialis m., tendons

Abductor digiti minimi m.

Fig. 416. Anterior aspect of the right forearm. Deep layer after exposure of the flexor digitorum profundus muscle.

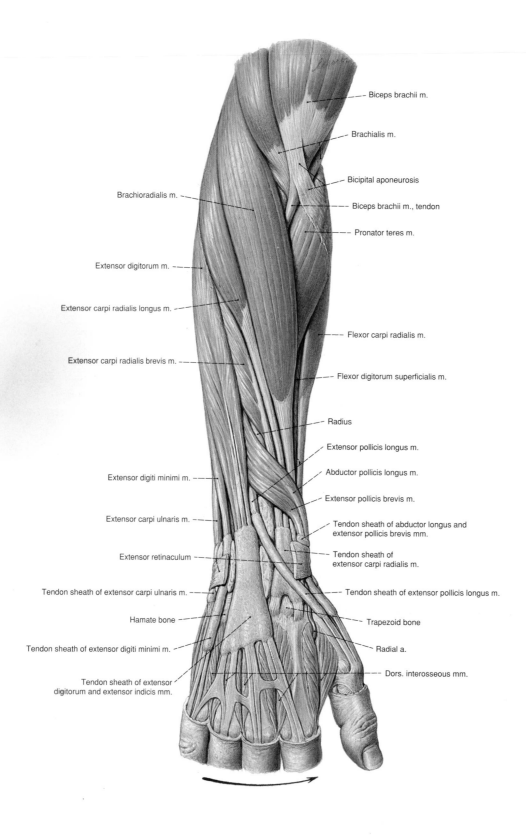

Biceps brachii m.

Brachialis m.

Bicipital aponeurosis

Brachioradialis m.

Biceps brachii m., tendon

Pronator teres m.

Extensor digitorum m.

Extensor carpi radialis longus m.

Flexor carpi radialis m.

Extensor carpi radialis brevis m.

Flexor digitorum superficialis m.

Radius

Extensor pollicis longus m.

Abductor pollicis longus m.

Extensor digiti minimi m.

Extensor pollicis brevis m.

Extensor carpi ulnaris m.

Tendon sheath of abductor longus and
extensor pollicis brevis mm.

Extensor retinaculum

Tendon sheath of
extensor carpi radialis m.

Tendon sheath of extensor carpi ulnaris m.

Tendon sheath of extensor pollicis longus m.

Hamate bone

Trapezoid bone

Tendon sheath of extensor digiti minimi m.

Radial a.

Tendon sheath of extensor
digitorum and extensor indicis mm.

Dors. interosseous mm.

Fig. 417. The forearm, pronated.

Extensor Muscles of the Forearm, Superficial, Radial Group

Name	Origin	Insertion	Innervation	Function
Brachioradialis muscle	Lateral supracondylar ridge of humerus; lateral intermuscular septum	Lateral side of base of styloid process of radius	Radial nerve (5th, 6th cervical nerves)	Flexes forearm; pronates forearm if flexed; supinates, if forearm is maximally pronated
Extensor carpi radialis longus muscle	Distal 1/3 of lateral supracondylar ridge of humerus and lateral intermuscular septum	Dorsal surface of base of 2nd metacarpal bone	Radial nerve (6th, 7th cervical nerves)	Extends (mainly extensor carpi radialis brevis); abducts (radially flexes) the hand; ext. carpi radialis longus supinates extended forearm (pronates when forearm flexed) and flexes elbow joint
Extensor carpi radialis brevis muscle	Lateral epicondyle of humerus; radial collateral ligament of elbow joint	Dorsal surface of base of 3rd metacarpal bone	same as above	

Extensor Muscles of the Forearm, Superficial, Posterior Group

Name	Origin	Insertion	Innervation	Function
Extensor digitorum muscle	Lateral epicondyle of humerus; antebrachial fascia	By 4 tendons to middle and distal phalanges of the fingers	Deep branch of radial nerve (6th, 7th, 8th cervical nerves)	Extends the fingers (proximal phalanges, in particular); indirectly extends the entire hand
Extensor digiti minimi muscle	same as above	Joins extensor digitorum tendon to little finger	same as above	Extends little finger; aids in extending hand
Extensor carpi ulnaris muscle (Separated from the 2 above by an intermuscular septum)	Lateral epicondyle of humerus; antebrachial fascia	Base of 5th metacarpal, dorsal surface	same as above	Extends (slightly) and adducts hand (ulnar flexes)

Extensor Muscles of the Forearm, Deep, Ulnar Group (Figs. 420, 421)

Name	Origin	Insertion	Innervation	Function
Extensor pollicis longus muscle	Posterior surface of ulna; interosseous membrane	Base of distal phalanx of thumb	Deep branch of radial nerve (6th, 7th, 8th cervical nerve)	Extend thumb and index finger; the former helps to extend the hand
Extensor indicis muscle	same as above	Extensor hood of the index finger	same as above	

Extensor Muscles of the Forearm, Deep, Radial (Oblique) Group (Fig. 417)

Name	Origin	Insertion	Innervation	Function
Abductor pollicis longus muscle	Posterior surfaces of ulna and radius; interosseous membrane	Base of 1st metacarpal of thumb, radial side	Deep branch of radial nerve (6th, 7th cervical nerves)	Abducts thumb; abducts hand (radially flex); former assists in supination; latter extends and abducts the proximal phalanx of thumb
Extensor pollicis brevis muscle	Posterior surface of radius; interosseous membrane	Base of proximal palanx of thumb	same as above	

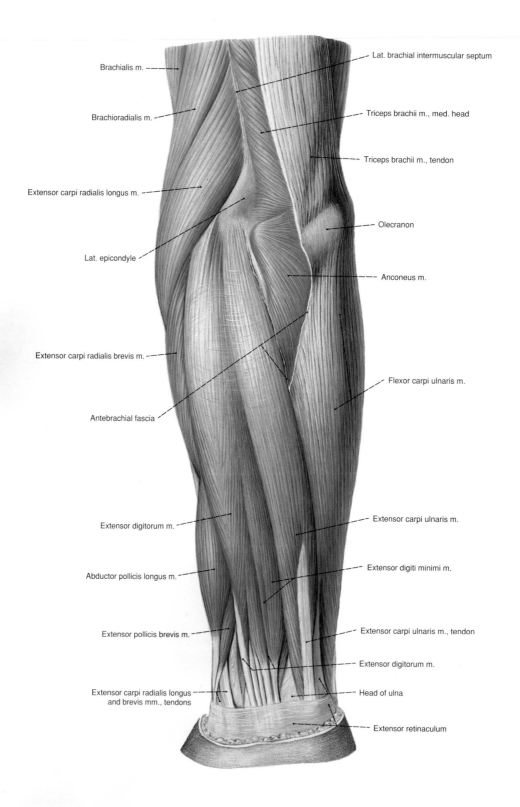

Brachialis m.

Brachioradialis m.

Extensor carpi radialis longus m.

Lat. epicondyle

Extensor carpi radialis brevis m.

Antebrachial fascia

Extensor digitorum m.

Abductor pollicis longus m.

Extensor pollicis brevis m.

Extensor carpi radialis longus and brevis mm., tendons

Lat. brachial intermuscular septum

Triceps brachii m., med. head

Triceps brachii m., tendon

Olecranon

Anconeus m.

Flexor carpi ulnaris m.

Extensor carpi ulnaris m.

Extensor digiti minimi m.

Extensor carpi ulnaris m., tendon

Extensor digitorum m.

Head of ulna

Extensor retinaculum

Fig. 418. Superficial muscles of the posterior aspect of the left forearm and of the distal portion of the arm.

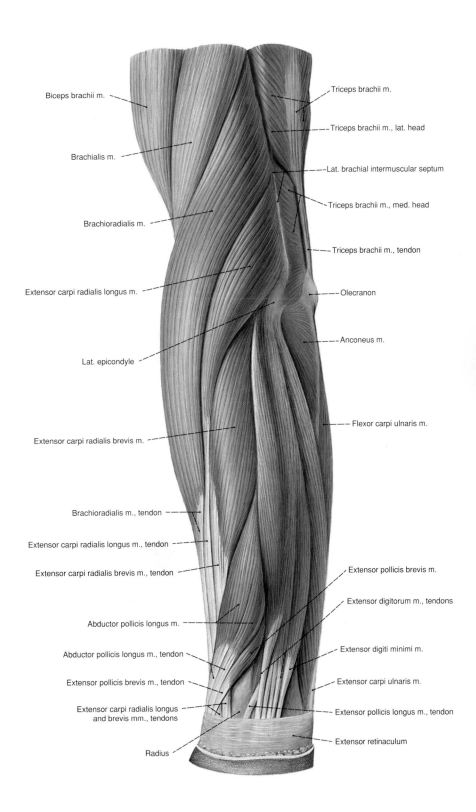

Biceps brachii m.

Brachialis m.

Brachioradialis m.

Extensor carpi radialis longus m.

Lat. epicondyle

Extensor carpi radialis brevis m.

Brachioradialis m., tendon

Extensor carpi radialis longus m., tendon

Extensor carpi radialis brevis m., tendon

Abductor pollicis longus m.

Abductor pollicis longus m., tendon

Extensor pollicis brevis m., tendon

Extensor carpi radialis longus and brevis mm., tendons

Radius

Triceps brachii m.

Triceps brachii m., lat. head

Lat. brachial intermuscular septum

Triceps brachii m., med. head

Triceps brachii m., tendon

Olecranon

Anconeus m.

Flexor carpi ulnaris m.

Extensor pollicis brevis m.

Extensor digitorum m., tendons

Extensor digiti minimi m.

Extensor carpi ulnaris m.

Extensor pollicis longus m., tendon

Extensor retinaculum

Fig. 419. Superficial muscles of the left forearm and the distal arm, viewed from the lateral (radial) side.

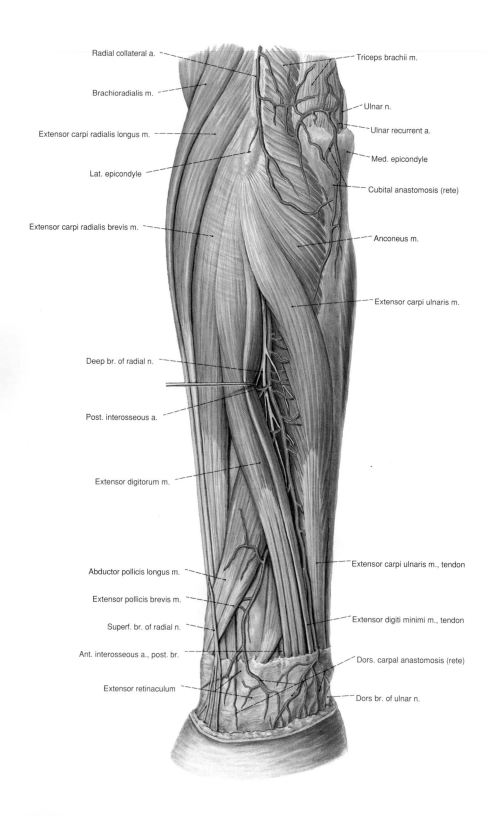

Radial collateral a.

Brachioradialis m.

Extensor carpi radialis longus m.

Lat. epicondyle

Extensor carpi radialis brevis m.

Deep br. of radial n.

Post. interosseous a.

Extensor digitorum m.

Abductor pollicis longus m.

Extensor pollicis brevis m.

Superf. br. of radial n.

Ant. interosseous a., post. br.

Extensor retinaculum

Triceps brachii m.

Ulnar n.

Ulnar recurrent a.

Med. epicondyle

Cubital anastomosis (rete)

Anconeus m.

Extensor carpi ulnaris m.

Extensor carpi ulnaris m., tendon

Extensor digiti minimi m., tendon

Dors. carpal anastomosis (rete)

Dors br. of ulnar n.

Fig. 420. Superficial layer of nerves and arteries of the posterior aspect of the left forearm.

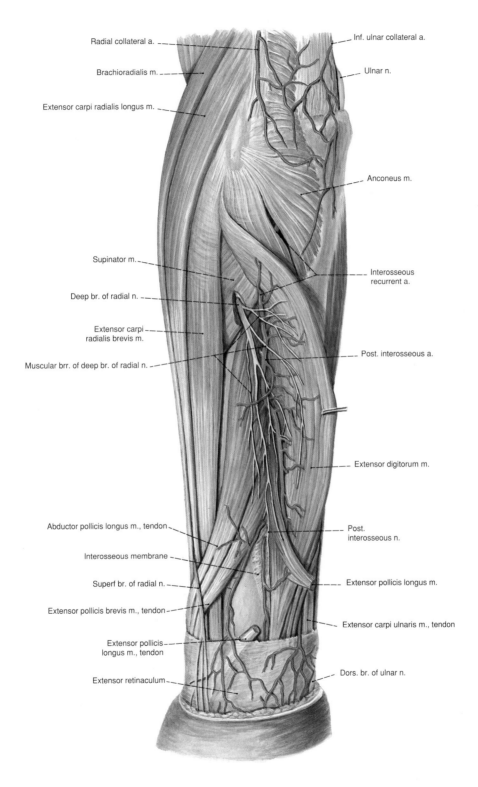

Radial collateral a.

Brachioradialis m.

Extensor carpi radialis longus m.

Supinator m.

Deep br. of radial n.

Extensor carpi
radialis brevis m.

Muscular brr. of deep br. of radial n.

Abductor pollicis longus m., tendon

Interosseous membrane

Superf br. of radial n.

Extensor pollicis brevis m., tendon

Extensor pollicis
longus m., tendon

Extensor retinaculum

Inf. ulnar collateral a.

Ulnar n.

Anconeus m.

Interosseous
recurrent a.

Post. interosseous a.

Extensor digitorum m.

Post.
interosseous n.

Extensor pollicis longus m.

Extensor carpi ulnaris m., tendon

Dors. br. of ulnar n.

Fig. 421. Deep layer of nerves and arteries of the posterior aspect of the left forearm. The extensor digitorum and digiti minimi muscles have been retracted toward the ulnar side, the extensor pollicis longus has been sectioned, and the supinator muscle has been slit along the deep branch of the radial nerve.

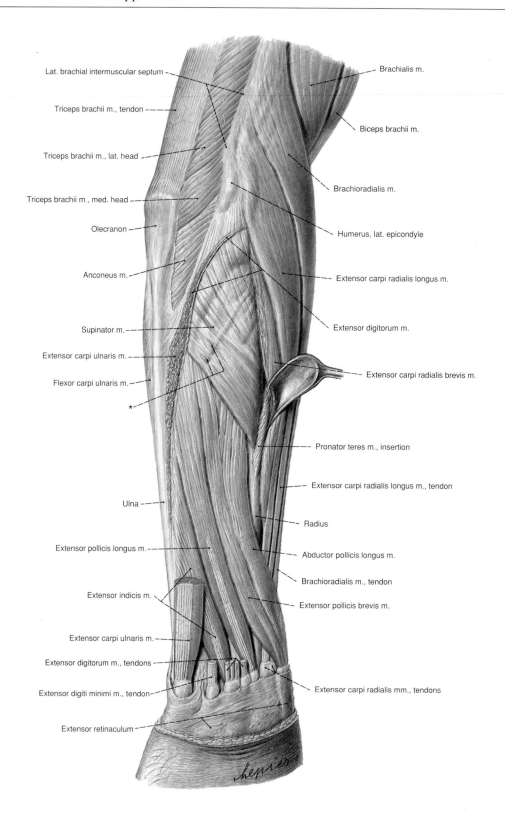

Lat. brachial intermuscular septum

Triceps brachii m., tendon

Triceps brachii m., lat. head

Triceps brachii m., med. head

Olecranon

Anconeus m.

Supinator m.

Extensor carpi ulnaris m.

Flexor carpi ulnaris m.

*

Ulna

Extensor pollicis longus m.

Extensor indicis m.

Extensor carpi ulnaris m.

Extensor digitorum m., tendons

Extensor digiti minimi m., tendon

Extensor retinaculum

Brachialis m.

Biceps brachii m.

Brachioradialis m.

Humerus, lat. epicondyle

Extensor carpi radialis longus m.

Extensor digitorum m.

Extensor carpi radialis brevis m.

Pronator teres m., insertion

Extensor carpi radialis longus m., tendon

Radius

Abductor pollicis longus m.

Brachioradialis m., tendon

Extensor pollicis brevis m.

Extensor carpi radialis mm., tendons

Fig. 422. Muscles of the posterior aspect of the right forearm after removal of the overlying (ulnar) extensor muscles.

* Points of penetration of the deep and posterior interosseous branches of the radial nerve

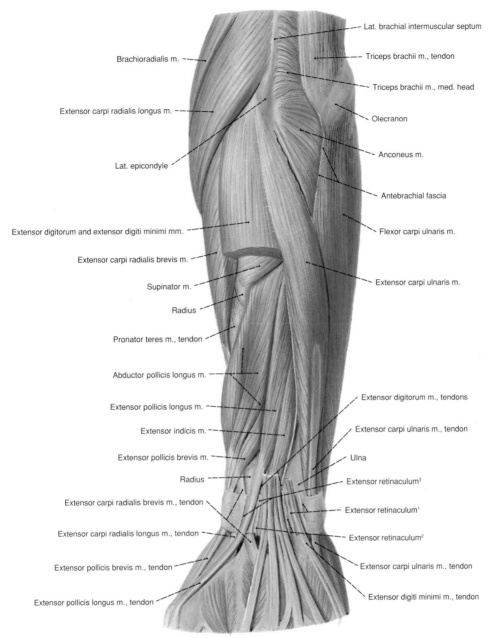

Brachioradialis m.

Extensor carpi radialis longus m.

Lat. epicondyle

Extensor digitorum and extensor digiti minimi mm.

Extensor carpi radialis brevis m.

Supinator m.

Radius

Pronator teres m., tendon

Abductor pollicis longus m.

Extensor pollicis longus m.

Extensor indicis m.

Extensor pollicis brevis m.

Radius

Extensor carpi radialis brevis m., tendon

Extensor carpi radialis longus m., tendon

Extensor pollicis brevis m., tendon

Extensor pollicis longus m., tendon

Lat. brachial intermuscular septum

Triceps brachii m., tendon

Triceps brachii m., med. head

Olecranon

Anconeus m.

Antebrachial fascia

Flexor carpi ulnaris m.

Extensor carpi ulnaris m.

Extensor digitorum m., tendons

Extensor carpi ulnaris m., tendon

Ulna

Extensor retinaculum[3]

Extensor retinaculum[1]

Extensor retinaculum[2]

Extensor carpi ulnaris m., tendon

Extensor digiti minimi m., tendon

Fig. 423. Muscles of the posterior aspect of the left forearm. The extensor digitorum and extensor digiti minimi muscles have been sectioned. The tendon compartments of the extensor retinaculum have been partially opened.

1 = Tendon compartment of extensor digiti minimi muscle
2 = Tendon compartment of extensor digitorum and extensor indicis muscles
3 = Tendon compartment for extensor pollicis longus muscle

Extensor Muscles of the Forearm, Deep Group, Supinator Muscle (Figs. 422, 425)

Name	Origin	Insertion	Innervation	Function
Supinator muscle Two layers (superficial and deep) between which the deep branch of radial nerve lies	Lateral epicondyle of humerus; radial collateral ligament; annular ligament of radius; supinator crest of ulna	Anterior surface and border, lateral surface, and posterior border of radius, proximal and distal to tuberosity of radius	Deep branch of radial nerve (6th cervical nerve)	Supinates hand and forearm

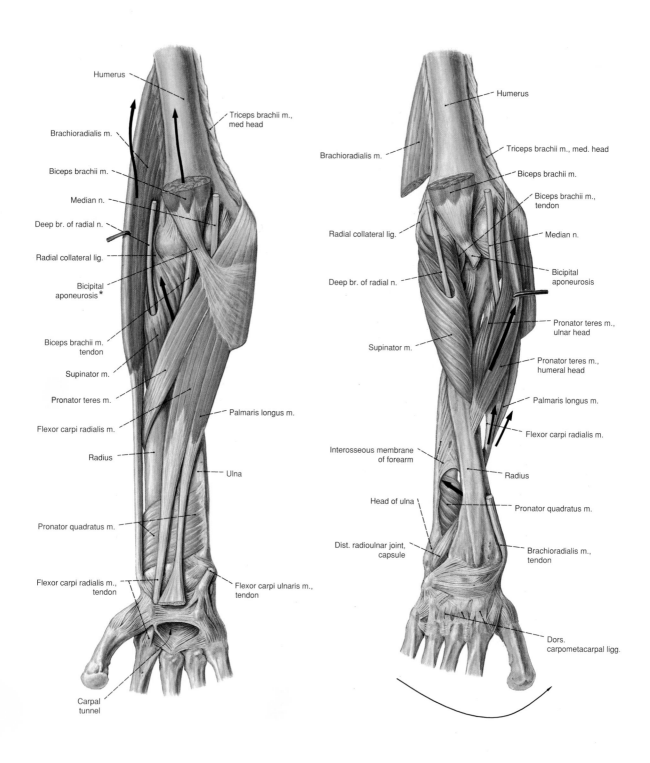

Humerus

Brachioradialis m.

Biceps brachii m.

Median n.

Deep br. of radial n.

Radial collateral lig.

Bicipital
aponeurosis *

Biceps brachii m.
tendon

Supinator m.

Pronator teres m.

Flexor carpi radialis m.

Radius

Pronator quadratus m.

Flexor carpi radialis m.,
tendon

Carpal
tunnel

Triceps brachii m.,
med head

Palmaris longus m.

Ulna

Flexor carpi ulnaris m.,
tendon

Humerus

Brachioradialis m.

Radial collateral lig.

Deep br. of radial n.

Supinator m.

Interosseous membrane
of forearm

Head of ulna

Dist. radioulnar joint,
capsule

Triceps brachii m., med. head

Biceps brachii m.

Biceps brachii m.,
tendon

Median n.

Bicipital
aponeurosis

Pronator teres m.,
ulnar head

Pronator teres m.,
humeral head

Palmaris longus m.

Flexor carpi radialis m.

Radius

Pronator quadratus m.

Brachioradialis m.,
tendon

Dors.
carpometacarpal ligg.

Fig. 424. Right forearm in supine position. The biceps brachii muscle is a supinator of the forearm when the forearm is partially flexed.

* Lacertus fibrosus

Fig. 425. The pronated forearm showing the deep muscles of pronation and supination. The brachialis muscle may supinate the forearm from extreme pronation or pronate from extreme supination.

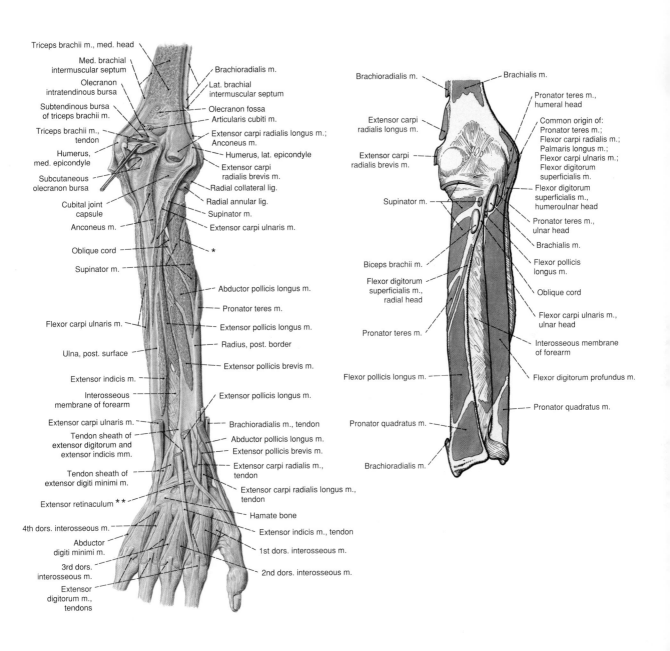

Triceps brachii m., med. head
Med. brachial intermuscular septum
Olecranon intratendinous bursa
Subtendinous bursa of triceps brachii m.
Triceps brachii m., tendon
Humerus, med. epicondyle
Subcutaneous olecranon bursa
Cubital joint capsule
Anconeus m.
Oblique cord
Supinator m.
Flexor carpi ulnaris m.
Ulna, post. surface
Extensor indicis m.
Interosseous membrane of forearm
Extensor carpi ulnaris m.
Tendon sheath of extensor digitorum and extensor indicis mm.
Tendon sheath of extensor digiti minimi m.
Extensor retinaculum **
4th dors. interosseous m.
Abductor digiti minimi m.
3rd dors. interosseous m.
Extensor digitorum m., tendons

Brachioradialis m.
Lat. brachial intermuscular septum
Olecranon fossa
Articularis cubiti m.
Extensor carpi radialis longus m.; Anconeus m.
Humerus, lat. epicondyle
Extensor carpi radialis brevis m.
Radial collateral lig.
Radial annular lig.
Supinator m.
Extensor carpi ulnaris m.
*
Abductor pollicis longus m.
Pronator teres m.
Extensor pollicis longus m.
Radius, post. border
Extensor pollicis brevis m.
Extensor pollicis longus m.
Brachioradialis m., tendon
Abductor pollicis longus m.
Extensor pollicis brevis m.
Extensor carpi radialis m., tendon
Extensor carpi radialis longus m., tendon
Hamate bone
Extensor indicis m., tendon
1st dors. interosseous m.
2nd dors. interosseous m.

Brachioradialis m.
Extensor carpi radialis longus m.
Extensor carpi radialis brevis m.
Supinator m.
Biceps brachii m.
Flexor digitorum superficialis m., radial head
Pronator teres m.
Flexor pollicis longus m.
Pronator quadratus m.
Brachioradialis m.

Brachialis m.
Pronator teres m., humeral head
Common origin of:
Pronator teres m.;
Flexor carpi radialis m.;
Palmaris longus m.;
Flexor carpi ulnaris m.;
Flexor digitorum superficialis m.
Flexor digitorum superficialis m., humeroulnar head
Pronator teres m., ulnar head
Brachialis m.
Flexor pollicis longus m.
Oblique cord
Flexor carpi ulnaris m., ulnar head
Interosseous membrane of forearm
Flexor digitorum profundus m.
Pronator quadratus m.

Fig. 426. Posterior forearm and dorsum of the right hand. The muscles have been extensively removed to expose their sites of attachment.

Fig. 427. Muscle attachments to the radius, ulna and distal end of the humerus. Right arm, anterior aspect.

   * Groove for the deep branch of the radial nerve
** Traditionally: Dorsal carpal ligament

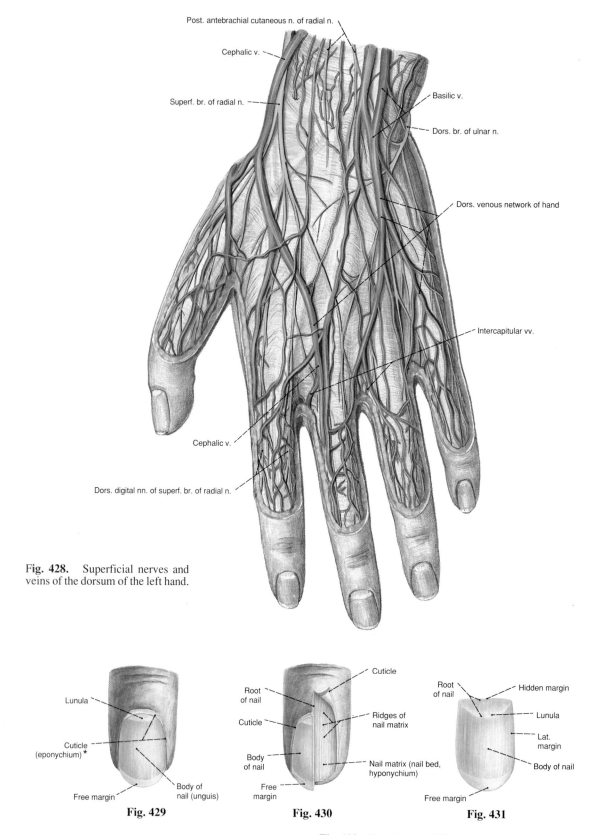

Post. antebrachial cutaneous n. of radial n.

Cephalic v.

Superf. br. of radial n.

Basilic v.

Dors. br. of ulnar n.

Dors. venous network of hand

Intercapitular vv.

Cephalic v.

Dors. digital nn. of superf. br. of radial n.

Fig. 428. Superficial nerves and veins of the dorsum of the left hand.

Lunula

Cuticle (eponychium) *

Free margin

Body of nail (unguis)

Fig. 429

Cuticle

Root of nail

Cuticle

Body of nail

Free margin

Ridges of nail matrix

Nail matrix (nail bed, hyponychium)

Fig. 430

Root of nail

Hidden margin

Lunula

Lat. margin

Body of nail

Free margin

Fig. 431

Fig. 430. Dorsal view of fingernail, sectioned longitudinally to expose the left half of the nail bed.

Fig. 431. Body of fingernail removed from the nail bed, dorsal view.

Fig. 429. Fingernail, dorsal view.

* Cuticle = epidermis

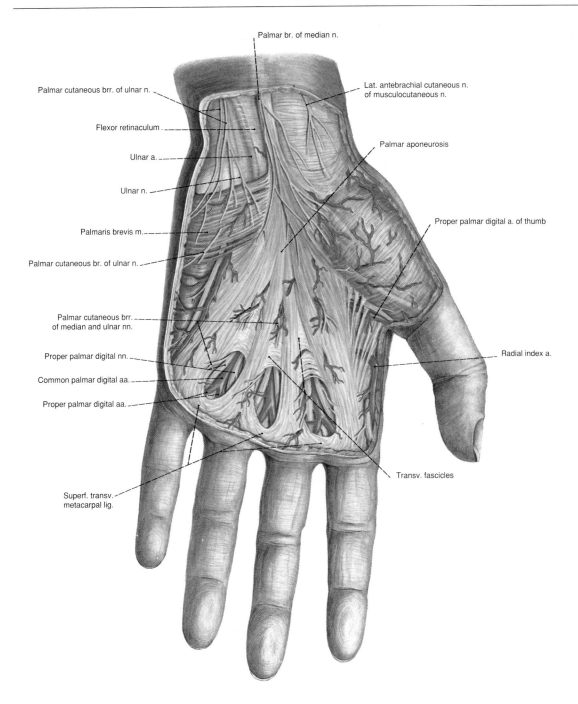

Palmar br. of median n.

Palmar cutaneous brr. of ulnar n.

Lat. antebrachial cutaneous n. of musculocutaneous n.

Flexor retinaculum

Palmar aponeurosis

Ulnar a.

Ulnar n.

Palmaris brevis m.

Proper palmar digital a. of thumb

Palmar cutaneous br. of ulnar n.

Palmar cutaneous brr. of median and ulnar nn.

Proper palmar digital nn.

Radial index a.

Common palmar digital aa.

Proper palmar digital aa.

Transv. fascicles

Superf. transv. metacarpal lig.

Fig. 432. Superficial nerves and arteries of the palm of the left hand.

Palmaris Brevis Muscle (Fig. 434)

Name	Origin	Insertion	Innervation	Function
Palmaris brevis muscle Cutaneous muscle, numerous separated fasciculi	Palmar aponeurosis, medial margin (occasionally trapezium bone)	Skin over medial border of palm	Ulnar nerve, superficial branch (8th cervical, 1st thoracic nerves)	Tenses skin of medial (ulnar) side of palm

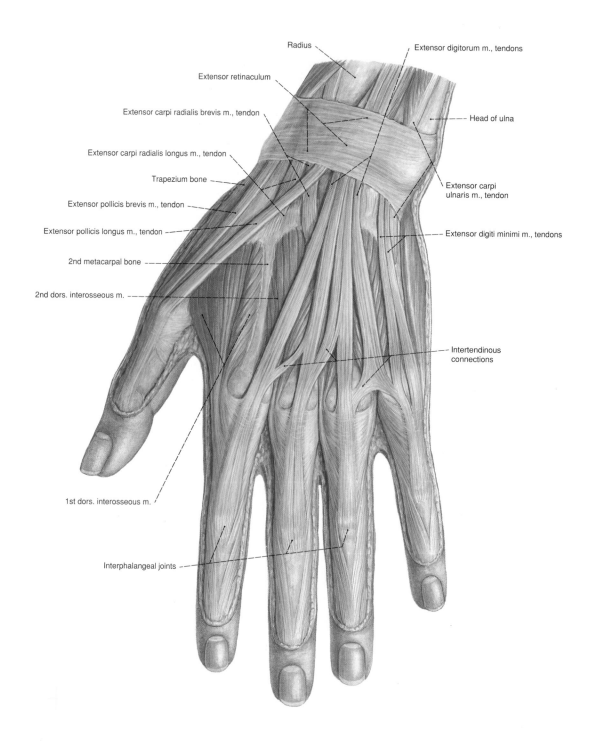

Radius

Extensor digitorum m., tendons

Extensor retinaculum

Extensor carpi radialis brevis m., tendon

Head of ulna

Extensor carpi radialis longus m., tendon

Trapezium bone

Extensor carpi ulnaris m., tendon

Extensor pollicis brevis m., tendon

Extensor pollicis longus m., tendon

Extensor digiti minimi m., tendons

2nd metacarpal bone

2nd dors. interosseous m.

Intertendinous connections

1st dors. interosseous m.

Interphalangeal joints

Fig. 433. Tendons of the dorsum of the left hand and the dorsal interosseous muscles.

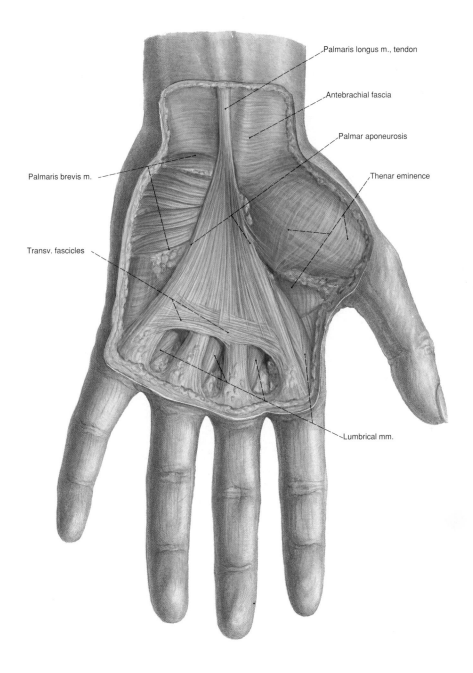

Palmaris longus m., tendon

Antebrachial fascia

Palmar aponeurosis

Thenar eminence

Palmaris brevis m.

Transv. fascicles

Lumbrical mm.

Fig. 434. The palmar aponeurosis and palmaris brevis muscle after removal of the skin of the palm of the left hand.

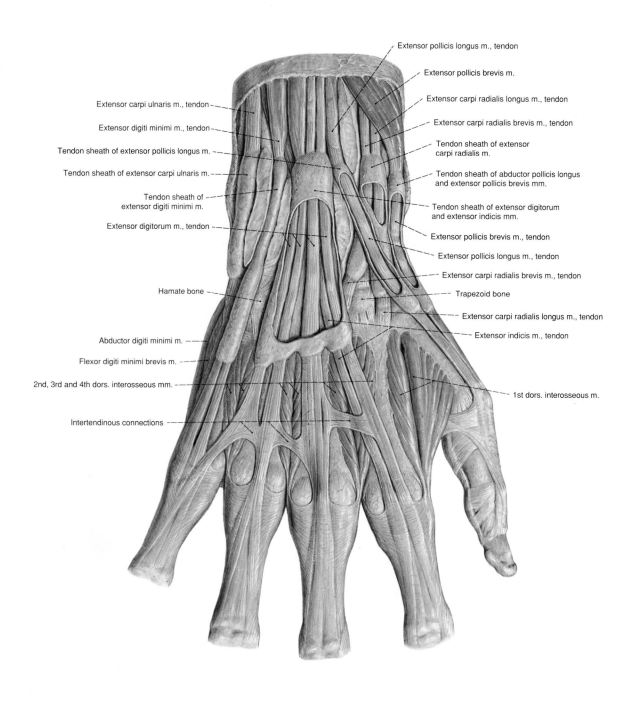

Extensor pollicis longus m., tendon

Extensor pollicis brevis m.

Extensor carpi radialis longus m., tendon

Extensor carpi radialis brevis m., tendon

Tendon sheath of extensor carpi radialis m.

Tendon sheath of abductor pollicis longus and extensor pollicis brevis mm.

Tendon sheath of extensor digitorum and extensor indicis mm.

Extensor pollicis brevis m., tendon

Extensor pollicis longus m., tendon

Extensor carpi radialis brevis m., tendon

Trapezoid bone

Extensor carpi radialis longus m., tendon

Extensor indicis m., tendon

1st dors. interosseous m.

Extensor carpi ulnaris m., tendon

Extensor digiti minimi m., tendon

Tendon sheath of extensor pollicis longus m.

Tendon sheath of extensor carpi ulnaris m.

Tendon sheath of extensor digiti minimi m.

Extensor digitorum m., tendon

Hamate bone

Abductor digiti minimi m.

Flexor digiti minimi brevis m.

2nd, 3rd and 4th dors. interosseous mm.

Intertendinous connections

Fig. 435. The dorsum of the right hand showing the tendon sheaths of the extensor muscles.

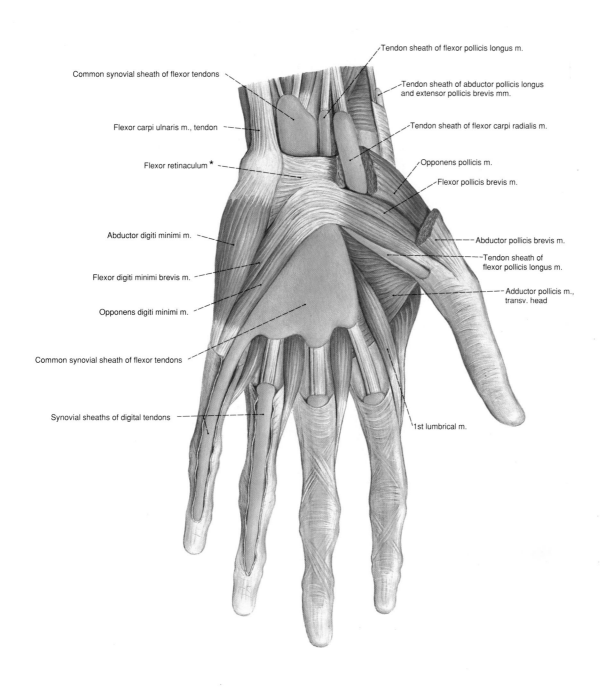

Tendon sheath of flexor pollicis longus m.

Common synovial sheath of flexor tendons

Tendon sheath of abductor pollicis longus and extensor pollicis brevis mm.

Flexor carpi ulnaris m., tendon

Tendon sheath of flexor carpi radialis m.

Flexor retinaculum *

Opponens pollicis m.

Flexor pollicis brevis m.

Abductor digiti minimi m.

Abductor pollicis brevis m.

Tendon sheath of flexor pollicis longus m.

Flexor digiti minimi brevis m.

Opponens digiti minimi m.

Adductor pollicis m., transv. head

Common synovial sheath of flexor tendons

Synovial sheaths of digital tendons

1st lumbrical m.

Fig. 436. Synovial sheaths of the tendons of the palm of the left hand. The tendon sheath of the flexor carpi radialis muscle has been exposed by cutting the muscles of the thenar eminence. The tendon sheaths of digits 4 and 5 have been split. Tendon sheaths have been colored blue. In vivo they contain a clear, colorless lubricating fluid.

* Transverse carpal ligament

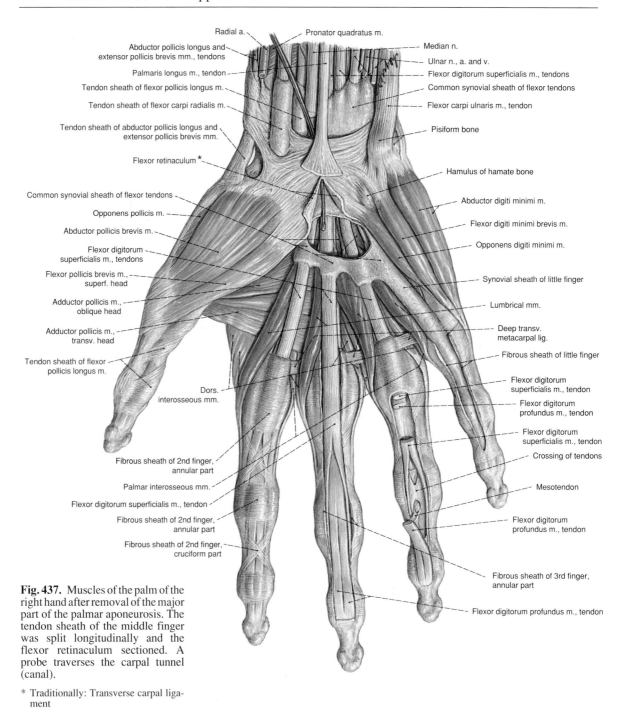

Radial a.

Pronator quadratus m.

Abductor pollicis longus and
extensor pollicis brevis mm., tendons

Median n.

Ulnar n., a. and v.

Palmaris longus m., tendon

Flexor digitorum superficialis m., tendons

Tendon sheath of flexor pollicis longus m.

Common synovial sheath of flexor tendons

Tendon sheath of flexor carpi radialis m.

Flexor carpi ulnaris m., tendon

Tendon sheath of abductor pollicis longus and
extensor pollicis brevis mm.

Pisiform bone

Flexor retinaculum *

Hamulus of hamate bone

Common synovial sheath of flexor tendons

Abductor digiti minimi m.

Opponens pollicis m.

Flexor digiti minimi brevis m.

Abductor pollicis brevis m.

Opponens digiti minimi m.

Flexor digitorum
superficialis m., tendons

Synovial sheath of little finger

Flexor pollicis brevis m.,
superf. head

Lumbrical mm.

Adductor pollicis m.,
oblique head

Deep transv.
metacarpal lig.

Adductor pollicis m.,
transv. head

Fibrous sheath of little finger

Tendon sheath of flexor
pollicis longus m.

Flexor digitorum
superficialis m., tendon

Flexor digitorum
profundus m., tendon

Dors.
interosseous mm.

Flexor digitorum
superficialis m., tendon

Crossing of tendons

Fibrous sheath of 2nd finger,
annular part

Mesotendon

Palmar interosseous mm.

Flexor digitorum superficialis m., tendon

Flexor digitorum
profundus m., tendon

Fibrous sheath of 2nd finger,
annular part

Fibrous sheath of 2nd finger,
cruciform part

Fibrous sheath of 3rd finger,
annular part

Fig. 437. Muscles of the palm of the
right hand after removal of the major
part of the palmar aponeurosis. The
tendon sheath of the middle finger
was split longitudinally and the
flexor retinaculum sectioned. A
probe traverses the carpal tunnel
(canal).

Flexor digitorum profundus m., tendon

* Traditionally: Transverse carpal ligament

Thenar Muscles (Figs. 437, 438). Continued on the next page.

Name	Origin	Insertion	Innervation	Function
1. Abductor pollicis brevis muscle	Flexor retinaculum; tuberosity of scaphoid; ridge of trapezium	Radial side of base of proximal phalanx of thumb; radial sesamoid bone	Median nerve (8th cervical, 1st thoracic nerves)	Abducts thumb, assists in opposition
2. Opponens pollicis muscle	Flexor retinaculum; ridge of trapezium	Entire length of radial border of metacarpal of thumb	Median nerve (8th cervical, 1st thoracic nerves)	Opposes thumb to other fingers, abducts, flexes, rotates first metacarpal

Fig. 438. The deep muscles of the palm of the right hand and the pronator quadratus muscle. The flexor retinaculum was severed. The tendons of the flexor digitorum superficialis muscle of each finger were sectioned at their intersection with the tendons of the flexor digitorum profundus flexor digitorum profundus to display the lumbrical muscles. The muscles of the thenar and hypothenar eminences were partially removed.

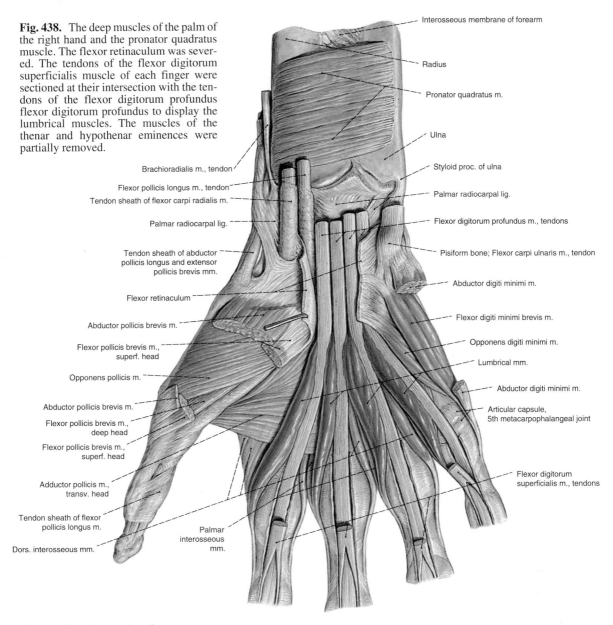

Brachioradialis m., tendon

Flexor pollicis longus m., tendon
Tendon sheath of flexor carpi radialis m.

Palmar radiocarpal lig.

Tendon sheath of abductor pollicis longus and extensor pollicis brevis mm.

Flexor retinaculum

Abductor pollicis brevis m.

Flexor pollicis brevis m., superf. head

Opponens pollicis m.

Abductor pollicis brevis m.

Flexor pollicis brevis m., deep head

Flexor pollicis brevis m., superf. head

Adductor pollicis m., transv. head

Tendon sheath of flexor pollicis longus m.

Dors. interosseous mm.

Palmar interosseous mm.

Interosseous membrane of forearm

Radius

Pronator quadratus m.

Ulna

Styloid proc. of ulna

Palmar radiocarpal lig.

Flexor digitorum profundus m., tendons

Pisiform bone; Flexor carpi ulnaris m., tendon

Abductor digiti minimi m.

Flexor digiti minimi brevis m.

Opponens digiti minimi m.

Lumbrical mm.

Abductor digiti minimi m.

Articular capsule, 5th metacarpophalangeal joint

Flexor digitorum superficialis m., tendons

Thenar Muscles, continued
The superficial muscle layer of the thenar eminence is formed by the abductor pollicis brevis, opponens pollicis and the superficial head of the flexor pollicis brevis muscles, while the deep layer is formed by the deep head of the flexor pollicis brevis and the adductor pollicis muscles.

Name	Origin	Insertion	Innervation	Function
3. Flexor pollicis brevis muscle *Superficial head* *Deep head*	Distal border of flexor retinaculum; distal part of tubercle of trapezium. Trapezoid and capitate bones	Radial side of base of proximal phalanx of thumb with radial sesamoid bone in its tendon	Superficial head: Median nerve; Deep head: Ulnar nerve, deep palmer branch (8th cervical, 1st thoracic nerves)	Flexes proximal phalanx of thumb; assists in opposition and adduction
4. Adductor pollicis muscle *Oblique head*	Capitate bone; bases of 2nd and 3rd metacarpals, intercarpal ligaments, sheath of flexor carpi radialis	Medial (ulnar) side of base of proximal phalanx of thumb, sesamoid bone	Ulnar nerve, deep palmar branch (8th cervical, 1st thoracic nerves)	Adducts the thumb; aids in opposition
Transverse head	Palmar surface of 3rd metacarpal			

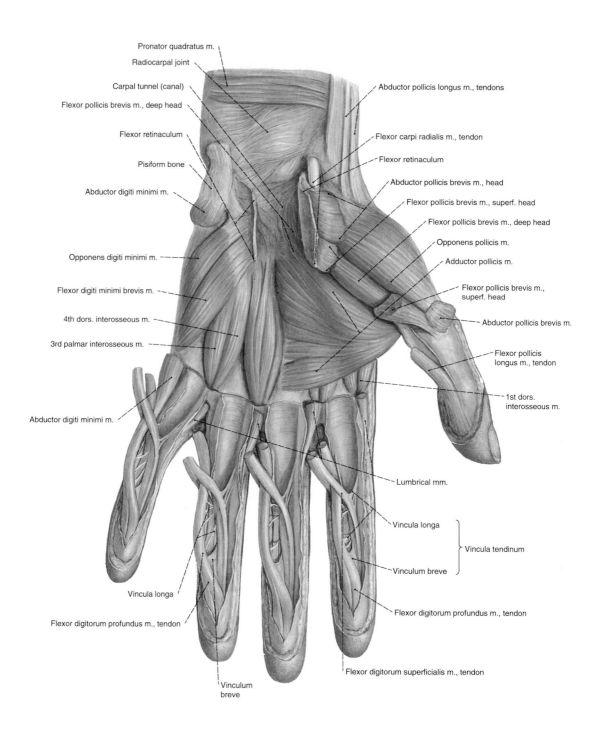

Pronator quadratus m.

Radiocarpal joint

Carpal tunnel (canal)

Flexor pollicis brevis m., deep head

Flexor retinaculum

Pisiform bone

Abductor digiti minimi m.

Opponens digiti minimi m.

Flexor digiti minimi brevis m.

4th dors. interosseous m.

3rd palmar interosseous m.

Abductor digiti minimi m.

Vincula longa

Flexor digitorum profundus m., tendon

Vinculum breve

Abductor pollicis longus m., tendons

Flexor carpi radialis m., tendon

Flexor retinaculum

Abductor pollicis brevis m., head

Flexor pollicis brevis m., superf. head

Flexor pollicis brevis m., deep head

Opponens pollicis m.

Adductor pollicis m.

Flexor pollicis brevis m., superf. head

Abductor pollicis brevis m.

Flexor pollicis longus m., tendon

1st dors. interosseous m.

Lumbrical mm.

Vincula longa

Vinculum breve

Vincula tendinum

Flexor digitorum profundus m., tendon

Flexor digitorum superficialis m., tendon

Fig. 439. Deep layer of the palm of the left hand with the carpal tunnel opened. The flexor tendons of the fingers, median nerve and the blood vessels have been removed, the digital tendon sheaths split open, and the abductor pollicis brevis, flexor pollicis brevis (superficial head) and abductor digiti minimi muscles sectioned.

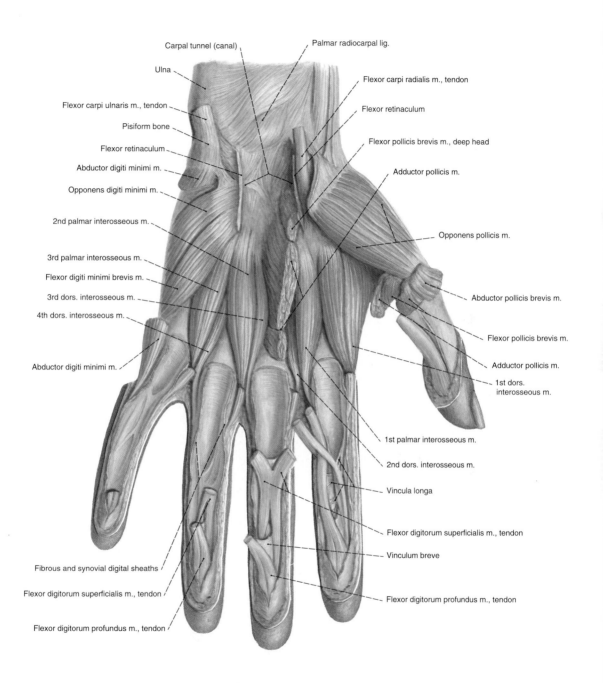

Carpal tunnel (canal)

Ulna

Flexor carpi ulnaris m., tendon

Pisiform bone

Flexor retinaculum

Abductor digiti minimi m.

Opponens digiti minimi m.

2nd palmar interosseous m.

3rd palmar interosseous m.

Flexor digiti minimi brevis m.

3rd dors. interosseous m.

4th dors. interosseous m.

Abductor digiti minimi m.

Fibrous and synovial digital sheaths

Flexor digitorum superficialis m., tendon

Flexor digitorum profundus m., tendon

Palmar radiocarpal lig.

Flexor carpi radialis m., tendon

Flexor retinaculum

Flexor pollicis brevis m., deep head

Adductor pollicis m.

Opponens pollicis m.

Abductor pollicis brevis m.

Flexor pollicis brevis m.

Adductor pollicis m.

1st dors. interosseous m.

1st palmar interosseous m.

2nd dors. interosseous m.

Vincula longa

Flexor digitorum superficialis m., tendon

Vinculum breve

Flexor digitorum profundus m., tendon

Fig. 440. Deepest layer of the palm of the left hand. The same dissection as in Fig. 439, but in addition, the flexor tendons of the fingers have been cut near their insertions and the adductor pollicis muscle as well as the deep head of the flexor pollicis brevis muscle were sectioned.

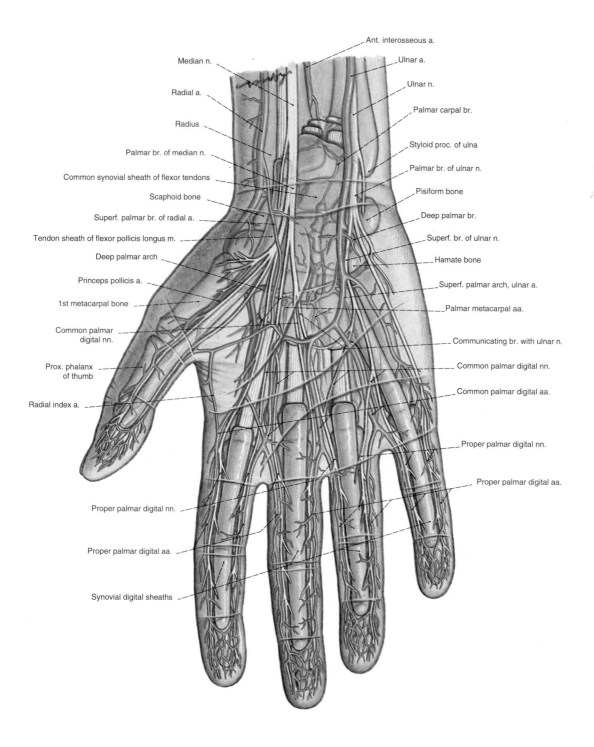

Fig. 441. Schematic representation of the tendon sheaths in the wrist, palm and fingers, and the arterial and nerve supply of the palm of the right hand.

Fig. 442. Middle layer of nerves and arteries of the palm of the left hand showing the superficial palmar arch. The palmar aponeurosis has been removed, and the abductor pollicis brevis muscle split along the superficial palmar branch of the radial artery.

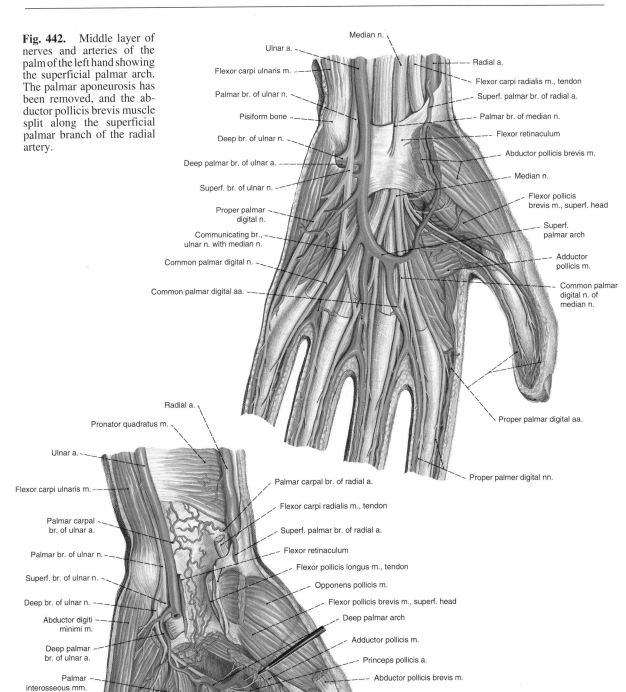

Fig. 443. Deep layer of nerves and arteries of the palm of the left hand showing the deep palmar arch. The abductor pollicis brevis, adductor pollicis and flexor digiti minimi muscles have been sectioned, and the flexor tendons, the median nerve, the superficial branch of the ulnar nerve and the superficial palmar arch removed.

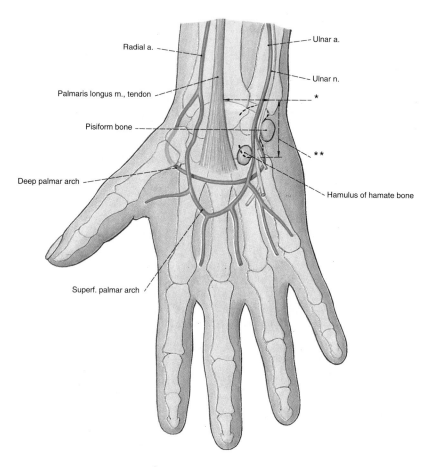

Radial a.

Ulnar a.

Ulnar n.

Palmaris longus m., tendon

*

Pisiform bone

**

Deep palmar arch

Hamulus of hamate bone

Superf. palmar arch

Fig. 444. The ulnar tunnel (clinically: Guyon's interstitial space) in the hypothenar eminence of the right hand: position, dimensions and transversing structures. The dashed lines mark the proximal and distal openings of the ulnar tunnel. (From: Dr. H.-M. Schmidt, Anatomical Institute of the University of Bonn.)

  * Width: radial 15.3 (8.6-24.1)
               ulnar 15.4 (6.6-25.9) (mm)
** Length: proximal 15.1 (9.0-20.1)
               distal 14.8 (9.3-19.2) (mm)

The ulnar tunnel of the hypothenar eminence (clinically: Guyon's interstitial space) has considerable practical significance. It lies between the pisiform bone and the hamulus of the hamate bone in the hypothenar compartment of the hand. Its proximal opening or hiatus is bounded by the antebrachial fascia (superficial part of the flexor retinaculum, known clinically as the palmar carpal ligament) and the flexor retinaculum. Together with the pisohamate and pisometacarpal ligaments, the retinaculum contributes to the floor of the interstitial space. Its roof consists of the palmar ulnocarpal ligament, connective tissue extensions of the flexor carpi ulnaris muscle and the antebrachial fascia, adipose tissue, and the palmaris brevis muscle.

The variable distal hiatus is covered by the tendinous origins of the abductor digiti minimi and the flexor digiti minimi muscles in a crescent shape. The superficial and deep branches of the ulnar nerve course in the interstitial space. The superficial and deep branches of the ulnar nerve are particularly vulnerable. They can be compressed by ganglia (hygromas or nodes = gelatinous enlargements near joints and tendons of the wrist), tumors, fractures, dislocations and sprained muscles. Pain in the wrist joint, radiating to the forearm and to the fourth and fifth fingers, hyp- and paresthesia in these fingers as well as weakness and atrophy of the muscles of the hand supplied by the ulnar nerve are characteristic of the ulnar tunnel syndrome.

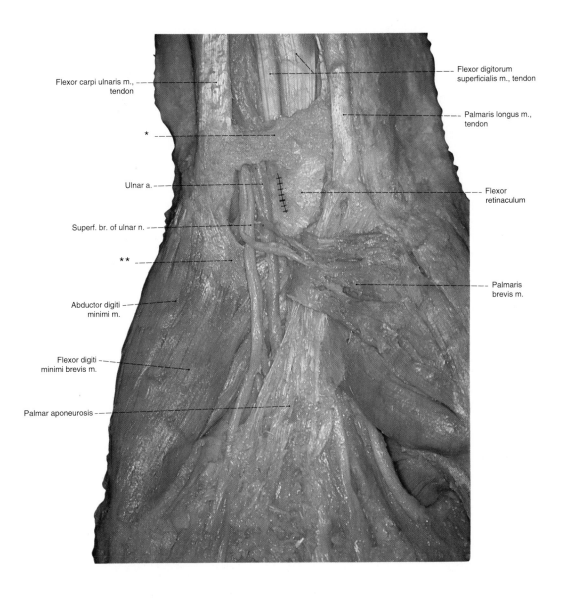

Flexor carpi ulnaris m., tendon

Flexor digitorum superficialis m., tendon

Palmaris longus m., tendon

*

Ulnar a.

Flexor retinaculum

Superf. br. of ulnar n.

**

Palmaris brevis m.

Abductor digiti minimi m.

Flexor digiti minimi brevis m.

Palmar aponeurosis

Fig. 445. The ulnar tunnel, left hand, palmar view: openings and traversing structures. (From: Dr. H.-M. SCHMIDT, Anatomical Institute of the University of Bonn.)

  * Palmar carpal (volar) ligament
** Tendinous arch at the origin of the hypothenar muscles and the distal hiatus of GUYON's interstitial space.

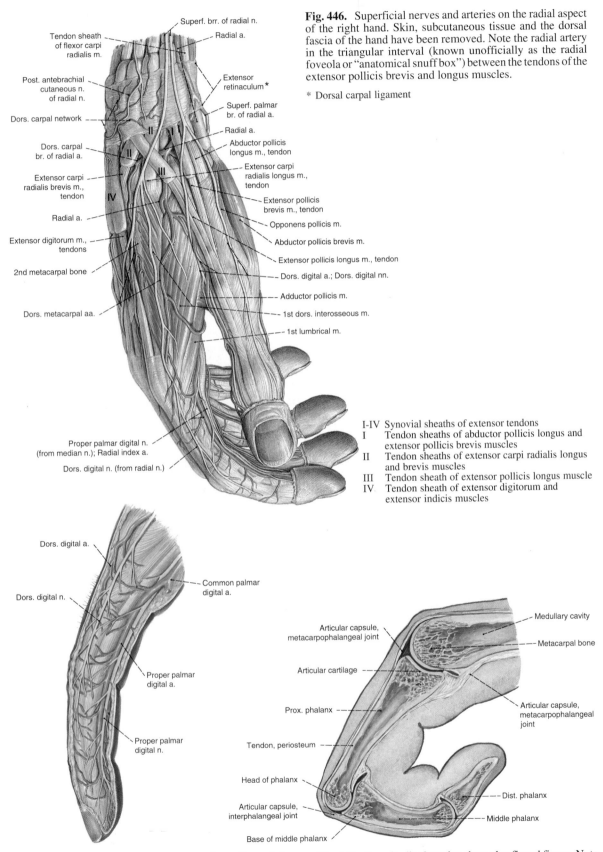

Tendon sheath of flexor carpi radialis m.

Post. antebrachial cutaneous n. of radial n.

Dors. carpal network

Dors. carpal br. of radial a.

Extensor carpi radialis brevis m., tendon

Radial a.

Extensor digitorum m., tendons

2nd metacarpal bone

Dors. metacarpal aa.

Proper palmar digital n. (from median n.); Radial index a.

Dors. digital n. (from radial n.)

Superf. brr. of radial n.

Radial a.

Extensor retinaculum*

Superf. palmar br. of radial a.

Radial a.

Abductor pollicis longus m., tendon

Extensor carpi radialis longus m., tendon

Extensor pollicis brevis m., tendon

Opponens pollicis m.

Abductor pollicis brevis m.

Extensor pollicis longus m., tendon

Dors. digital a.; Dors. digital nn.

Adductor pollicis m.

1st dors. interosseous m.

1st lumbrical m.

Fig. 446. Superficial nerves and arteries on the radial aspect of the right hand. Skin, subcutaneous tissue and the dorsal fascia of the hand have been removed. Note the radial artery in the triangular interval (known unofficially as the radial foveola or "anatomical snuff box") between the tendons of the extensor pollicis brevis and longus muscles.

* Dorsal carpal ligament

I-IV Synovial sheaths of extensor tendons
I   Tendon sheaths of abductor pollicis longus and extensor pollicis brevis muscles
II   Tendon sheaths of extensor carpi radialis longus and brevis muscles
III Tendon sheath of extensor pollicis longus muscle
IV Tendon sheath of extensor digitorum and extensor indicis muscles

Dors. digital a.

Dors. digital n.

Common palmar digital a.

Proper palmar digital a.

Proper palmar digital n.

Articular capsule, metacarpophalangeal joint

Articular cartilage

Prox. phalanx

Tendon, periosteum

Head of phalanx

Articular capsule, interphalangeal joint

Base of middle phalanx

Medullary cavity

Metacarpal bone

Articular capsule, metacarpophalangeal joint

Dist. phalanx

Middle phalanx

Fig. 447. Nerves and arteries of the index finger. Lateral view.

Fig. 448. Longitudinal section through a flexed finger. Note the position of the flexion creases in relation to the corresponding joints.

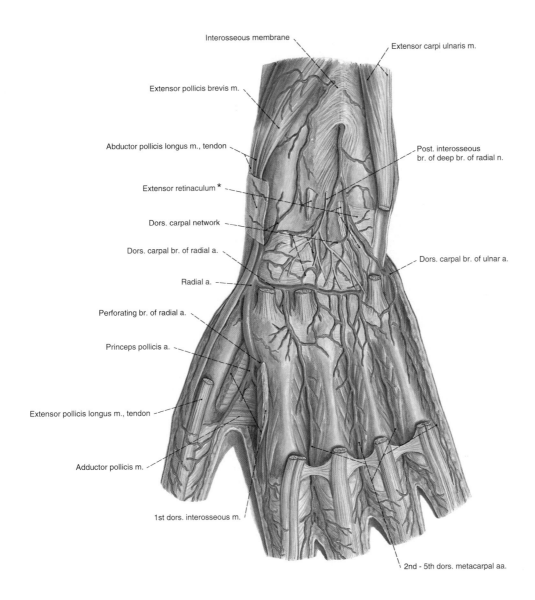

Interosseous membrane

Extensor carpi ulnaris m.

Extensor pollicis brevis m.

Abductor pollicis longus m., tendon

Post. interosseous
br. of deep br. of radial n.

Extensor retinaculum *

Dors. carpal network

Dors. carpal br. of radial a.

Dors. carpal br. of ulnar a.

Radial a.

Perforating br. of radial a.

Princeps pollicis a.

Extensor pollicis longus m., tendon

Adductor pollicis m.

1st dors. interosseous m.

2nd - 5th dors. metacarpal aa.

Fig. 449. Arteries of the dorsum of the left hand and distal end of the extensor compartment of the forearm. The extensor tendons with the exception of those of the abductor pollicis longus and extensor pollicis brevis muscles have been cut and partially removed; the extensor retinaculum has been opened and partially removed; the first dorsal interosseous muscle has been sectioned.

* Dorsal carpal ligament

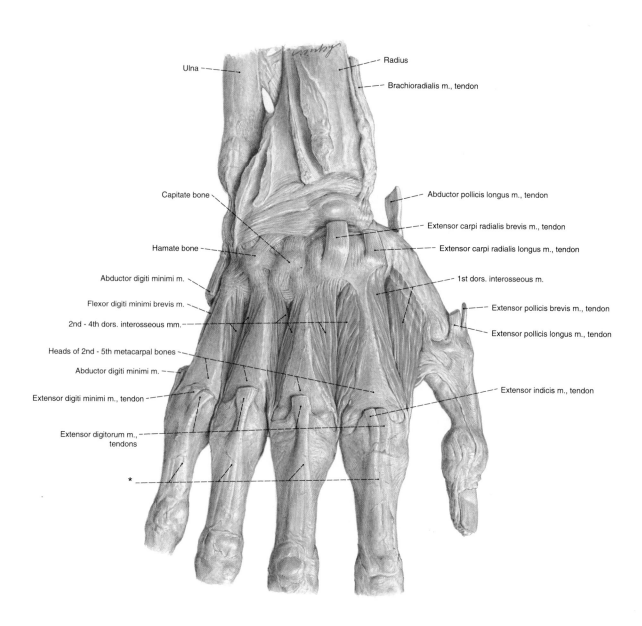

Ulna

Radius

Brachioradialis m., tendon

Capitate bone

Abductor pollicis longus m., tendon

Extensor carpi radialis brevis m., tendon

Hamate bone

Extensor carpi radialis longus m., tendon

Abductor digiti minimi m.

1st dors. interosseous m.

Flexor digiti minimi brevis m.

Extensor pollicis brevis m., tendon

2nd - 4th dors. interosseous mm.

Extensor pollicis longus m., tendon

Heads of 2nd - 5th metacarpal bones

Abductor digiti minimi m.

Extensor digiti minimi m., tendon

Extensor indicis m., tendon

Extensor digitorum m.,
tendons

*

Fig. 450. Dorsum of the right hand showing the interosseous
muscles and the extensor hoods of the fingers.

* Extensor hood or expansion (extensor aponeurosis)

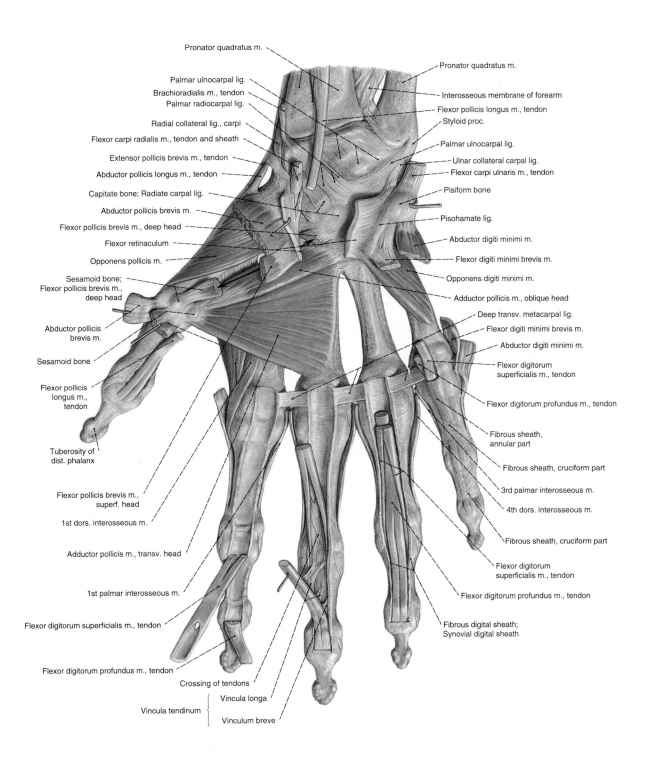

Pronator quadratus m.

Palmar ulnocarpal lig.

Brachioradialis m., tendon

Palmar radiocarpal lig.

Radial collateral lig., carpi

Flexor carpi radialis m., tendon and sheath

Extensor pollicis brevis m., tendon

Abductor pollicis longus m., tendon

Capitate bone; Radiate carpal lig.

Abductor pollicis brevis m.

Flexor pollicis brevis m., deep head

Flexor retinaculum

Opponens pollicis m.

Sesamoid bone;
Flexor pollicis brevis m.,
deep head

Abductor pollicis
brevis m.

Sesamoid bone

Flexor pollicis
longus m.,
tendon

Tuberosity of
dist. phalanx

Flexor pollicis brevis m.,
superf. head

1st dors. interosseous m.

Adductor pollicis m., transv. head

1st palmar interosseous m.

Flexor digitorum superficialis m., tendon

Flexor digitorum profundus m., tendon

Crossing of tendons

Vincula longa

Vincula tendinum

Vinculum breve

Pronator quadratus m.

Interosseous membrane of forearm

Flexor pollicis longus m., tendon

Styloid proc.

Palmar ulnocarpal lig.

Ulnar collateral carpal lig.

Flexor carpi ulnaris m., tendon

Pisiform bone

Pisohamate lig.

Abductor digiti minimi m.

Flexor digiti minimi brevis m.

Opponens digiti minimi m.

Adductor pollicis m., oblique head

Deep transv. metacarpal lig.

Flexor digiti minimi brevis m.

Abductor digiti minimi m.

Flexor digitorum
superficialis m., tendon

Flexor digitorum profundus m., tendon

Fibrous sheath,
annular part

Fibrous sheath, cruciform part

3rd palmar interosseous m.

4th dors. interosseous m.

Fibrous sheath, cruciform part

Flexor digitorum
superficialis m., tendon

Flexor digitorum profundus m., tendon

Fibrous digital sheath;
Synovial digital sheath

Fig. 451. Deepest layer of the musculature of the palm of the right hand. The carpal canal was opened by cutting the flexor retinaculum. The origin and insertion of the pronator quadratus muscle were sectioned at the distal end of the ulna and radius, respectively. The tendon of the flexor pollicis longus muscle is partially retained. After opening the tendon sheaths, the flexor tendons of the fingers were dissected to their insertions. An arrow connects the insertion and origin of the deep head of flexor pollicis brevis muscle.

Fig. 452. Diagram of the dorsal interosseous muscles and the extensor hoods of the fingers of the left hand. Dorsal view. ▼

Fig. 453. Diagram of the palmar intereosseous muscles of the left hand. Palmar view. ▼

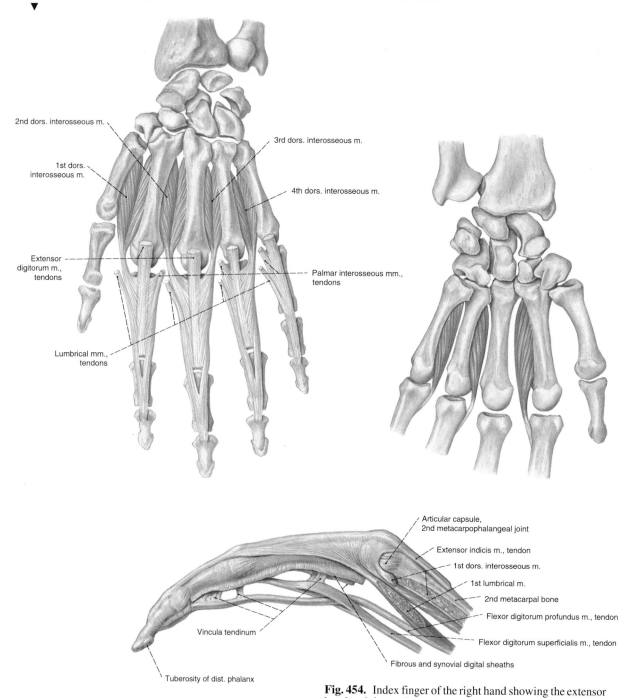

2nd dors. interosseous m.

3rd dors. interosseous m.

1st dors. interosseous m.

4th dors. interosseous m.

Extensor digitorum m., tendons

Palmar interosseous mm., tendons

Lumbrical mm., tendons

Articular capsule, 2nd metacarpophalangeal joint

Extensor indicis m., tendon

1st dors. interosseous m.

1st lumbrical m.

2nd metacarpal bone

Flexor digitorum profundus m., tendon

Vincula tendinum

Flexor digitorum superficialis m., tendon

Fibrous and synovial digital sheaths

Tuberosity of dist. phalanx

Fig. 454. Index finger of the right hand showing the extensor hood and the tendons of its flexor and extensor muscles.

Lumbrical Muscles

Name	Origin	Insertion	Innervation	Function
Lumbrical muscles	Tendons of flexor digitorum profundus muscle; 1st and 2nd have one head; 3rd and 4th, 2 heads	Extensor head of 2nd to 5th digits	1st and 2nd lumbricals: median nerve (8th cervical and 1st thoracic nerves); 3rd and 4th lumbricals: ulnar nerve (8th cervical and 1st thoracic nerves)	Flex the metacarpophalangeal joints; extend the two distal phalanges

Fig. 455. Diagram of the lumbrical muscles of the left hand. Palmar view.
▼

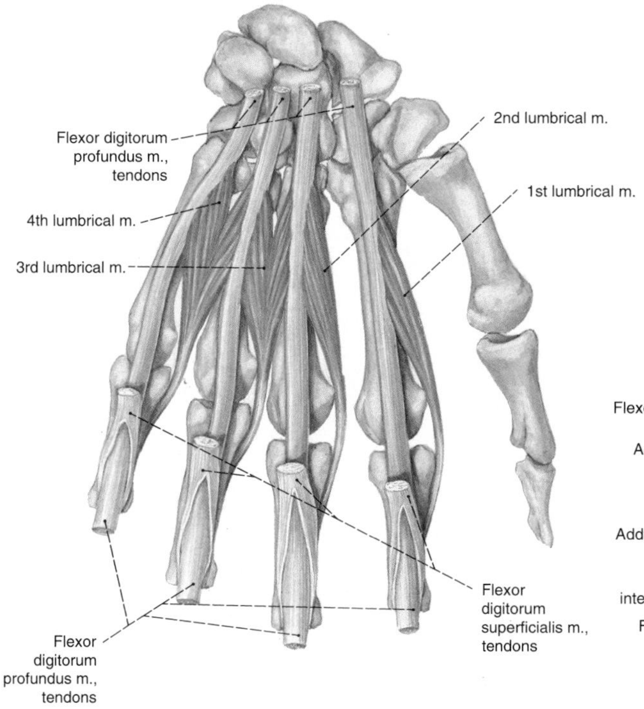

Fig. 456. Muscle origins and insertions on the palmar aspect of the right hand. The tendon compartments of the radiocarpal joint are designated by proximally directed arrows.
▼

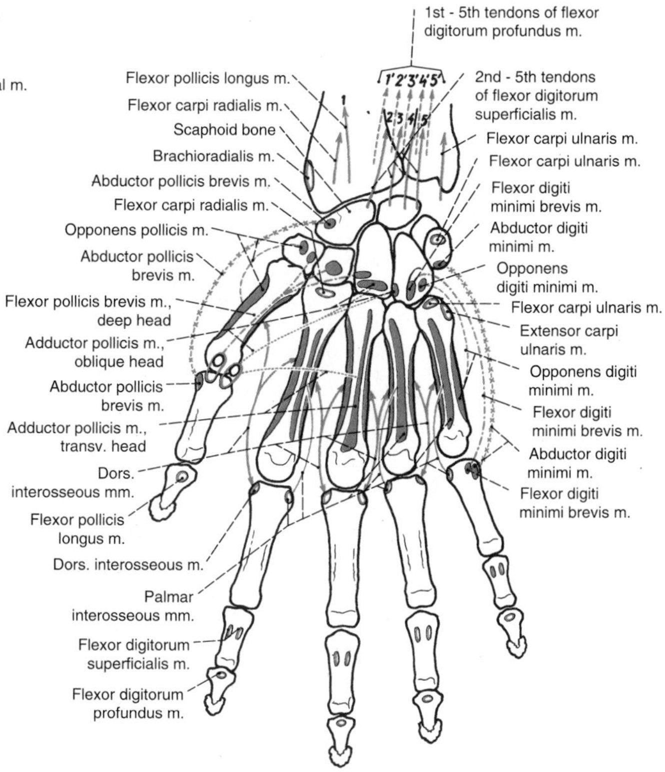

Hypothenar Muscles

Name	Origin	Insertion	Innervation	Function
Abductor digiti minimi muscle	Pisiform bone and tendon of flexor carpi ulnaris m.	Base of proximal phalanx of little finger, ulnar side of dorsal aponeurosis of little finger	Ulnar nerve, deep branch (8th cervical, 1st thoracic nerves)	As their names indicate: abduct and flex little finger; opponens digiti minimi muscle also rotates 5th metacarpal and brings it out of plane of palm to meet the thumb and cup the hand
Flexor digiti minimi brevis muscle (variable)	Flexor retinaculum; hamulus of hamate bone	Ulnar side of base of proximal phalanx of little finger		
Opponens digiti minimi muscle	Flexor retinaculum; hamulus of hamate bone	Ulnar margin of 5th metacarpal bone		
Palmar is brevis muscle	(see page 255)			

Interosseous Muscles (Figs. 452, 453)

Name	Origin	Insertion	Innervation	Function
Dorsal interosseous muscles	By two heads from 1st-5th metacarpal bones	Extensor hood of 2nd-4th fingers	Ulnar nerve, deep palmar brach (8th cervical, 1st thoracic nerves)	Dorsal interosseous muscles abduct; palmar interosseous adduct the fingers. All flex 2nd-5th fingers at metacarpophalangeal joints and extend the two distal phalanges at the corresponding interphalangeal joints
Palmar interosseous muscles	By one head from 2nd-5th metacarpal bones	Extensor hood of 2nd-5th fingers		

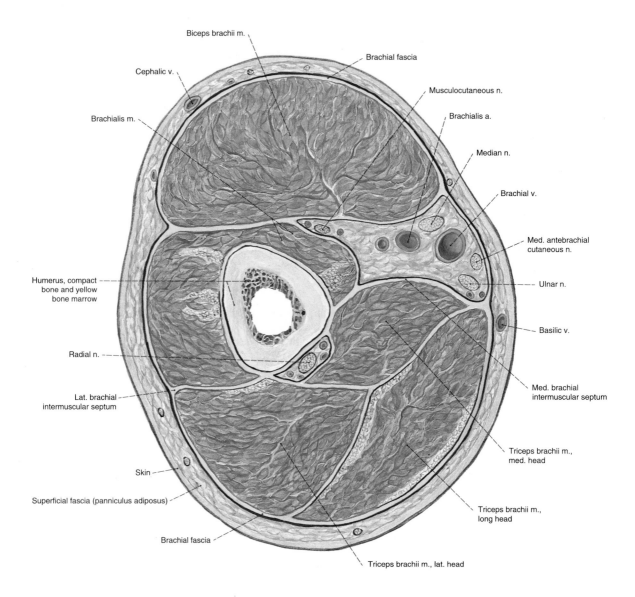

Fig. 457. Cross section through the arm at the level of the groove for the radial nerve in the middle third of the left arm. Distal surface of section.

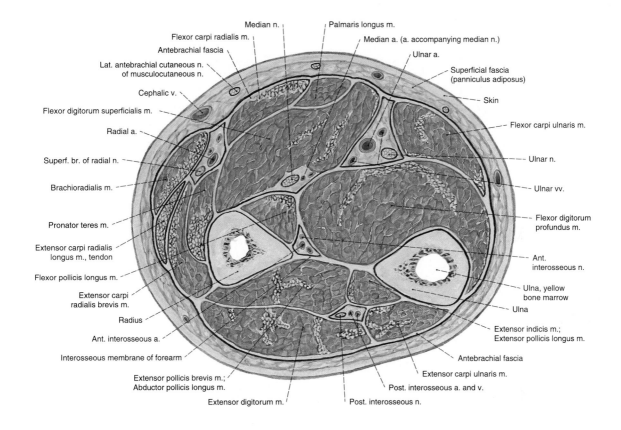

Fig. 458. Cross section through the middle third of the left forearm. Distal surface of section.

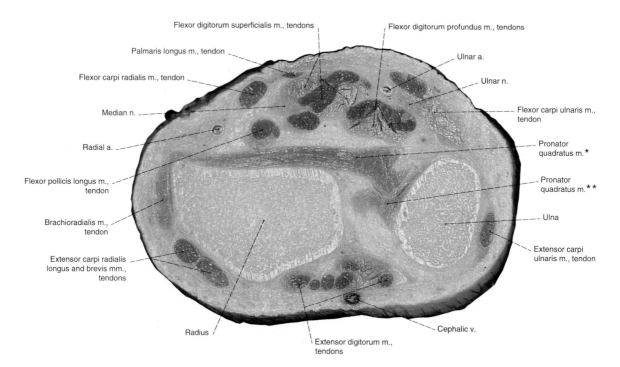

Fig. 459. Cross section through the left forearm proximal to the distal radioulnar joint. Proximal surface of section.

  * Superficial part of the pronator quadratus muscle
** Deep part of the pronator quadratus muscle

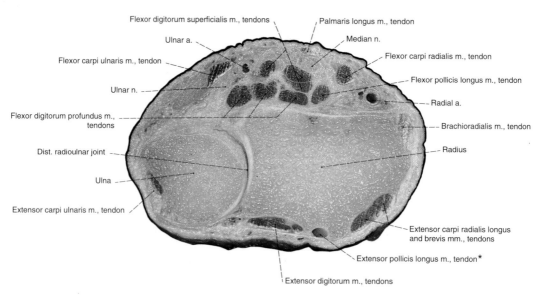

Fig. 460. Cross section through the left forearm at the level of the distal radioulnar joint. Distal surface of section. (From: Prof. Dr. H.-M. SCHMIDT, Anatomical Institute of the University of Bonn.)

* Tendon has been removed

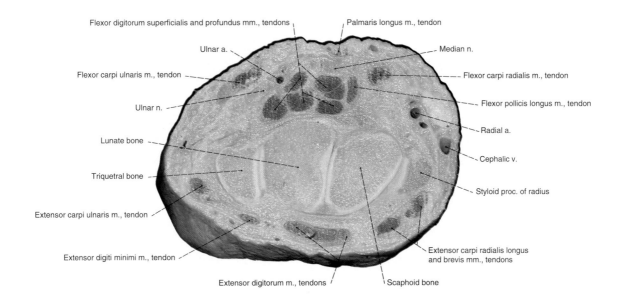

Flexor digitorum superficialis and profundus mm., tendons

Palmaris longus m., tendon

Ulnar a.

Median n.

Flexor carpi ulnaris m., tendon

Flexor carpi radialis m., tendon

Ulnar n.

Flexor pollicis longus m., tendon

Radial a.

Lunate bone

Cephalic v.

Triquetral bone

Styloid proc. of radius

Extensor carpi ulnaris m., tendon

Extensor digiti minimi m., tendon

Extensor carpi radialis longus
and brevis mm., tendons

Extensor digitorum m., tendons

Scaphoid bone

Fig. 461. Cross section through the left hand at the level of the proximal wrist bones. Distal surface of section.

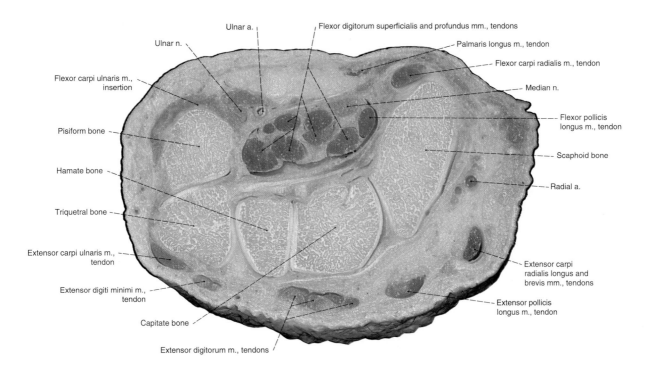

Ulnar a.

Flexor digitorum superficialis and profundus mm., tendons

Ulnar n.

Palmaris longus m., tendon

Flexor carpi radialis m., tendon

Flexor carpi ulnaris m., insertion

Median n.

Pisiform bone

Flexor pollicis longus m., tendon

Hamate bone

Scaphoid bone

Triquetral bone

Radial a.

Extensor carpi ulnaris m., tendon

Extensor digiti minimi m., tendon

Extensor carpi radialis longus and brevis mm., tendons

Capitate bone

Extensor pollicis longus m., tendon

Extensor digitorum m., tendons

Fig. 462. Cross section through the left hand at the level of the pisiform bone. Distal surface of section. (From: Dr. H.-M. SCHMIDT, Anatomical Institute of the University of Bonn.)

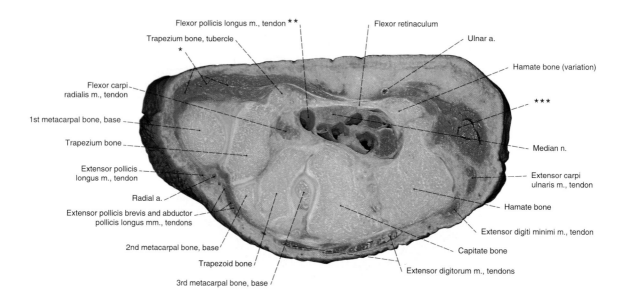

Fig. 463. Cross section through the left hand at the level of the hamulus of the hamate bone. Proximal surface of section.

  * Thenar muscles
 ** Tendon has been removed
*** Hypothenar muscles

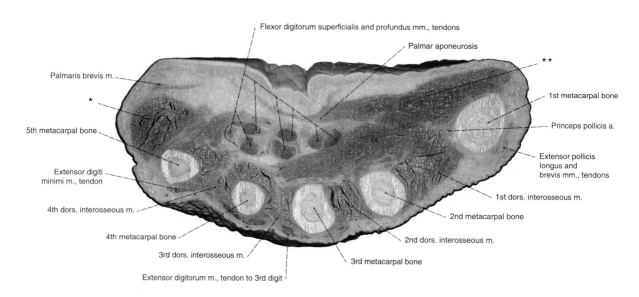

Fig. 464. Cross section through the left hand at midshaft level of the metacarpal bones. Distal surface of section. (From: Dr. H.-M. SCHMIDT, Anatomical Institute of the University of Bonn.)

 * Hypothenar muscles
** Thenar muscles

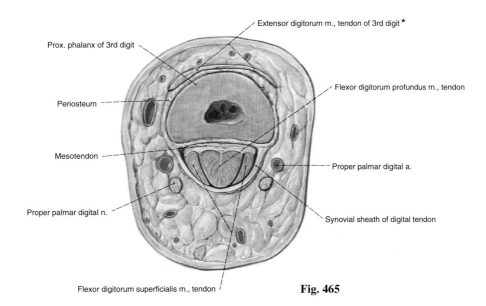

Extensor digitorum m., tendon of 3rd digit *

Prox. phalanx of 3rd digit

Flexor digitorum profundus m., tendon

Periosteum

Mesotendon

Proper palmar digital a.

Proper palmar digital n.

Synovial sheath of digital tendon

Flexor digitorum superficialis m., tendon

Fig. 465

Figs. 465 and 466. Cross sections through the middle finger at the level of the proximal (Fig. 465) and the middle (Fig. 466) phalanges. Note the location of the arteries and nerves lateral to the flexor tendons at these levels (important for anesthesia and hemostasis).

* Transition of the tendon into extensor hood (extensor aponeurosis)

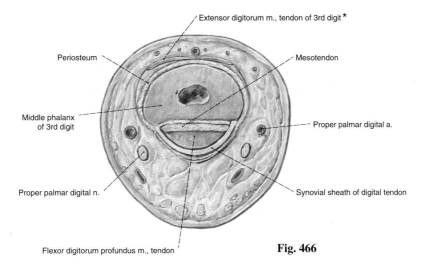

Extensor digitorum m., tendon of 3rd digit *

Periosteum

Mesotendon

Middle phalanx of 3rd digit

Proper palmar digital a.

Proper palmar digital n.

Synovial sheath of digital tendon

Flexor digitorum profundus m., tendon

Fig. 466

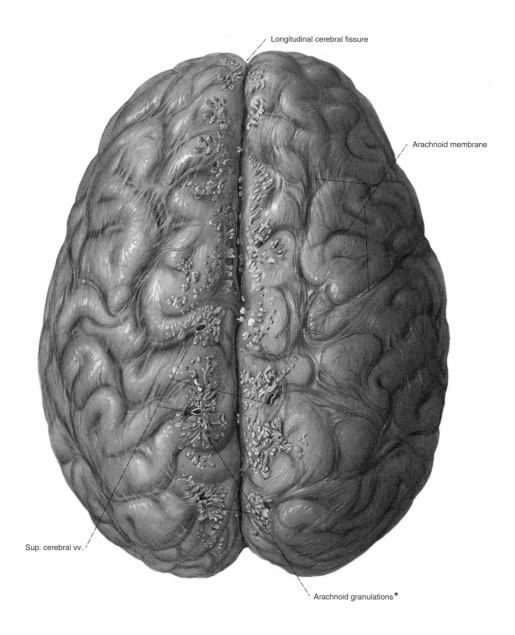

Longitudinal cerebral fissure

Arachnoid membrane

Sup. cerebral vv.

Arachnoid granulations *

Fig. 467. Cerebral hemispheres with the arachnoid and pia
mater (leptomeninges) viewed from above.

* Pacchionian bodies

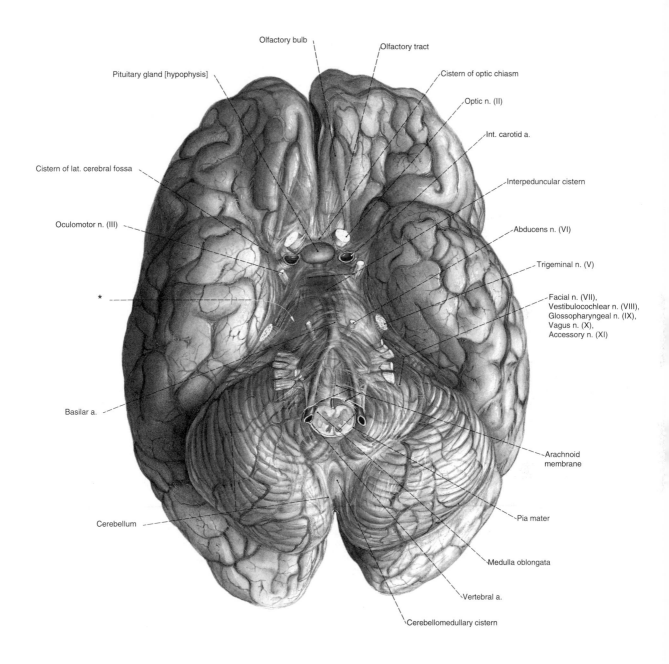

Olfactory bulb

Olfactory tract

Pituitary gland [hypophysis]

Cistern of optic chiasm

Optic n. (II)

Int. carotid a.

Cistern of lat. cerebral fossa

Interpeduncular cistern

Oculomotor n. (III)

Abducens n. (VI)

Trigeminal n. (V)

*

Facial n. (VII),
Vestibulocochlear n. (VIII),
Glossopharyngeal n. (IX),
Vagus n. (X),
Accessory n. (XI)

Basilar a.

Arachnoid
membrane

Cerebellum

Pia mater

Medulla oblongata

Vertebral a.

Cerebellomedullary cistern

Fig. 468. Inferior surface of the brain with the arachnoid and pia mater. Subarachnoid cisternae and the cranial nerves exiting from the base of the brain are visible.

* Cisterna ambiens

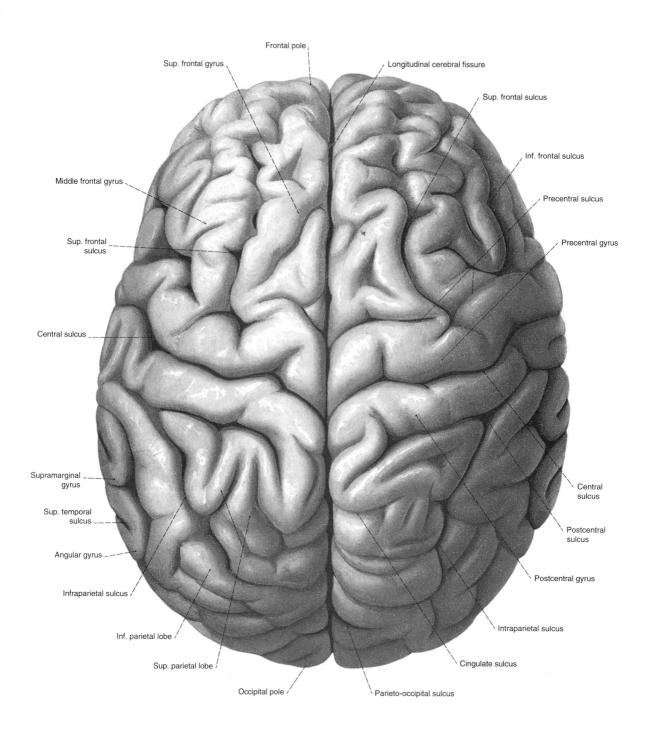

Fig. 469. Cerebral hemispheres after removal of the arachnoid and pia mater, viewed from above.

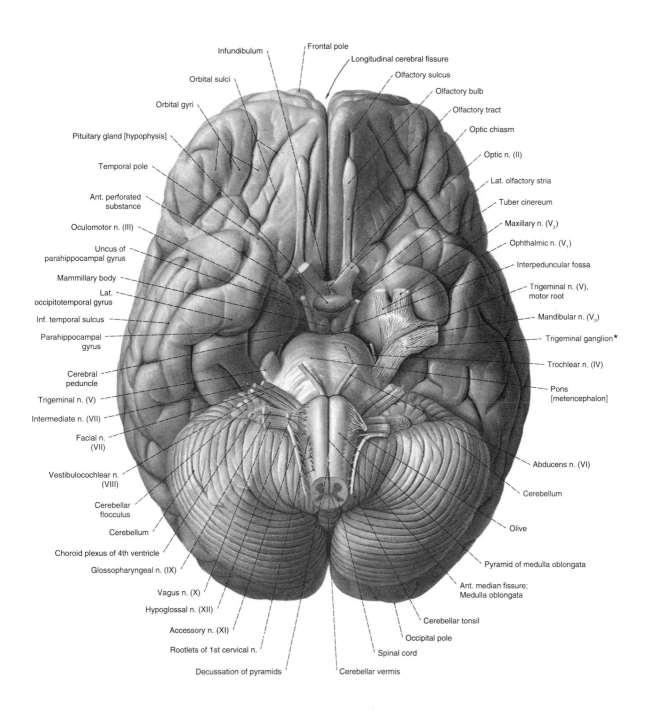

Infundibulum

Frontal pole

Orbital sulci

Longitudinal cerebral fissure

Orbital gyri

Olfactory sulcus

Pituitary gland [hypophysis]

Olfactory bulb

Temporal pole

Olfactory tract

Ant. perforated
substance

Optic chiasm

Oculomotor n. (III)

Optic n. (II)

Uncus of
parahippocampal gyrus

Lat. olfactory stria

Mammillary body

Tuber cinereum

Lat.
occipitotemporal gyrus

Maxillary n. (V₂)

Inf. temporal sulcus

Ophthalmic n. (V₁)

Parahippocampal
gyrus

Interpeduncular fossa

Cerebral
peduncle

Trigeminal n. (V),
motor root

Trigeminal n. (V)

Mandibular n. (V₃)

Intermediate n. (VII)

Trigeminal ganglion*

Facial n.
(VII)

Trochlear n. (IV)

Vestibulocochlear n.
(VIII)

Pons
[metencephalon]

Cerebellar
flocculus

Abducens n. (VI)

Cerebellum

Cerebellum

Choroid plexus of 4th ventricle

Olive

Glossopharyngeal n. (IX)

Pyramid of medulla oblongata

Vagus n. (X)

Ant. median fissure;
Medulla oblongata

Hypoglossal n. (XII)

Accessory n. (XI)

Cerebellar tonsil

Rootlets of 1st cervical n.

Occipital pole

Spinal cord

Decussation of pyramids

Cerebellar vermis

Fig. 470. The cranial nerves on the inferior surface of the brain. The cerebrum is shown in pale red, the cerebellum in yellow, and the brain stem and cranial nerves in gray. The trigeminal ganglion is retained on the left. The pituitary gland is slightly deflected to expose the infundibulum.

* Semilunar or GASSERIAN ganglion

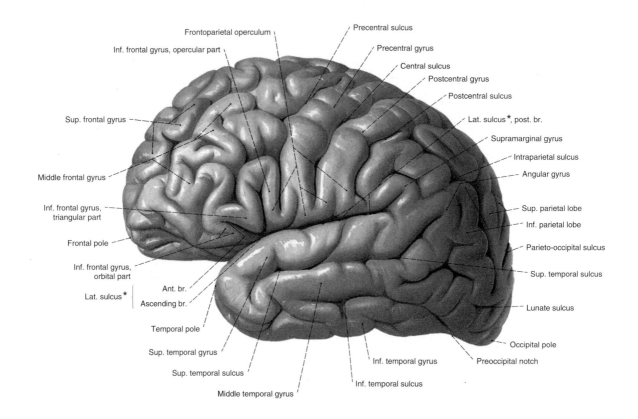

Frontoparietal operculum
Inf. frontal gyrus, opercular part
Precentral sulcus
Precentral gyrus
Central sulcus
Postcentral gyrus
Postcentral sulcus
Lat. sulcus *, post. br.
Supramarginal gyrus
Intraparietal sulcus
Angular gyrus
Sup. frontal gyrus
Middle frontal gyrus
Sup. parietal lobe
Inf. parietal lobe
Inf. frontal gyrus, triangular part
Parieto-occipital sulcus
Frontal pole
Sup. temporal sulcus
Inf. frontal gyrus, orbital part
Lat. sulcus *
Ant. br.
Ascending br.
Lunate sulcus
Temporal pole
Sup. temporal gyrus
Occipital pole
Preoccipital notch
Sup. temporal sulcus
Inf. temporal gyrus
Middle temporal gyrus
Inf. temporal sulcus

Fig. 471. Lobes, sulci and gyri of the left cerebral hemisphere after removal of the arachnoid and pia mater. Lateral view.

* Sylvian sulcus

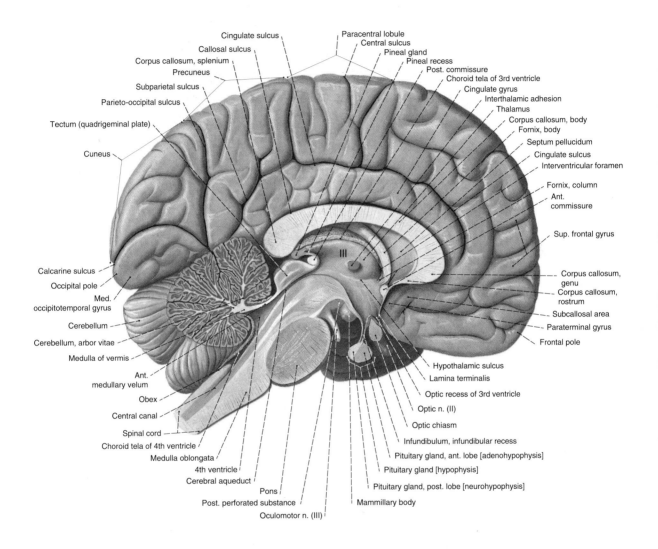

Cingulate sulcus
Callosal sulcus
Corpus callosum, splenium
Precuneus
Subparietal sulcus
Parieto-occipital sulcus
Tectum (quadrigeminal plate)
Cuneus
Calcarine sulcus
Occipital pole
Med. occipitotemporal gyrus
Cerebellum
Cerebellum, arbor vitae
Medulla of vermis
Ant. medullary velum
Obex
Central canal
Spinal cord
Choroid tela of 4th ventricle
Medulla oblongata
4th ventricle
Cerebral aqueduct
Pons
Post. perforated substance
Oculomotor n. (III)

Paracentral lobule
Central sulcus
Pineal gland
Pineal recess
Post. commissure
Choroid tela of 3rd ventricle
Cingulate gyrus
Interthalamic adhesion
Thalamus
Corpus callosum, body
Fornix, body
Septum pellucidum
Cingulate sulcus
Interventricular foramen
Fornix, column
Ant. commissure
Sup. frontal gyrus
Corpus callosum, genu
Corpus callosum, rostrum
Subcallosal area
Paraterminal gyrus
Frontal pole

III

Hypothalamic sulcus
Lamina terminalis
Optic recess of 3rd ventricle
Optic n. (II)
Optic chiasm
Infundibulum, infundibular recess
Pituitary gland, ant. lobe [adenohypophysis]
Pituitary gland [hypophysis]
Pituitary gland, post. lobe [neurohypophysis]
Mammillary body

Fig. 472. Midsagittal section through the brain. Medial aspect of the left half of the brain.
III = third ventricle

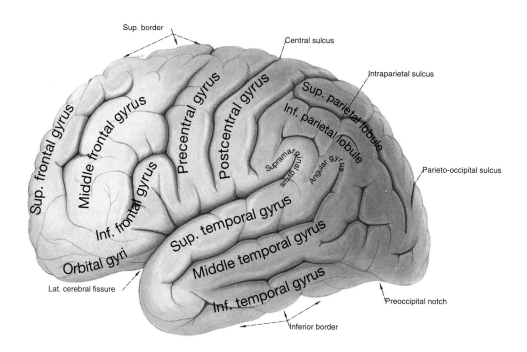

Fig. 473. Sulci and gyri of the lateral surface of the left cerebral hemisphere.

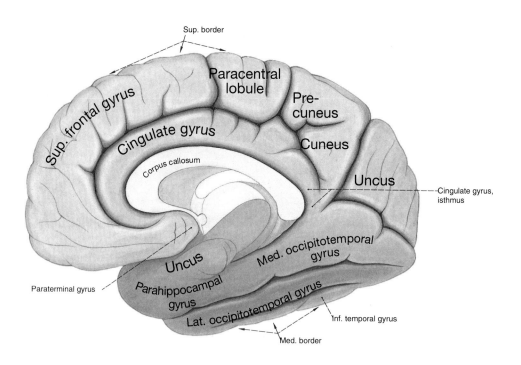

Fig. 474. Sulci and gyri of the medial surface of the right cerebral hemisphere. The brain has been hemisected in the median plane and the brain stem and cerebellum have been removed.

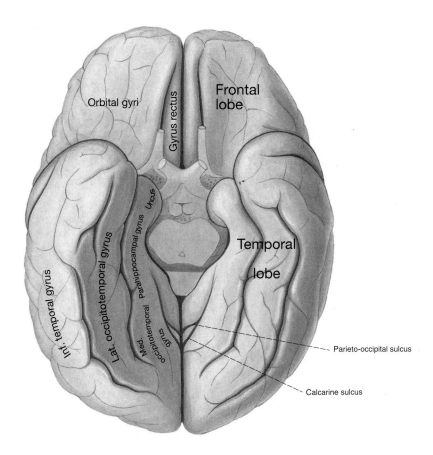

Orbital gyri

Gyrus rectus

Frontal lobe

Uncus

Inf. temporal gyrus

Lat. occipitotemporal gyrus

Parahippocampal gyrus

Med. occipitotemporal gyrus

Temporal lobe

Parieto-occipital sulcus

Calcarine sulcus

Fig. 475. Sulci and gyri of the inferior surface of the cerebral hemispheres. Brain stem and cerebellum have been removed.

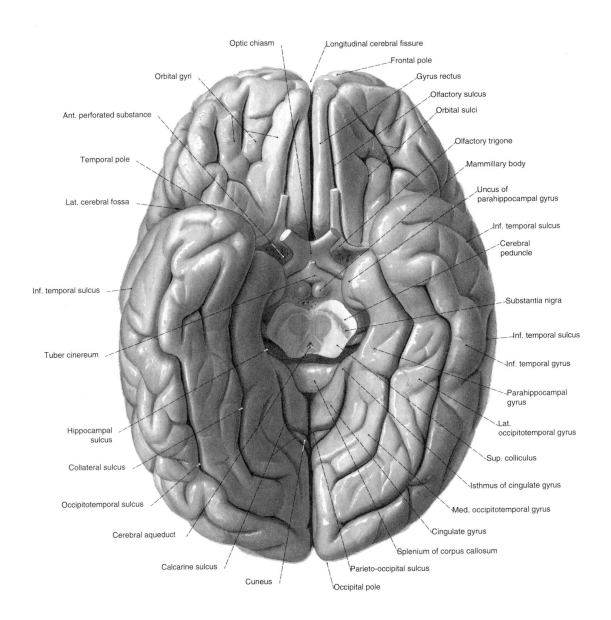

Optic chiasm

Longitudinal cerebral fissure

Frontal pole

Orbital gyri

Gyrus rectus

Olfactory sulcus

Ant. perforated substance

Orbital sulci

Olfactory trigone

Temporal pole

Mammillary body

Uncus of parahippocampal gyrus

Lat. cerebral fossa

Inf. temporal sulcus

Cerebral peduncle

Inf. temporal sulcus

Substantia nigra

Tuber cinereum

Inf. temporal sulcus

Inf. temporal gyrus

Parahippocampal gyrus

Hippocampal sulcus

Lat. occipitotemporal gyrus

Collateral sulcus

Sup. colliculus

Occipitotemporal sulcus

Isthmus of cingulate gyrus

Cerebral aqueduct

Med. occipitotemporal gyrus

Cingulate gyrus

Calcarine sulcus

Splenium of corpus callosum

Cuneus

Parieto-occipital sulcus

Occipital pole

Fig. 476. Cerebral hemispheres after removal of the arachnoid and pia mater. Basal view with brain stem and cerebellum removed by sectioning across the midbrain. The cut surface of the midbrain shows the red nuclei situated bilaterally beneath the substantia nigra.

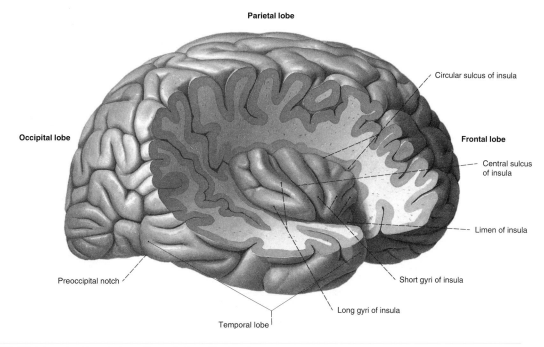

Fig. 477. Right cerebral hemisphere. Lateral view. The insula has been exposed by removing the frontal, frontoparietal and temporal opercula.

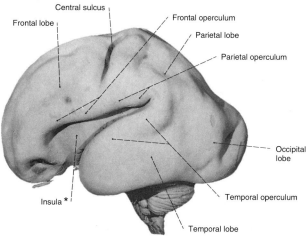

Fig. 478. Brain of a human fetus of 20 cm crown-rump length (CRL), natural size (HOCHSTETTER Collection); fronto-occipital diameter 7 cm. Gyri have not yet developed, only the primary sulci, i.e., the central and lateral cerebral sulci, are formed. The opercula do not yet cover the insula; thus a smooth insula, without gyri and sulci, is visible deep within the lateral sulcus.

* REIL's insula

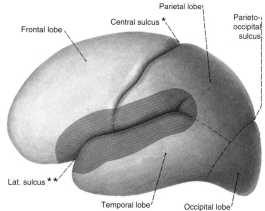

Fig. 479. The lobes of the telencephalon. The region of the frontal, frontoparietal and temporal opercula is colored lavender.

  * Fissure of ROLANDO
** SYLVIAN fissure

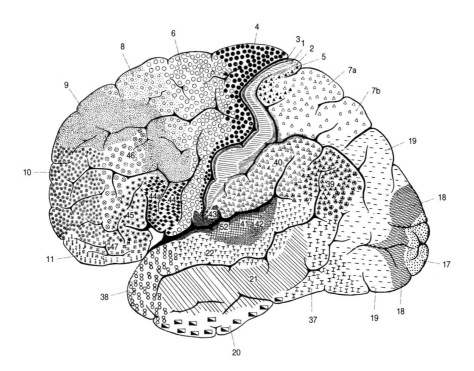

Fig. 480

Figs. 480 and 481. The cortical areas of the lateral (upper) and medial (lower) surfaces of the cerebral hemispheres described by K. BRODMANN on the basis of cytoarchitectonic features. The areas are numbered and designated by different symbols.

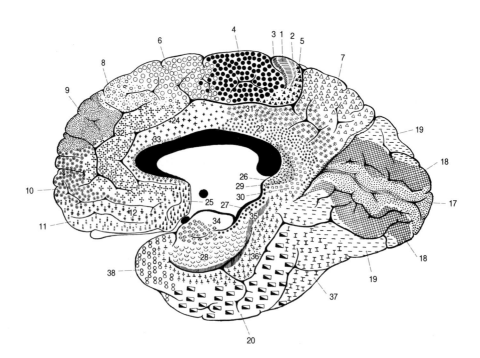

Fig. 481

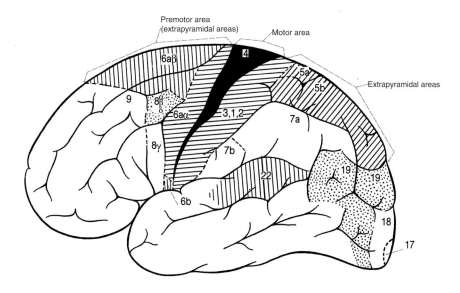

Fig. 482

Figs. 482 and 483. The motor cortical areas on the lateral (upper) and medial (lower) surfaces of the cerebral hemispheres.

The motor area is black, the extrapyramidal areas are hatched, and the eye fields are stippled; SR, central sulcus (ROLANDO); SPO, parieto-occipital sulcus; SF, cingulate sulcus. (From: O. FOERSTER, O. BUMKE, Handbook of Neurology, Vol. VI, Springer, Berlin, 1940.)

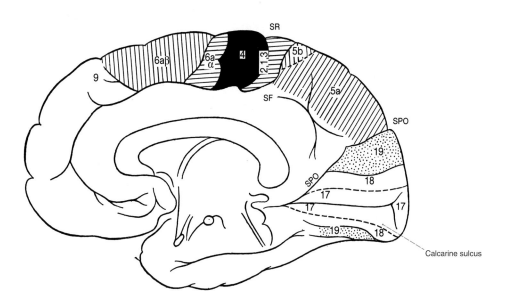

Fig. 483

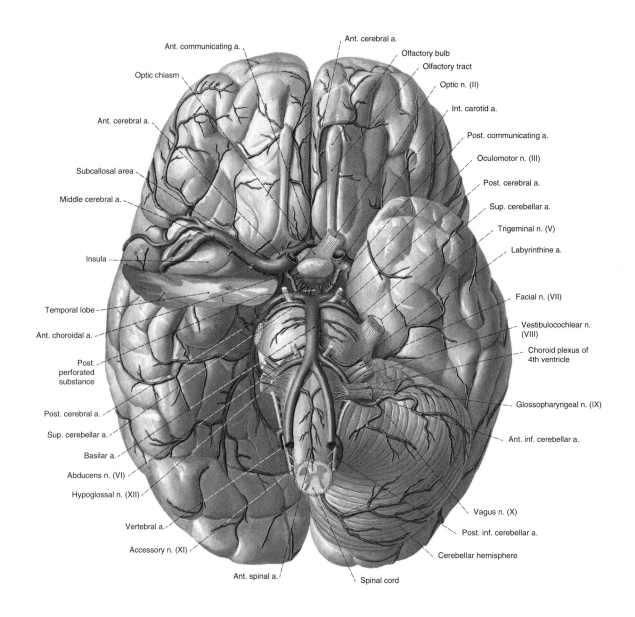

Ant. communicating a.

Optic chiasm

Ant. cerbral a.

Subcallosal area

Middle cerebral a.

Insula

Temporal lobe

Ant. choroidal a.

Post. perforated substance

Post. cerebral a.

Sup. cerebellar a.

Basilar a.

Abducens n. (VI)

Hypoglossal n. (XII)

Vertebral a.

Accessory n. (XI)

Ant. spinal a.

Ant. cerebral a.

Olfactory bulb

Olfactory tract

Optic n. (II)

Int. carotid a.

Post. communicating a.

Oculomotor n. (III)

Post. cerebral a.

Sup. cerebellar a.

Trigeminal n. (V)

Labyrinthine a.

Facial n. (VII)

Vestibulocochlear n. (VIII)

Choroid plexus of 4th ventricle

Glossopharyngeal n. (IX)

Ant. inf. cerebellar a.

Vagus n. (X)

Post. inf. cerebellar a.

Cerebellar hemisphere

Spinal cord

Fig. 484.   Arteries at the base of the brain, arterial circle of
WILLIS. The optic nerve, the anterior portion of the temporal
lobe, and the cerebellar hemisphere of the right side have been
removed.

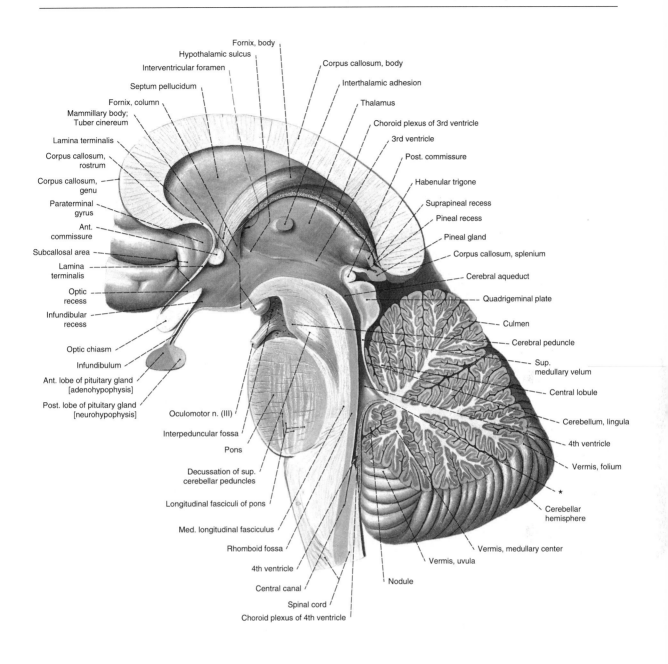

Fornix, body
Hypothalamic sulcus
Interventricular foramen
Septum pellucidum
Fornix, column
Mammillary body;
Tuber cinereum
Lamina terminalis
Corpus callosum,
rostrum
Corpus callosum,
genu
Paraterminal
gyrus
Ant.
commissure
Subcallosal area
Lamina
terminalis
Optic
recess
Infundibular
recess
Optic chiasm
Infundibulum
Ant. lobe of pituitary gland
[adenohypophysis]
Post. lobe of pituitary gland
[neurohypophysis]
Oculomotor n. (III)
Interpeduncular fossa
Pons
Decussation of sup.
cerebellar peduncles
Longitudinal fasciculi of pons
Med. longitudinal fasciculus
Rhomboid fossa
4th ventricle
Central canal
Spinal cord
Choroid plexus of 4th ventricle

Corpus callosum, body
Interthalamic adhesion
Thalamus
Choroid plexus of 3rd ventricle
3rd ventricle
Post. commissure
Habenular trigone
Suprapineal recess
Pineal recess
Pineal gland
Corpus callosum, splenium
Cerebral aqueduct
Quadrigeminal plate
Culmen
Cerebral peduncle
Sup.
medullary velum
Central lobule
Cerebellum, lingula
4th ventricle
Vermis, folium
*
Cerebellar
hemisphere
Vermis, medullary center
Vermis, uvula
Nodule

Fig. 485. Midsagittal section through the brain stem. Medial surface of the right half. The walls of the third and fourth ventricles and of the cerebral aqueduct are shown in yellow.

* Fastigium

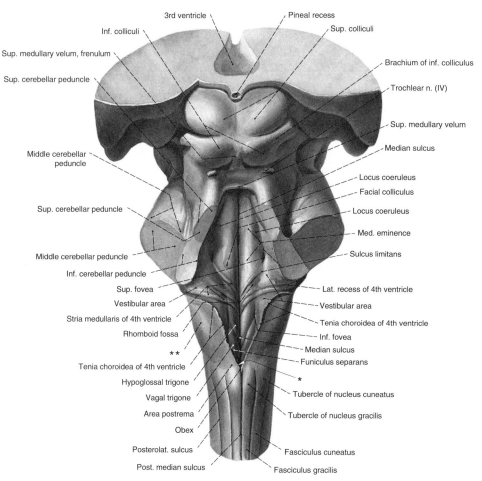

Fig. 486. Floor of the fourth ventricle (rhomboid fossa). Dorsal view of the quadrigeminal plate. The cerebellum and pineal gland have been removed.

* Calamus scriptorius
** Tuberculum cinereum

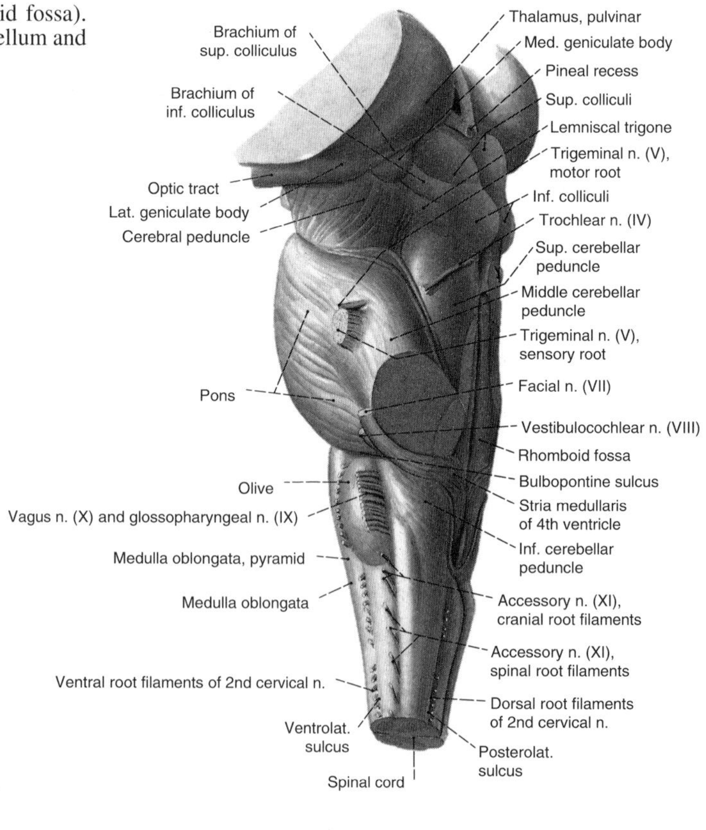

Fig. 487. Brain stem viewed from left and dorsal.

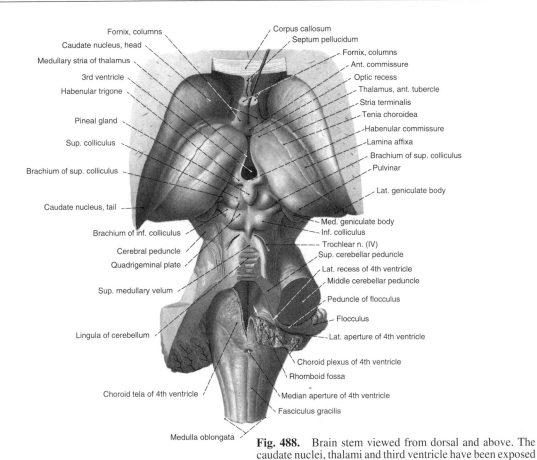

Fornix, columns
Caudate nucleus, head
Medullary stria of thalamus
3rd ventricle
Habenular trigone
Pineal gland
Sup. colliculus
Brachium of sup. colliculus
Caudate nucleus, tail
Brachium of inf. colliculus
Cerebral peduncle
Quadrigeminal plate
Sup. medullary velum
Lingula of cerebellum
Choroid tela of 4th ventricle
Medulla oblongata

Corpus callosum
Septum pellucidum
Fornix, columns
Ant. commissure
Optic recess
Thalamus, ant. tubercle
Stria terminalis
Tenia choroidea
Habenular commissure
Lamina affixa
Brachium of sup. colliculus
Pulvinar
Lat. geniculate body
Med. geniculate body
Inf. colliculus
Trochlear n. (IV)
Sup. cerebellar peduncle
Lat. recess of 4th ventricle
Middle cerebellar peduncle
Peduncle of flocculus
Flocculus
Lat. aperture of 4th ventricle
Choroid plexus of 4th ventricle
Rhomboid fossa
Median aperture of 4th ventricle
Fasciculus gracilis

Fig. 488. Brain stem viewed from dorsal and above. The caudate nuclei, thalami and third ventricle have been exposed by removal of the corpus callosum, the fornix, and the tela choroidea of the third ventricle. The cerebellum has been mostly removed. The tela choroidea of the fourth ventricle has been split in the midline and reflected on the right side.

3rd ventricle; Post. commissure; Suprapineal recess
Habenular trigone
Brachium of sup. colliculus
Brachium of inf. colliculus
Lemniscal trigone
Sup. medullary velum
Rhomboid fossa, med. eminence

Stria terminalis
Tenia choroidea
Thalamus, pulvinar
Pineal gland
Med. and lat. geniculate bodies
Sup. colliculus
Inf. colliculus
Trochlear n. (IV)
Sup. cerebellar peduncle
Middle cerebellar peduncle

Fig. 489. The midbrain (pink), quadrigeminal (tectal) plate and pineal gland in a dorsal view. The medial and lateral geniculate bodies of the metathalamus are colored green.

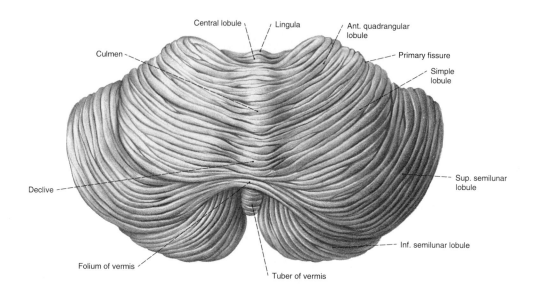

Fig. 490. The superior surface of the cerebellum.

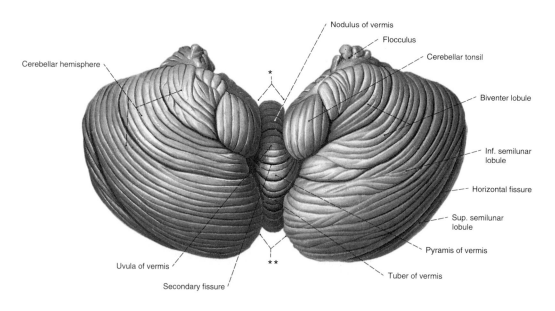

Fig. 491. The inferior surface of the cerebellum. The secondary fissure separates the uvula of the vermis from the pyramids.

 * Anterior cerebellar incisure
** Posterior cerebellar incisure

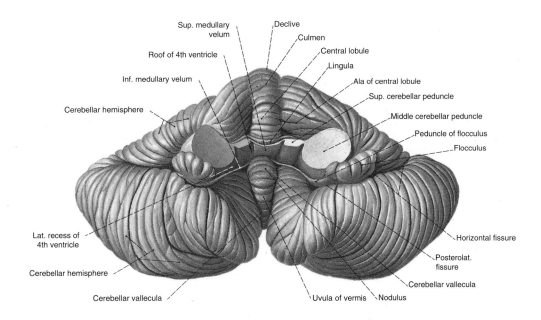

Fig. 492. The cerebellum viewed from its anterior aspect. The cerebellar peduncles have been cut.

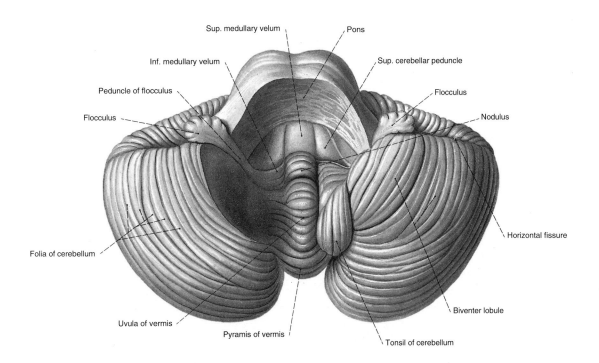

Fig. 493. Inferior cerebellar surface after removal of the right cerebellar tonsil and a part of the biventer lobule. The pons has been sectioned transversely. View from below onto the roof of the fourth ventricle.

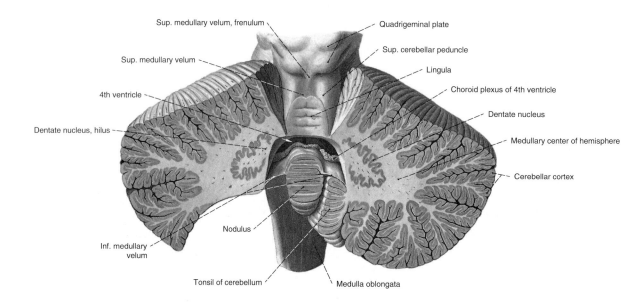

Sup. medullary velum, frenulum

Quadrigeminal plate

Sup. medullary velum

Sup. cerebellar peduncle

4th ventricle

Lingula

Choroid plexus of 4th ventricle

Dentate nucleus

Dentate nucleus, hilus

Medullary center of hemisphere

Cerebellar cortex

Nodulus

Inf. medullary velum

Tonsil of cerebellum

Medulla oblongata

Fig. 494. Transverse section through the cerebellum exposing the fourth ventricle.

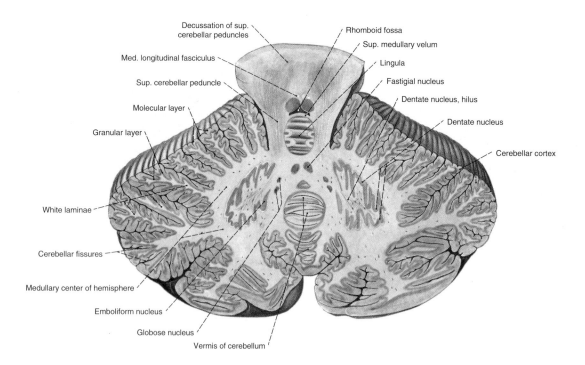

Decussation of sup. cerebellar peduncles

Rhomboid fossa

Med. longitudinal fasciculus

Sup. medullary velum

Sup. cerebellar peduncle

Lingula

Molecular layer

Fastigial nucleus

Granular layer

Dentate nucleus, hilus

Dentate nucleus

Cerebellar cortex

White laminae

Cerebellar fissures

Medullary center of hemisphere

Emboliform nucleus

Globose nucleus

Vermis of cerebellum

Fig. 495. Section through the cerebellum in the direction of the superior cerebellar peduncles showing the cerebellar nuclei.

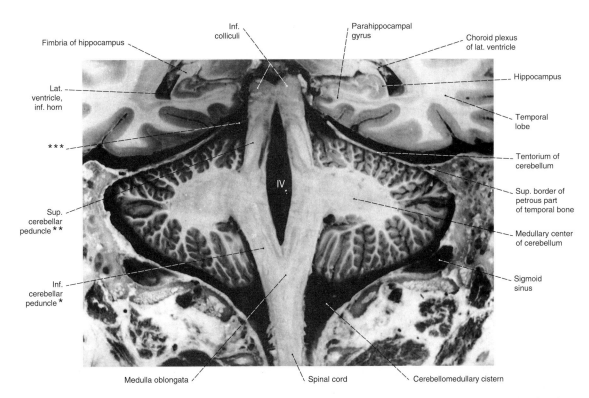

Fimbria of hippocampus

Inf. colliculi

Parahippocampal gyrus

Choroid plexus of lat. ventricle

Hippocampus

Lat. ventricle, inf. horn

Temporal lobe

***

Tentorium of cerebellum

Sup. cerebellar peduncle **

Sup. border of petrous part of temporal bone

Medullary center of cerebellum

Inf. cerebellar peduncle *

Sigmoid sinus

IV

Medulla oblongata

Spinal cord

Cerebellomedullary cistern

* Restiform body, medullocerebellar crus
** Brachium conjunctivum, cerebellocerebral crus
*** Cisterna ambiens

Fig. 496. Frontal section through the hindbrain [rhomben-cephalon] at the level of the infratentorial space. Note the position of the midbrain in the tentorial incisure. IV = fourth ventricle.

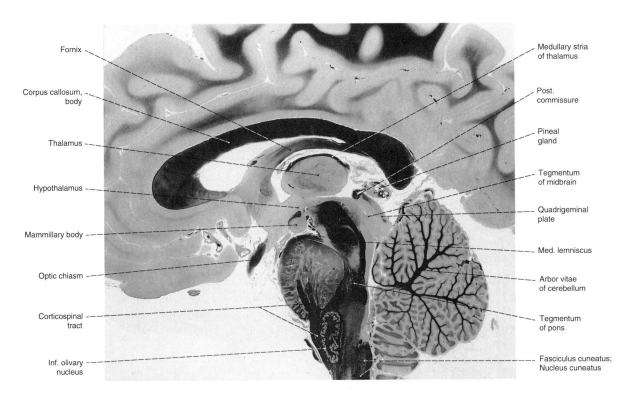

Fornix

Medullary stria of thalamus

Corpus callosum, body

Post. commissure

Thalamus

Pineal gland

Hypothalamus

Tegmentum of midbrain

Mammillary body

Quadrigeminal plate

Optic chiasm

Med. lemniscus

Corticospinal tract

Arbor vitae of cerebellum

Tegmentum of pons

Inf. olivary nucleus

Fasciculus cuneatus; Nucleus cuneatus

Fig. 497. Paramedian sagittal section through the brain stem. WEIGERT's myelin stain (black myelinated fibers), compare with Fig. 472. The myelinated fiber tracts are easily recognized as black regions, e.g., corpus callosum, medulla of cerebellum, pyramidal tract, optic chiasm, etc. The cortical regions and subcortical nuclei are lightly stained, e.g., dorsal thalamus, mamillary body, quadrigeminal (tectal) plate.

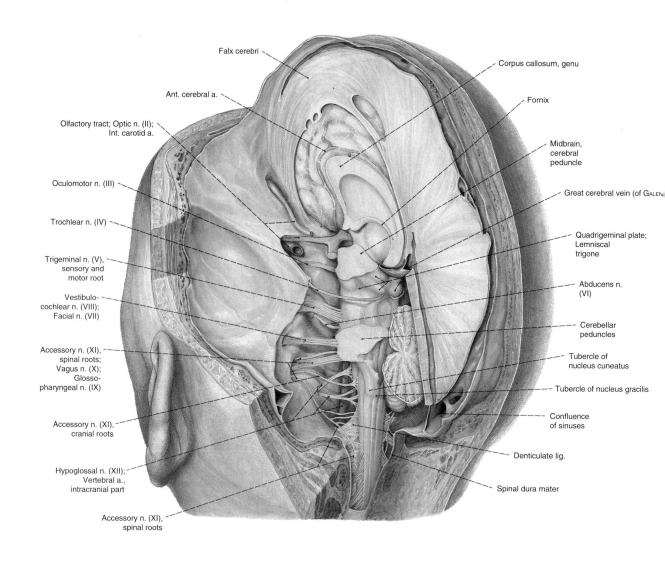

Falx cerebri

Corpus callosum, genu

Ant. cerebral a.

Fornix

Olfactory tract; Optic n. (II);
Int. carotid a.

Midbrain,
cerebral
peduncle

Oculomotor n. (III)

Great cerebral vein (of GALEN)

Trochlear n. (IV)

Quadrigeminal plate;
Lemniscal
trigone

Trigeminal n. (V),
sensory and
motor root

Abducens n.
(VI)

Vestibulo-
cochlear n. (VIII);
Facial n. (VII)

Cerebellar
peduncles

Accessory n. (XI),
spinal roots;
Vagus n. (X);
Glosso-
pharyngeal n. (IX)

Tubercle of
nucleus cuneatus

Tubercle of nucleus gracilis

Accessory n. (XI),
cranial roots

Confluence
of sinuses

Denticulate lig.

Hypoglossal n. (XII);
Vertebral a.,
intracranial part

Spinal dura mater

Accessory n. (XI),
spinal roots

Fig. 498.   The brain in situ after removal of the left cerebral
and cerebellar hemispheres, showing the brain stem, falx
cerebri, and intracranial portions of the cranial nerves. (From:
PERNKOPF: Atlas of Topographic and Applied Human
Anatomy, Vol. 1, 3rd edition [W. PLATZER, Ed.], Urban &
Schwarzenberg, Munich-Vienna-Baltimore, 1987.)

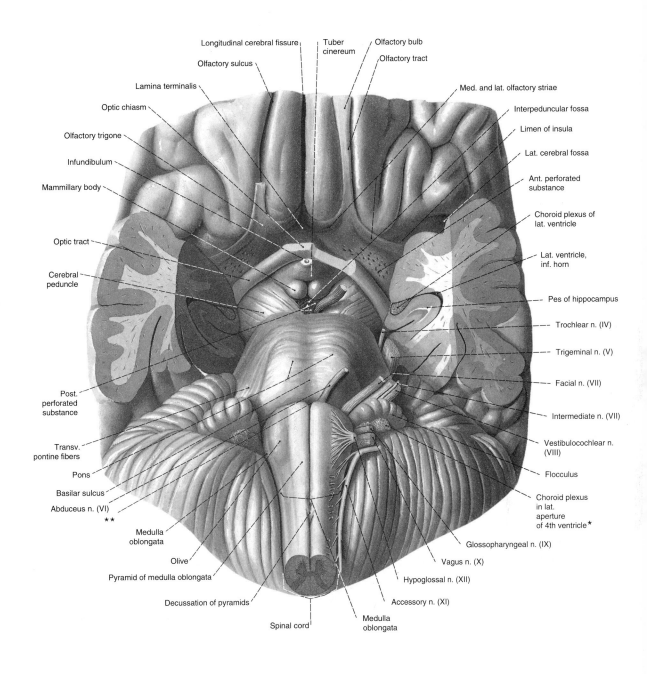

Longitudinal cerebral fissure

Olfactory sulcus

Lamina terminalis

Optic chiasm

Olfactory trigone

Infundibulum

Mammillary body

Optic tract

Cerebral peduncle

Post. perforated substance

Transv. pontine fibers

Pons

Basilar sulcus

Abduceus n. (VI)
**

Medulla oblongata

Olive

Pyramid of medulla oblongata

Decussation of pyramids

Spinal cord

Tuber cinereum

Olfactory bulb

Olfactory tract

Med. and lat. olfactory striae

Interpeduncular fossa

Limen of insula

Lat. cerebral fossa

Ant. perforated substance

Choroid plexus of lat. ventricle

Lat. ventricle, inf. horn

Pes of hippocampus

Trochlear n. (IV)

Trigeminal n. (V)

Facial n. (VII)

Intermediate n. (VII)

Vestibulocochlear n. (VIII)

Flocculus

Choroid plexus in lat. aperture of 4th ventricle *

Glossopharyngeal n. (IX)

Vagus n. (X)

Hypoglossal n. (XII)

Accessory n. (XI)

Medulla oblongata

Fig. 499. Basal view of the brain stem and adjacent parts of the brain. The temporal poles have been removed. The cranial nerves have been preserved on the left side of the specimen and removed on the right.

* Lateral aperture of fourth ventricle with choroid plexus
** Foramen cecum of medulla oblongata

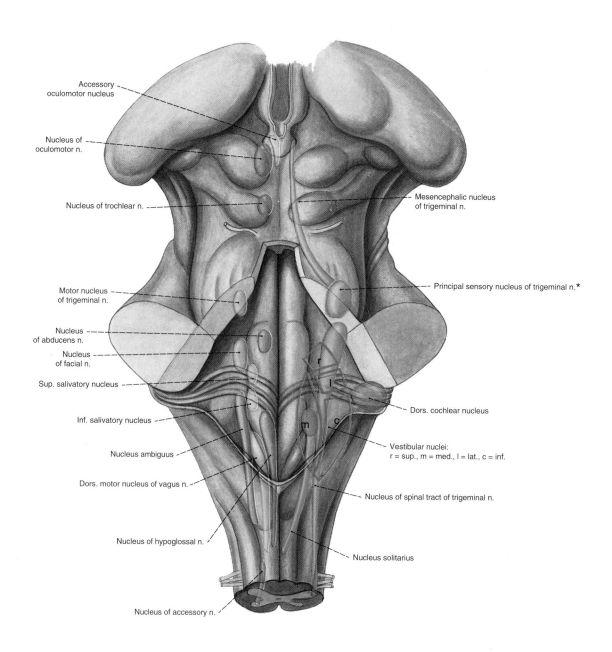

Fig. 500. The cranial nerve nuclei projected diagramatically onto a dorsal view of the brain stem. On the left are the motor nuclei and on the right, the sensory nuclei. (From: BENNING-HOFF: Macroscopic and Microscopic Human Anatomy, Vol. 3, 13/14th edition [W. Zenker, Ed.], Urban & Schwarzenberg, Munich-Vienna-Baltimore, 1985.)

* Pontine nucleus of the trigeminal nerve

Red - General somatic efferent nuclear column
Yellow - General visceral efferent nuclear column
Violet - Special visceral efferent nuclear column
Light green - General visceral afferent nuclear group
 (nucleus solitarius)
Dark green - Special visceral afferent nucleus
Light blue - General somatic afferent nuclear column
Dark blue - Special somatic afferent nuclei

Labels in figure:
Accessory oculomotor nucleus
Nucleus of oculomotor n.
Nucleus of trochlear n.
Mesencephalic nucleus of trigeminal n.
Motor nucleus of trigeminal n.
Principal sensory nucleus of trigeminal n.*
Nucleus of abducens n.
Nucleus of facial n.
Sup. salivatory nucleus
Dors. cochlear nucleus
Inf. salivatory nucleus
Vestibular nuclei: r = sup., m = med., l = lat., c = inf.
Nucleus ambiguus
Dors. motor nucleus of vagus n.
Nucleus of spinal tract of trigeminal n.
Nucleus of hypoglossal n.
Nucleus solitarius
Nucleus of accessory n.

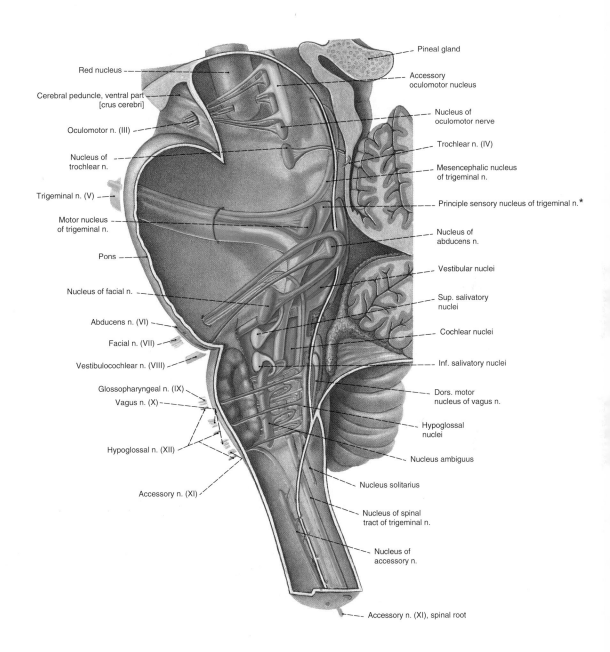

Red nucleus

Cerebral peduncle, ventral part [crus cerebri]

Oculomotor n. (III)

Nucleus of trochlear n.

Trigeminal n. (V)

Motor nucleus of trigeminal n.

Pons

Nucleus of facial n.

Abducens n. (VI)

Facial n. (VII)

Vestibulocochlear n. (VIII)

Glossopharyngeal n. (IX)

Vagus n. (X)

Hypoglossal n. (XII)

Accessory n. (XI)

Pineal gland

Accessory oculomotor nucleus

Nucleus of oculomotor nerve

Trochlear n. (IV)

Mesencephalic nucleus of trigeminal n.

Principle sensory nucleus of trigeminal n. *

Nucleus of abducens n.

Vestibular nuclei

Sup. salivatory nuclei

Cochlear nuclei

Inf. salivatory nuclei

Dors. motor nucleus of vagus n.

Hypoglossal nuclei

Nucleus ambiguus

Nucleus solitarius

Nucleus of spinal tract of trigeminal n.

Nucleus of accessory n.

Accessory n. (XI), spinal root

Fig. 501. Nuclei and intracerebral course of the cranial nerves (spatial diagram). View from the median sagittal plane onto the hollowed out shell of the right half of the midbrain [mesencephalon], pons [metencephalon] and medulla oblongata [bulb, myelencephalon]. (According to BRAUS/ELZE: Human Anatomy, Vol. 3, Springer, Berlin, 1932; from BENNINGHOFF: Macroscopic and Microscopic Human Anatomy, Vol. 3, 13/14th edition [W. Zenker, Ed.], Urban & Schwarzenberg, Munich-Vienna-Baltimore, 1985.)

Red - General somatic efferent nuclei
Yellow - General visceral efferent nuclei
Violet - Special visceral efferent nuclei
Pale green - General and special visceral afferent nuclei
Light blue - General somatic afferent nuclei
Dark blue - Vestibular and cochlear nuclei

* Pontine nucleus of the trigeminal nerve

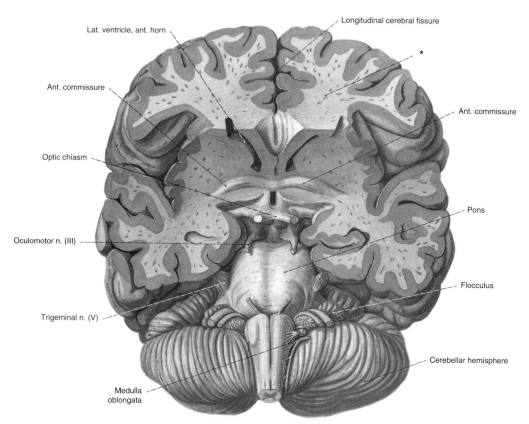

Fig. 502. Anterior commissure exposed by cutting a curved segment out of the base of the brain. The section shows only the middle portion of the commissure.

* Centrum semiovale

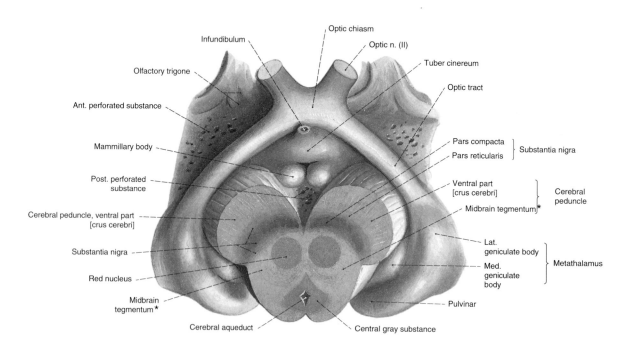

Fig. 503. Section through the midbrain. Caudal view of the cut surface, showing the optic tract, geniculate bodies, hypothalamus (compare with Fig. 499).

* Midbrain tegmentum = dorsal part of the cerebral peduncles. The tegmentum extends from the substantia nigra of each side dorsally to the level of the cerebral aqueduct.

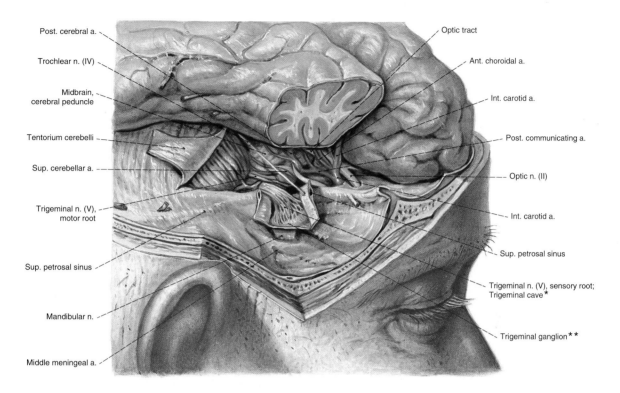

Post. cerebral a.

Trochlear n. (IV)

Midbrain, cerebral peduncle

Tentorium cerebelli

Sup. cerebellar a.

Trigeminal n. (V), motor root

Sup. petrosal sinus

Mandibular n.

Middle meningeal a.

Optic tract

Ant. choroidal a.

Int. carotid a.

Post. communicating a.

Optic n. (II)

Int. carotid a.

Sup. petrosal sinus

Trigeminal n. (V), sensory root; Trigeminal cave*

Trigeminal ganglion**

Fig. 504. The middle cranial fossa and the intracranial portion of the trigeminal nerve. The pocket of dura (cave trigeminal), which lodges the trigeminal ganglion and the distal part of the roots of the trigeminal nerve surrounded by arachnoid membrane and cerebrospinal fluid, has been opened. The superior petrosal sinus has been sectioned at the trigeminal impression and the temporal lobe has been elevated. (From: H. FERNER and R. KAUTZKY: Handbook of Neurosurgery, Springer, Heidelberg, 1959.)

  * Trigeminal cave of MECKEL
** Semilunar ganglion (GASSERIAN ganglion)

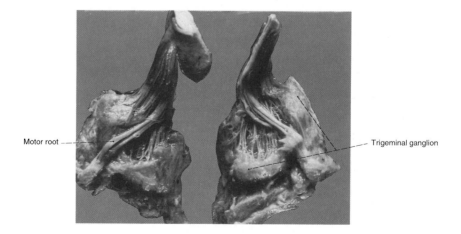

Motor root

Trigeminal ganglion

Fig. 505.   The left and right trigeminal ganglia with the sensory and the compact motor roots, viewed from the caudal aspect. (From: Dr. H. FERNER, Vienna.)

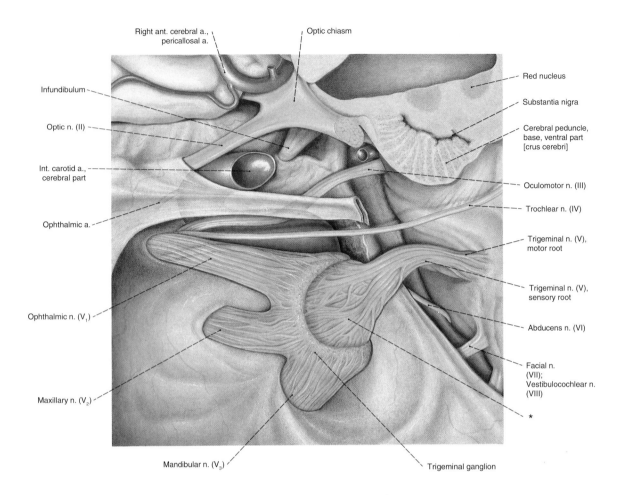

Right ant. cerebral a., pericallosal a.

Optic chiasm

Infundibulum

Optic n. (II)

Int. carotid a., cerebral part

Ophthalmic a.

Ophthalmic n. (V₁)

Maxillary n. (V₂)

Mandibular n. (V₃)

Red nucleus

Substantia nigra

Cerebral peduncle, base, ventral part [crus cerebri]

Oculomotor n. (III)

Trochlear n. (IV)

Trigeminal n. (V), motor root

Trigeminal n. (V), sensory root

Abducens n. (VI)

Facial n. (VII); Vestibulocochlear n. (VIII)

*

Trigeminal ganglion

Fig. 506. Topography of the region of the sella turcica and the intracranial portions of the left trigeminal nerve, viewed from lateral and above. The lateral dural wall of the cavernous sinus, the dura mater over the trigeminal ganglion and its branches, as well as the tentorium cerebelli have been removed. The oculomotor nerve (III) courses over the edge of the clivus. Note the position of the infundibulum and its relation to the optic chiasm. The third ventricle has been opened and the midbrain sectioned transversely behind the optic tract. (From: H. FERNER and R. KAUTZKY: Handbook of Neurosurgery, Vol. 1, Springer, Heidelberg, 1959.)

* Triangular portion of the trigeminal root

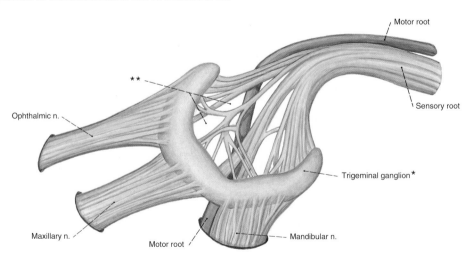

Motor root

Sensory root

**

Ophthalmic n.

Trigeminal ganglion *

Maxillary n.

Motor root

Mandibular n.

Fig. 507. The intracranial segments of the trigeminal nerve (V) (according to Prof. H. FERNER, Vienna).

* Semilunar or GASSERIAN ganglion
** Trigeminal cistern

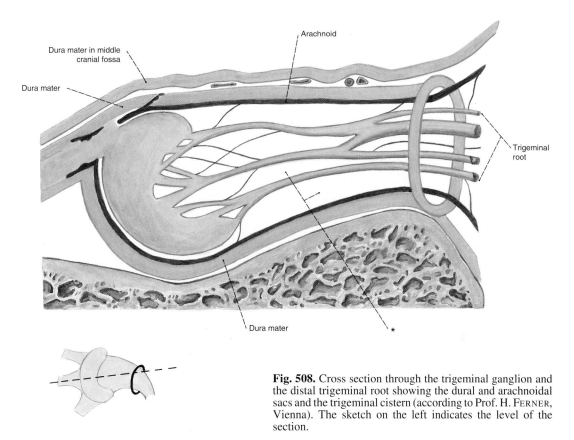

Fig. 508. Cross section through the trigeminal ganglion and the distal trigeminal root showing the dural and arachnoidal sacs and the trigeminal cistern (according to Prof. H. FERNER, Vienna). The sketch on the left indicates the level of the section.

* Trigeminal cistern

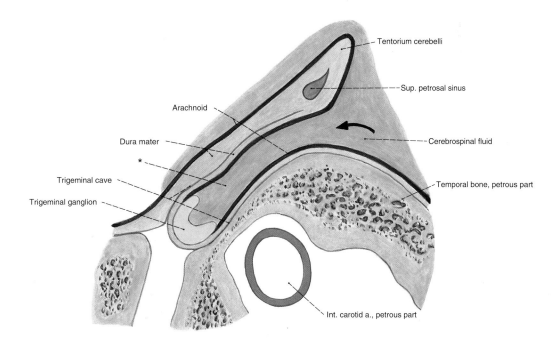

Fig. 509.   Schematic frontal section through the trigeminal cistern (*). Cerebrospinal fluid is indicated in green.

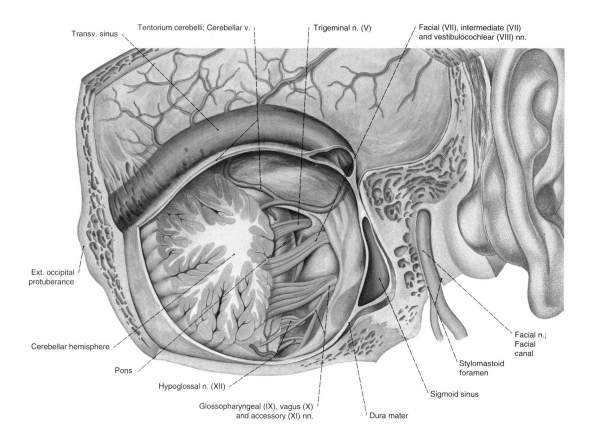

Transv. sinus

Tentorium cerebelli; Cerebellar v.

Trigeminal n. (V)

Facial (VII), intermediate (VII)
and vestibulocochlear (VIII) nn.

Ext. occipital
protuberance

Cerebellar hemisphere

Pons

Hypoglossal n. (XII)

Glossopharyngeal (IX), vagus (X)
and accessory (XI) nn.

Dura mater

Facial n.;
Facial
canal

Stylomastoid
foramen

Sigmoid sinus

Fig. 510. The subtentorial space after partial removal of the
right cerebellar hemisphere, viewed from the dorsolateral
aspect.

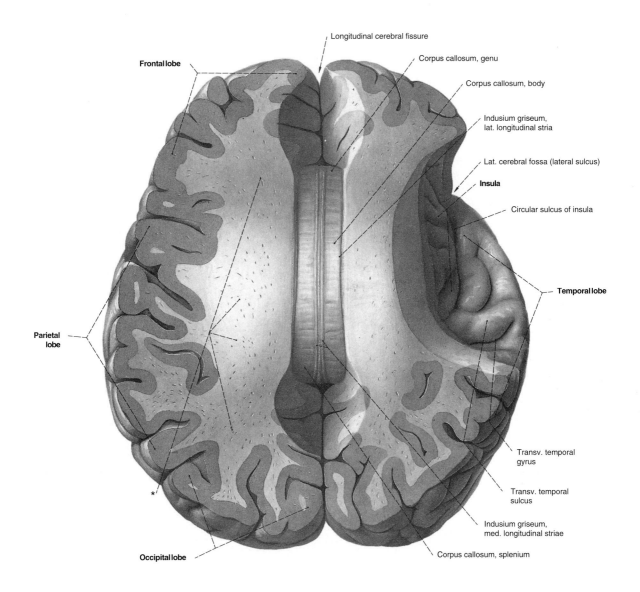

Frontal lobe

Longitudinal cerebral fissure

Corpus callosum, genu

Corpus callosum, body

Indusium griseum,
lat. longitudinal stria

Lat. cerebral fossa (lateral sulcus)

Insula

Circular sulcus of insula

Temporal lobe

**Parietal
lobe**

Transv. temporal
gyrus

Transv. temporal
sulcus

Indusium griseum,
med. longitudinal striae

*

Occipital lobe

Corpus callosum, splenium

Fig. 511.   Horizontal section through the cerebral hemispheres exposing the corpus callosum and the insula, viewed
from above.

* Centrum semiovale

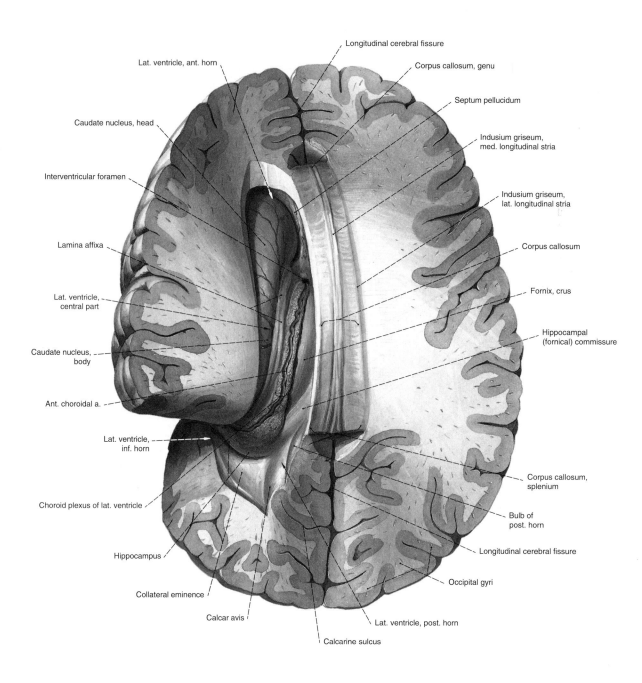

Longitudinal cerebral fissure

Lat. ventricle, ant. horn

Corpus callosum, genu

Septum pellucidum

Caudate nucleus, head

Indusium griseum,
med. longitudinal stria

Interventricular foramen

Indusium griseum,
lat. longitudinal stria

Corpus callosum

Lamina affixa

Fornix, crus

Lat. ventricle,
central part

Hippocampal
(fornical) commissure

Caudate nucleus,
body

Ant. choroidal a.

Lat. ventricle,
inf. horn

Corpus callosum,
splenium

Choroid plexus of lat. ventricle

Bulb of
post. horn

Longitudinal cerebral fissure

Hippocampus

Occipital gyri

Collateral eminence

Calcar avis

Lat. ventricle, post. horn

Calcarine sulcus

Fig. 512. Horizontal section through the cerebral hemispheres. View of the corpus callosum and the opened left
lateral ventricle from above and somewhat to the left.

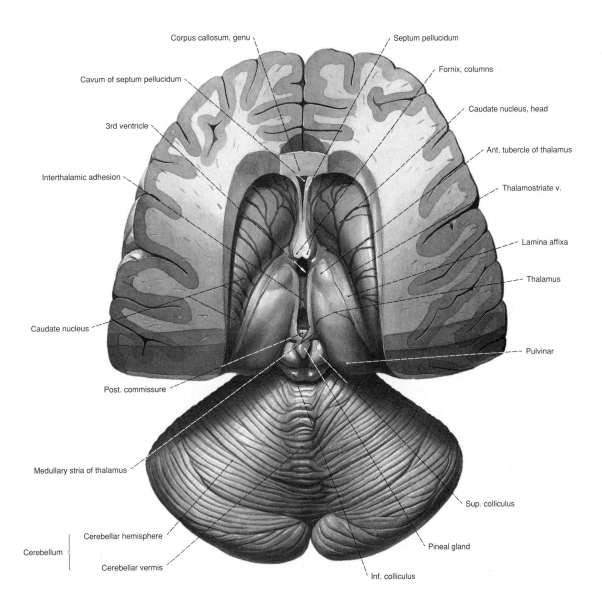

Corpus callosum, genu

Septum pellucidum

Cavum of septum pellucidum

Fornix, columns

Caudate nucleus, head

3rd ventricle

Ant. tubercle of thalamus

Interthalamic adhesion

Thalamostriate v.

Lamina affixa

Thalamus

Caudate nucleus

Pulvinar

Post. commissure

Medullary stria of thalamus

Sup. colliculus

Cerebellar hemisphere

Pineal gland

Cerebellum

Cerebellar vermis

Inf. colliculus

Fig. 513. Basal ganglia, third ventricle, quadrigeminal (tectal) plate, pineal gland and cerebellum exposed from above. The columns of the fornix, the choroid plexus of the third ventricle as well as the temporal and occipital lobes of the cerebral hemispheres have been removed. The lamina affixa is indicated in yellow.

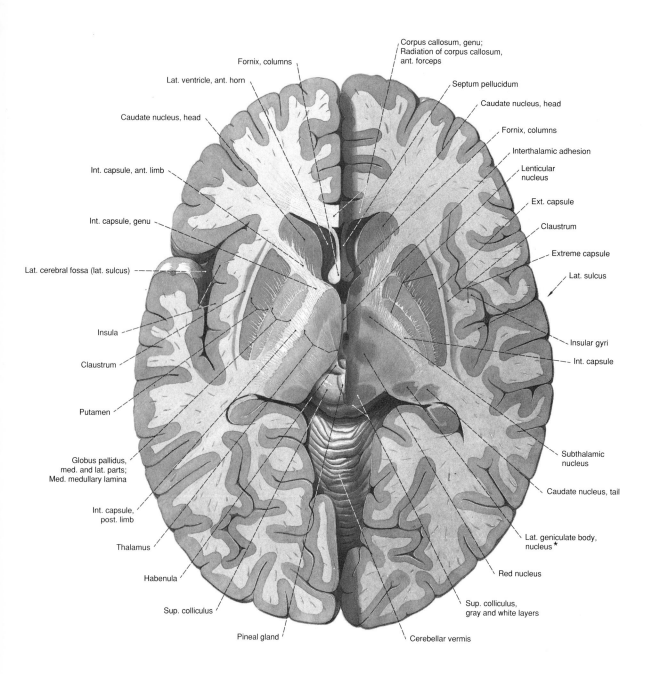

Corpus callosum, genu;
Radiation of corpus callosum,
ant. forceps

Fornix, columns

Lat. ventricle, ant. horn

Septum pellucidum

Caudate nucleus, head

Caudate nucleus, head

Fornix, columns

Interthalamic adhesion

Int. capsule, ant. limb

Lenticular
nucleus

Ext. capsule

Int. capsule, genu

Claustrum

Extreme capsule

Lat. cerebral fossa (lat. sulcus)

Lat. sulcus

Insula

Insular gyri

Claustrum

Int. capsule

Putamen

Globus pallidus,
med. and lat. parts;
Med. medullary lamina

Subthalamic
nucleus

Caudate nucleus, tail

Int. capsule,
post. limb

Thalamus

Lat. geniculate body,
nucleus *

Habenula

Red nucleus

Sup. colliculus

Sup. colliculus,
gray and white layers

Pineal gland

Cerebellar vermis

Fig. 514. Horizontal section through the cerebrum exposing
the basal ganglia and the internal capsule. On the left the plane
of section is at the level of the thalamus, on the right the section
is about 1 cm deeper at the level of the superior colliculi and
the subthalamic nucleus.

* Also known as the geniculate nucleus

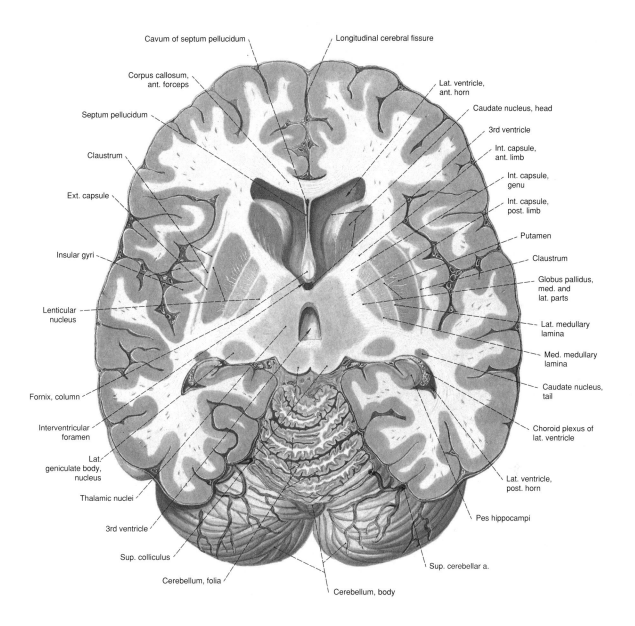

Cavum of septum pellucidum

Longitudinal cerebral fissure

Corpus callosum, ant. forceps

Lat. ventricle, ant. horn

Septum pellucidum

Caudate nucleus, head

3rd ventricle

Claustrum

Int. capsule, ant. limb

Int. capsule, genu

Ext. capsule

Int. capsule, post. limb

Putamen

Insular gyri

Claustrum

Globus pallidus, med. and lat. parts

Lenticular nucleus

Lat. medullary lamina

Med. medullary lamina

Caudate nucleus, tail

Fornix, column

Interventricular foramen

Choroid plexus of lat. ventricle

Lat. geniculate body, nucleus

Lat. ventricle, post. horn

Thalamic nuclei

Pes hippocampi

3rd ventricle

Sup. colliculus

Cerebellum, folia

Sup. cerebellar a.

Cerebellum, body

Fig. 515. Horizontal section through the cerebrum at the level of the basal ganglia and the internal capsule (cf. Figs. 517a and b).

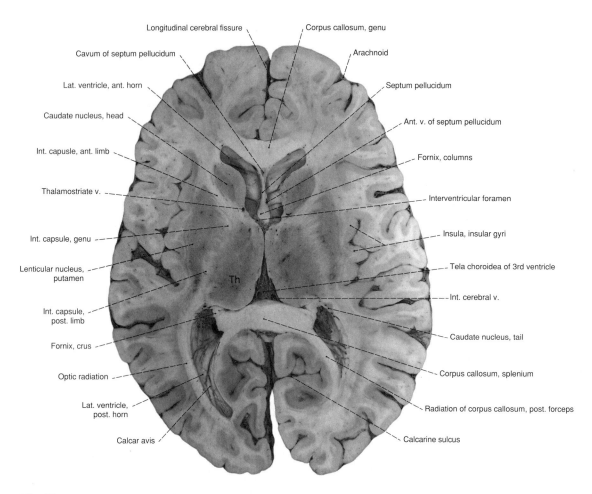

Longitudinal cerebral fissure

Cavum of septum pellucidum

Lat. ventricle, ant. horn

Caudate nucleus, head

Int. capsule, ant. limb

Thalamostriate v.

Int. capsule, genu

Lenticular nucleus, putamen

Int. capsule, post. limb

Fornix, crus

Optic radiation

Lat. ventricle, post. horn

Calcar avis

Corpus callosum, genu

Arachnoid

Septum pellucidum

Ant. v. of septum pellucidum

Fornix, columns

Interventricular foramen

Insula, insular gyri

Tela choroidea of 3rd ventricle

Int. cerebral v.

Caudate nucleus, tail

Corpus callosum, splenium

Radiation of corpus callosum, post. forceps

Calcarine sulcus

Th

Fig. 516a.   Horizontal section through the cerebrum at the level of the basal ganglia and the interventricular foramina. Th = thalamus

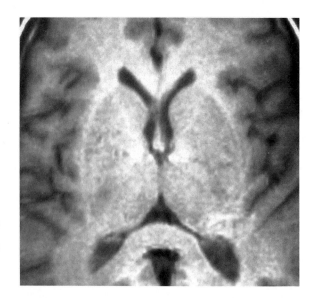

Fig. 516b.   Magnetic resonance (MR) image. Transverse section through the cerebrum at the level of Fig 516a. (From: Dr. H. FRIEDBURG, Radiology Clinic, Division of Diagnostic Radiology of the University Hospital, Freiburg i. Br.)

Refer to the explanatory text in Vol. II, page 146.

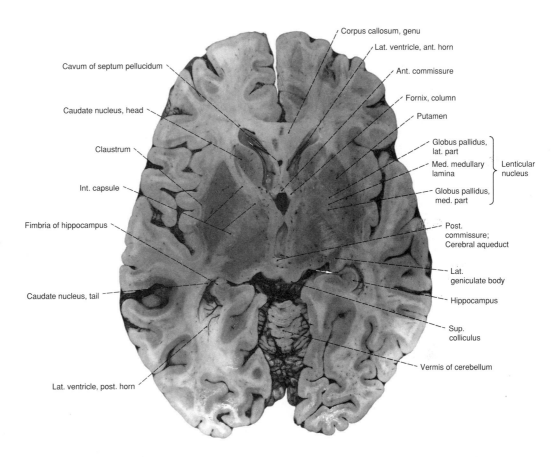

Corpus callosum, genu
Lat. ventricle, ant. horn
Ant. commissure
Fornix, column
Putamen
Globus pallidus, lat. part
Med. medullary lamina
Globus pallidus, med. part
Lenticular nucleus
Post. commissure; Cerebral aqueduct
Lat. geniculate body
Hippocampus
Sup. colliculus
Vermis of cerebellum

Cavum of septum pellucidum
Caudate nucleus, head
Claustrum
Int. capsule
Fimbria of hippocampus
Caudate nucleus, tail
Lat. ventricle, post. horn

Fig. 517a. Horizontal section through the cerebrum at the level of the internal capsule and the superior colliculi. Arrow = hippocampal sulcus, III = third ventricle, Th = thalamus

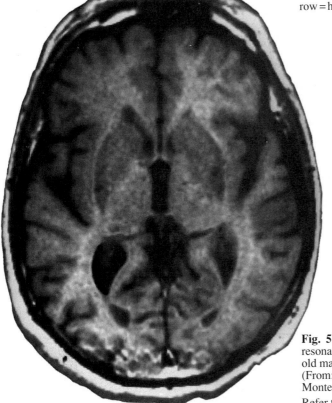

Fig. 517b. Magnetic resonance (MR) image (Magnetic resonance system, General Electric). Brain image (19-year-old male) corresponding to the plane of section in Fig. 517a. (From: Dr. M. T. McNamara, Princess Grace Hospital, Monte Carlo, Monaco.)

Refer to the explanatory text in Vol. II, page 146.

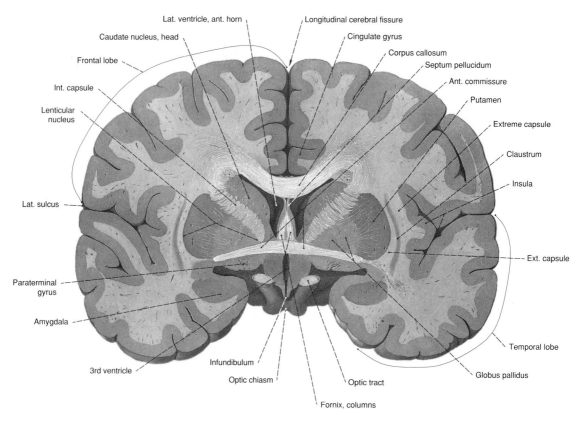

Fig. 518.   Frontal section through the telencephalon at the level of the anterior commissure.

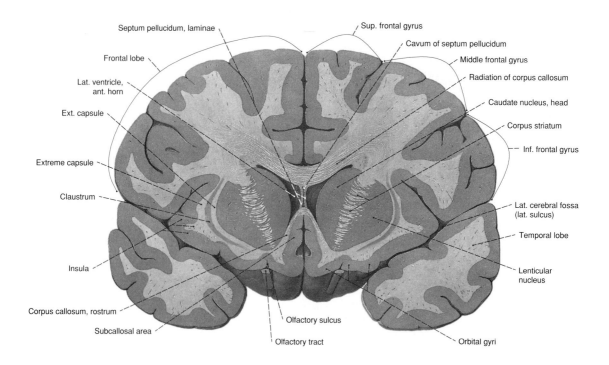

Fig. 519.   Frontal section through the telencephalon rostral to the thalamus. The anterior horns of the lateral ventricles do not possess a choroid plexus.

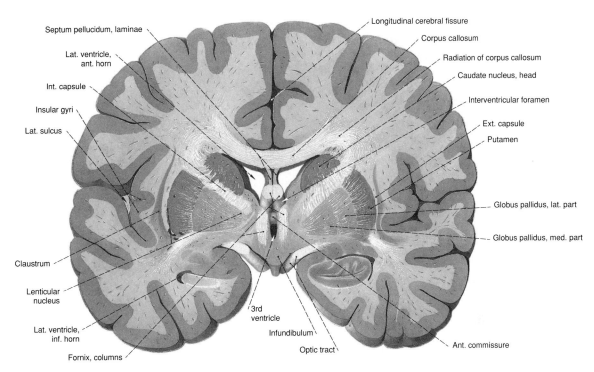

Septum pellucidum, laminae
Lat. ventricle, ant. horn
Int. capsule
Insular gyri
Lat. sulcus

Claustrum
Lenticular nucleus
Lat. ventricle, inf. horn
Fornix, columns

Longitudinal cerebral fissure
Corpus callosum
Radiation of corpus callosum
Caudate nucleus, head
Interventricular foramen
Ext. capsule
Putamen
Globus pallidus, lat. part
Globus pallidus, med. part

3rd ventricle
Infundibulum
Optic tract
Ant. commissure

Fig. 520. Frontal section through the diencephalon immediately behind the anterior commissure. View of rostral surface.

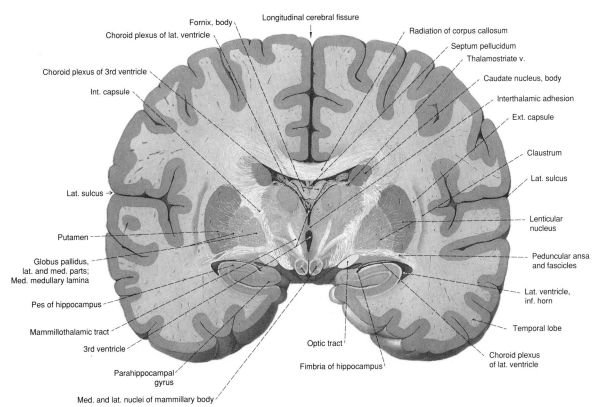

Fornix, body
Choroid plexus of lat. ventricle
Choroid plexus of 3rd ventricle
Int. capsule

Lat. sulcus

Putamen
Globus pallidus, lat. and med. parts; Med. medullary lamina
Pes of hippocampus
Mammillothalamic tract
3rd ventricle
Parahippocampal gyrus
Med. and lat. nuclei of mammillary body

Longitudinal cerebral fissure
Radiation of corpus callosum
Septum pellucidum
Thalamostriate v.
Caudate nucleus, body
Interthalamic adhesion
Ext. capsule
Claustrum
Lat. sulcus
Lenticular nucleus
Peduncular ansa and fascicles
Lat. ventricle, inf. horn
Temporal lobe
Choroid plexus of lat. ventricle

Optic tract
Fimbria of hippocampus

Fig. 521. Frontal section through the diencephalon at the level of the mammillary bodies and the thalamus. View of rostral surface.

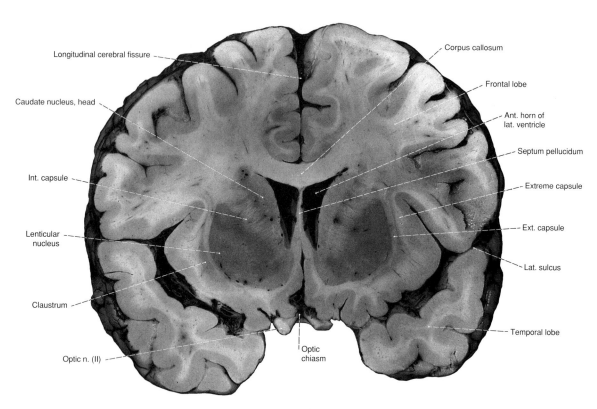

Fig. 522. Frontal section through the telencephalon in front of the optic chiasm. The anterior horn of the lateral ventricle lacks a choroid plexus. (From PERNKOPF: Atlas of Topographical and Applied Human Anatomy, Vol. 1, 2nd edition [H. FERNER, Ed.], Urban & Schwarzenberg, Baltimore-Munich, 1980.)

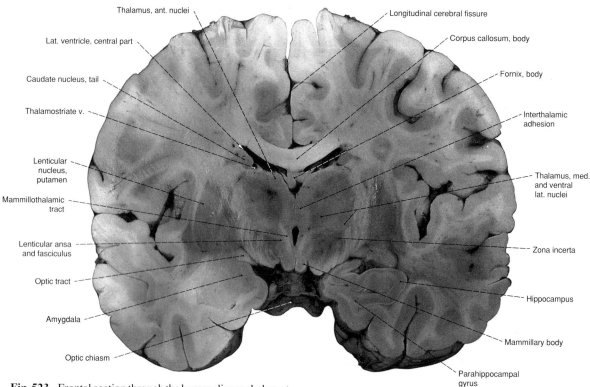

Fig. 523. Frontal section through the human diencephalon at the level of the mammillary bodies. The caudal surface of the section is shown.

Fig. 524. Cast of the ventricular system of the adult human brain viewed from the left side.

  * Foramen of MAGENDI
 ** Foramen of LUSCHKA
*** Fastigium

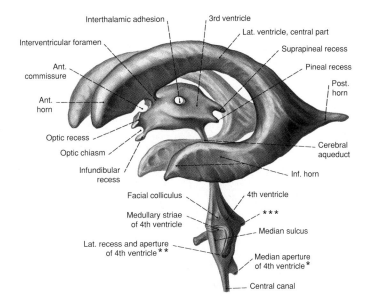

Interthalamic adhesion
3rd ventricle
Interventricular foramen
Lat. ventricle, central part
Suprapineal recess
Ant. commissure
Pineal recess
Post. horn
Ant. horn
Optic recess
Optic chiasm
Cerebral aqueduct
Infundibular recess
Inf. horn
Facial colliculus
4th ventricle
Medullary striae of 4th ventricle
***
Median sulcus
Lat. recess and aperture of 4th ventricle **
Median aperture of 4th ventricle *
Central canal

Note: Obstruction of the narrow passages of the ventricular system, i.e., interventricular foramina, cerebral aqueduct and median and lateral apertures of the fourth ventricle, may lead to obstructive or noncommunicating hydrocephalus.

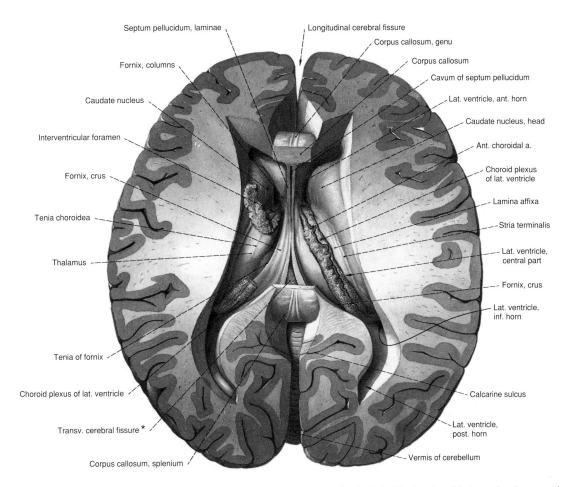

Septum pellucidum, laminae
Longitudinal cerebral fissure
Corpus callosum, genu
Fornix, columns
Corpus callosum
Cavum of septum pellucidum
Caudate nucleus
Lat. ventricle, ant. horn
Caudate nucleus, head
Interventricular foramen
Ant. choroidal a.
Choroid plexus of lat. ventricle
Fornix, crus
Lamina affixa
Tenia choroidea
Stria terminalis
Lat. ventricle, central part
Thalamus
Fornix, crus
Lat. ventricle, inf. horn
Tenia of fornix
Choroid plexus of lat. ventricle
Calcarine sulcus
Lat. ventricle, post. horn
Transv. cerebral fissure *
Vermis of cerebellum
Corpus callosum, splenium

Fig. 525. Horizontal section through the hemispheres. The lateral ventricles are opened from above. Fornix and septum pellucidum are visible after partial removal of the corpus callosum. On the left side the choroid plexus has been sectioned and reflected.
* Telodiencephalic fissure

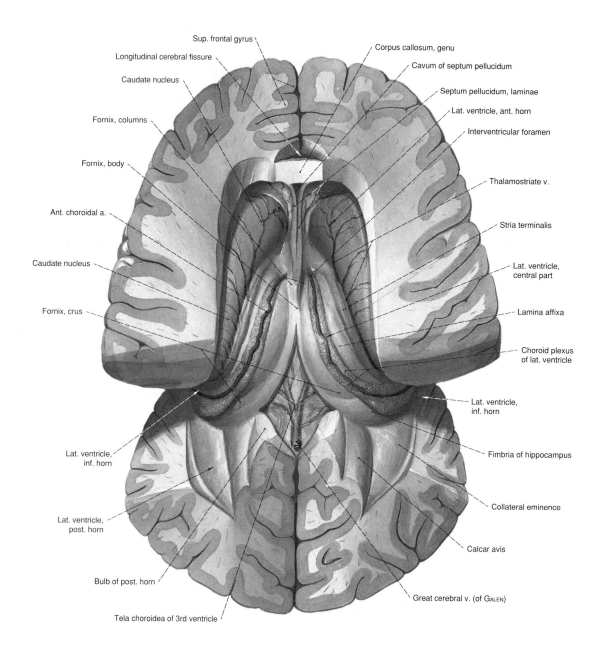

Sup. frontal gyrus

Longitudinal cerebral fissure

Caudate nucleus

Fornix, columns

Fornix, body

Ant. choroidal a.

Caudate nucleus

Fornix, crus

Lat. ventricle, inf. horn

Lat. ventricle, post. horn

Bulb of post. horn

Tela choroidea of 3rd ventricle

Corpus callosum, genu

Cavum of septum pellucidum

Septum pellucidum, laminae

Lat. ventricle, ant. horn

Interventricular foramen

Thalamostriate v.

Stria terminalis

Lat. ventricle, central part

Lamina affixa

Choroid plexus of lat. ventricle

Lat. ventricle, inf. horn

Fimbria of hippocampus

Collateral eminence

Calcar avis

Great cerebral v. (of Galen)

Fig. 526. The lateral ventricles opened from above. Most of the corpus callosum has been removed. The posterior horns of both lateral ventricles have been exposed.

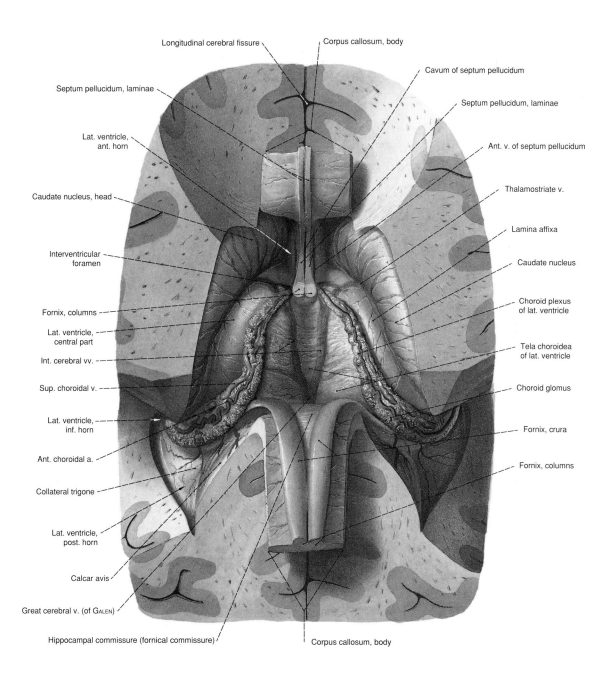

Longitudinal cerebral fissure

Corpus callosum, body

Cavum of septum pellucidum

Septum pellucidum, laminae

Septum pellucidum, laminae

Ant. v. of septum pellucidum

Lat. ventricle, ant. horn

Thalamostriate v.

Caudate nucleus, head

Lamina affixa

Caudate nucleus

Interventricular foramen

Choroid plexus of lat. ventricle

Fornix, columns

Tela choroidea of lat. ventricle

Lat. ventricle, central part

Int. cerebral vv.

Choroid glomus

Sup. choroidal v.

Lat. ventricle, inf. horn

Fornix, crura

Ant. choroidal a.

Fornix, columns

Collateral trigone

Lat. ventricle, post. horn

Calcar avis

Great cerebral v. (of GALEN)

Hippocampal commissure (fornical commissure)

Corpus callosum, body

Fig. 527. The lateral ventricles opened from above. Corpus callosum and fornix sectioned and reflected forward and backward.

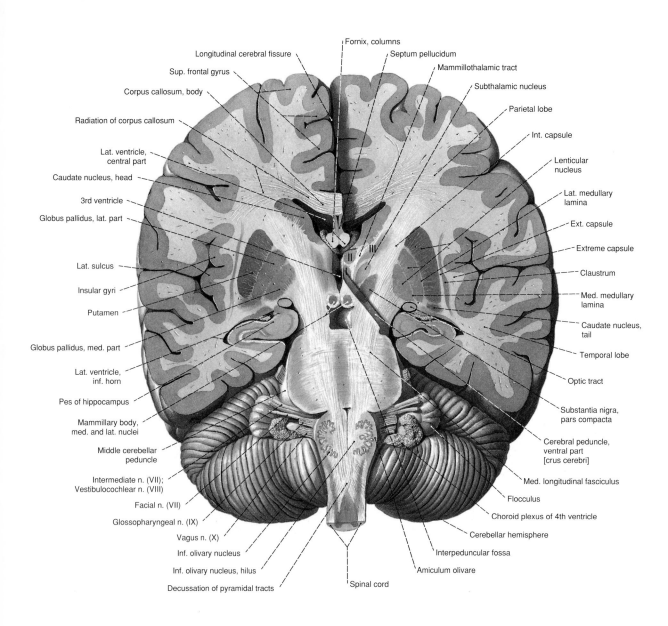

Fornix, columns
Longitudinal cerebral fissure
Sup. frontal gyrus
Corpus callosum, body
Radiation of corpus callosum
Lat. ventricle, central part
Caudate nucleus, head
3rd ventricle
Globus pallidus, lat. part
Lat. sulcus
Insular gyri
Putamen
Globus pallidus, med. part
Lat. ventricle, inf. horn
Pes of hippocampus
Mammillary body, med. and lat. nuclei
Middle cerebellar peduncle
Intermediate n. (VII); Vestibulocochlear n. (VIII)
Facial n. (VII)
Glossopharyngeal n. (IX)
Vagus n. (X)
Inf. olivary nucleus
Inf. olivary nucleus, hilus
Decussation of pyramidal tracts

Septum pellucidum
Mammillothalamic tract
Subthalamic nucleus
Parietal lobe
Int. capsule
Lenticular nucleus
Lat. medullary lamina
Ext. capsule
Extreme capsule
Claustrum
Med. medullary lamina
Caudate nucleus, tail
Temporal lobe
Optic tract
Substantia nigra, pars compacta
Cerebral peduncle, ventral part [crus cerebri]
Med. longitudinal fasciculus
Flocculus
Choroid plexus of 4th ventricle
Cerebellar hemisphere
Interpeduncular fossa
Amiculum olivare
Spinal cord

Fig. 528. Section through the brain parallel to the cerebral peduncles [crura cerebri], viewed from anterior. On the right side of the figure, the section reaches back to about the middle of the cerebral peduncle (oblique cut) and is somewhat posterior to the left side. I-III = thalamic nuclear groups: I = medial nuclei, II = anterior nuclei, III = ventral and lateral nuclei.

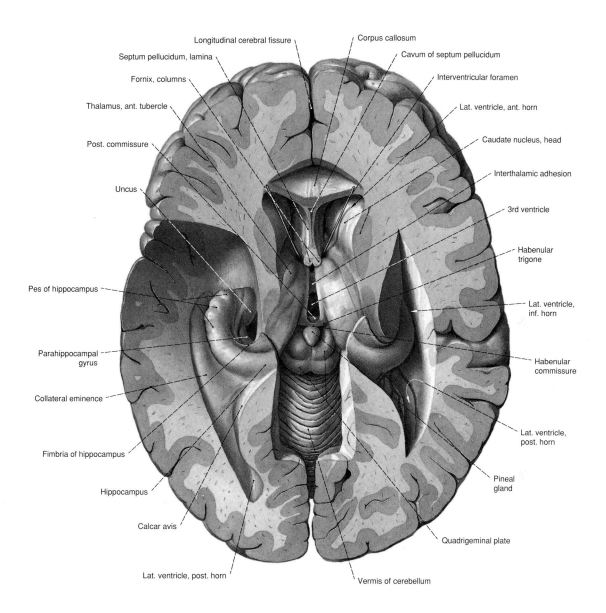

Longitudinal cerebral fissure

Septum pellucidum, lamina

Fornix, columns

Thalamus, ant. tubercle

Post. commissure

Uncus

Pes of hippocampus

Parahippocampal gyrus

Collateral eminence

Fimbria of hippocampus

Hippocampus

Calcar avis

Lat. ventricle, post. horn

Corpus callosum

Cavum of septum pellucidum

Interventricular foramen

Lat. ventricle, ant. horn

Caudate nucleus, head

Interthalamic adhesion

3rd ventricle

Habenular trigone

Lat. ventricle, inf. horn

Habenular commissure

Lat. ventricle, post. horn

Pineal gland

Quadrigeminal plate

Vermis of cerebellum

Fig. 529. Horizontal section through both hemispheres. Lateral ventricles and third ventricle are seen from above. The body and splenium of the corpus callosum, the columns of the fornix and the tela choroidea of the third ventricle have been removed. The left temporal lobe has been excavated down to the tips of the temporal (inferior) and the occipital (posterior) horns of the lateral ventricle. Probes are in the interventricular foramina.

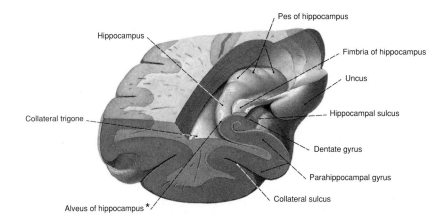

Hippocampus

Pes of hippocampus

Fimbria of hippocampus

Uncus

Hippocampal sulcus

Collateral trigone

Dentate gyrus

Parahippocampal gyrus

Collateral sulcus

Alveus of hippocampus *

Fig. 530. Anterior end of the temporal lobe after opening the temporal (inferior) horn of the lateral ventricle with a frontal section. View from behind and above.

* Thin layer of white substance on the hippocampus

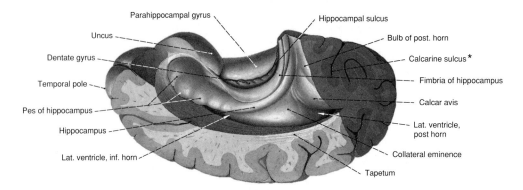

Parahippocampal gyrus

Hippocampal sulcus

Uncus

Bulb of post. horn

Dentate gyrus

Calcarine sulcus *

Temporal pole

Fimbria of hippocampus

Pes of hippocampus

Calcar avis

Hippocampus

Lat. ventricle, post horn

Lat. ventricle, inf. horn

Collateral eminence

Tapetum

Fig. 531. Occipital (posterior) and temporal (inferior) horns of the left lateral ventricle, opened laterally to expose the hippocampus.

* Deep sulcus under the cuneus in the region of the primary visual receiving area

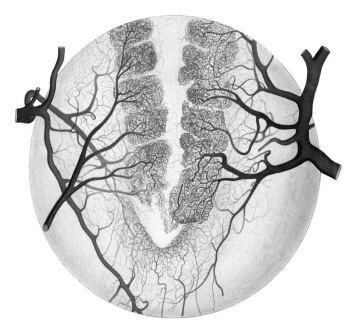

Fig. 532. Caudal section of the choroid plexus of the fourth ventricle, the vessels highlighted by injection of a dye.

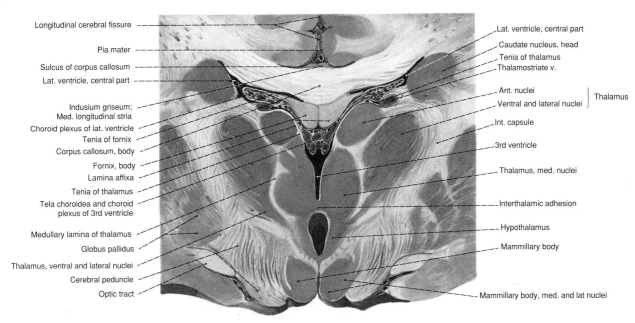

Longitudinal cerebral fissure

Pia mater

Sulcus of corpus callosum

Lat. ventricle, central part

Indusium griseum;
Med. longitudinal stria

Choroid plexus of lat. ventricle

Tenia of fornix

Corpus callosum, body

Fornix, body

Lamina affixa

Tenia of thalamus

Tela choroidea and choroid
plexus of 3rd ventricle

Medullary lamina of thalamus

Globus pallidus

Thalamus, ventral and lateral nuclei

Cerebral peduncle

Optic tract

Lat. ventricle, central part

Caudate nucleus, head

Tenia of thalamus

Thalamostriate v.

Ant. nuclei

Ventral and lateral nuclei } Thalamus

Int. capsule

3rd ventricle

Thalamus, med. nuclei

Interthalamic adhesion

Hypothalamus

Mammillary body

Mammillary body, med. and lat nuclei

Fig. 533. Frontal section through the lateral ventricles, third ventricle, corpus callosum, fornix and hypothalamus at the level of the mammillary bodies.

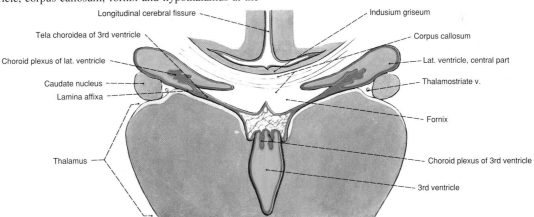

Longitudinal cerebral fissure

Tela choroidea of 3rd ventricle

Choroid plexus of lat. ventricle

Caudate nucleus

Lamina affixa

Thalamus

Indusium griseum

Corpus callosum

Lat. ventricle, central part

Thalamostriate v.

Fornix

Choroid plexus of 3rd ventricle

3rd ventricle

Fig. 534. Schematic frontal section through the central part of the lateral ventricle and the tela choroidea of the third ventricle. The pia mater and arachnoid are shown in red, the ependyma and epithelium of the choroid plexus in blue and the cerebrospinal fluid in green (cf. Fig. 533).

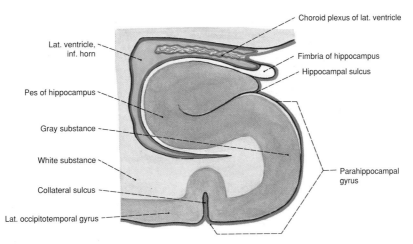

Lat. ventricle,
inf. horn

Pes of hippocampus

Gray substance

White substance

Collateral sulcus

Lat. occipitotemporal gyrus

Choroid plexus of lat. ventricle

Fimbria of hippocampus

Hippocampal sulcus

Parahippocampal
gyrus

Fig. 535. Schematic frontal, somewhat oblique section through the temporal (inferior) horn of the lateral ventricle. The pia mater and arachnoid are shown in red, the ependyma and epithelium of the choroid plexus in blue (cf. Fig. 528), and the cerebrospinal fluid in green.

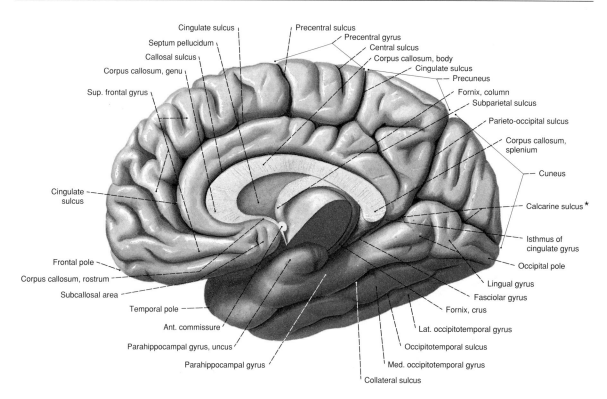

Fig. 536. Right cerebral hemisphere. The brain has been halved at the midline. The brain stem and cerebellum have been removed with an oblique section. View of the medial and inferior surfaces of the hemisphere.

* Sulcus under the cuneus in the region of the primary visual receiving area; joins with the descending parieto-occipital sulcus

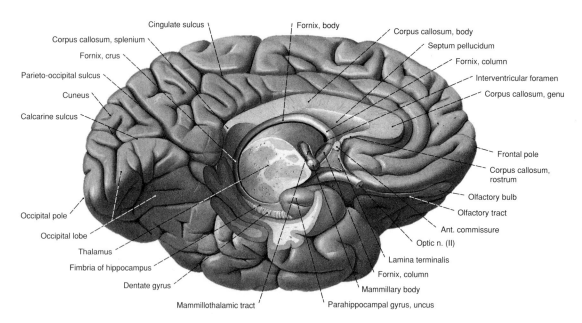

Fig. 537. The fornix exposed in its entirety. The brain has been halved in the midline and the brain stem and cerebellum removed with an oblique section. The parahippocampal gyrus has been removed to expose the hippocampal fimbria and the dentate gyrus. View of the medial and inferior surfaces of the left hemisphere.

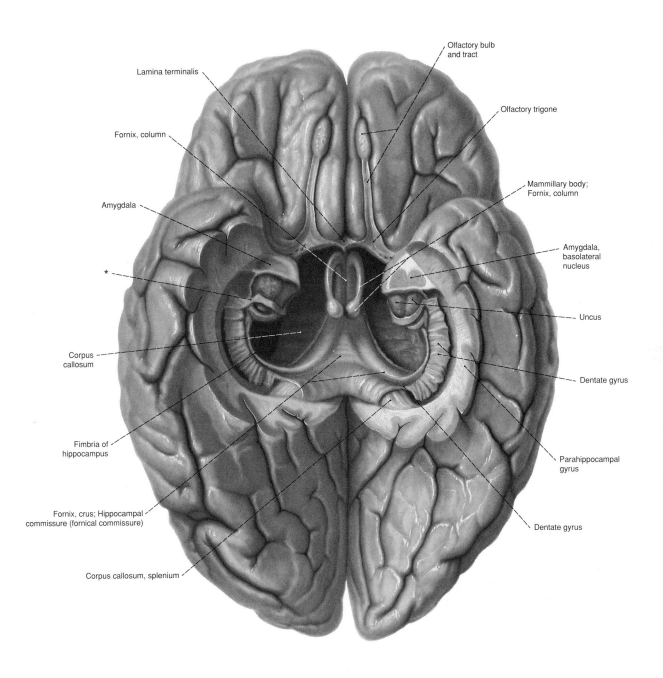

Lamina terminalis

Fornix, column

Amygdala

*

Corpus callosum

Fimbria of hippocampus

Fornix, crus; Hippocampal commissure (fornical commissure)

Corpus callosum, splenium

Olfactory bulb and tract

Olfactory trigone

Mammillary body; Fornix, column

Amygdala, basolateral nucleus

Uncus

Dentate gyrus

Parahippocampal gyrus

Dentate gyrus

Fig. 538. Inferior surface of the cerebrum. Dissection of the two fornices from the basal aspect. (From PERNKOPF: Atlas of Topographic and Applied Human Anatomy, Vol. 1, 3rd edition [W. PLATZER, Ed.], Urban & Schwarzenberg, Baltimore-Munich, 1989.)

* Uncal band

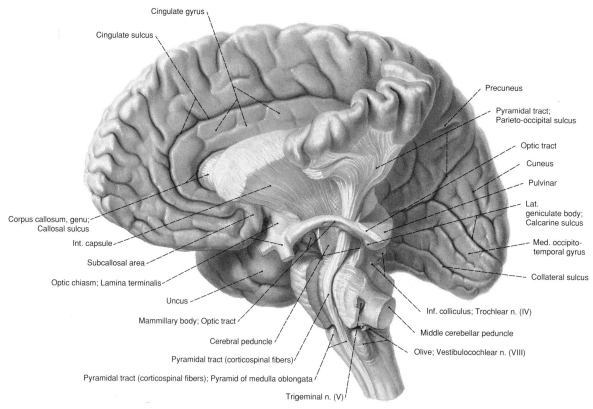

Cingulate gyrus

Cingulate sulcus

Precuneus

Pyramidal tract;
Parieto-occipital sulcus

Optic tract

Cuneus

Pulvinar

Lat.
geniculate body;
Calcarine sulcus

Med. occipito-
temporal gyrus

Collateral sulcus

Corpus callosum, genu;
Callosal sulcus

Int. capsule

Subcallosal area

Optic chiasm; Lamina terminalis

Uncus

Mammillary body; Optic tract

Cerebral peduncle

Pyramidal tract (corticospinal fibers)

Pyramidal tract (corticospinal fibers); Pyramid of medulla oblongata

Trigeminal n. (V)

Inf. colliculus; Trochlear n. (IV)

Middle cerebellar peduncle

Olive; Vestibulocochlear n. (VIII)

Fig. 539. Dissection of the left corticospinal (pyramidal)
fibers and their downward course through the brain stem.

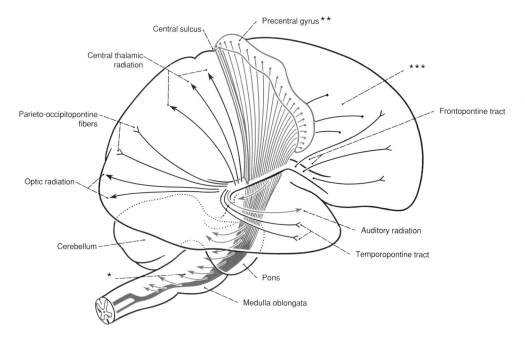

Central sulcus

Precentral gyrus **

Central thalamic
radiation

***

Parieto-occipitopontine
fibers

Frontopontine tract

Optic radiation

Auditory radiation

Temporopontine tract

Cerebellum

*

Pons

Medulla oblongata

Fig. 540. Course of the corticospinal (pyramidal) fibers
through the internal capsule and the brain stem.

* Arrows = fibers leaving the pyramidal tracts for the superior and
inferior colliculi, the pontine nuclei, the cerebellum and the
nuclei of the medulla oblongata. The pyramidal fibers continue
as the crossed lateral corticospinal tract and the uncrossed
anterior corticospinal tract.

** First order neurons of the corticospinal (pyramidal) fibers which
converge and occupy the anterior two thirds of the posterior limb
of the internal capsule (cf. Fig. 541).

*** Premotor pathways from the superior, medial and inferior frontal gyri (areas 6 and 8).

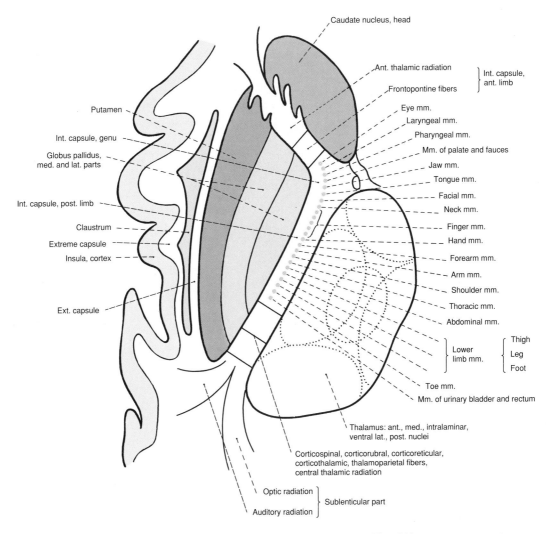

Caudate nucleus, head

Ant. thalamic radiation

Frontopontine fibers

Int. capsule, ant. limb

Eye mm.
Laryngeal mm.
Pharyngeal mm.
Mm. of palate and fauces
Jaw mm.
Tongue mm.
Facial mm.
Neck mm.
Finger mm.
Hand mm.
Forearm mm.
Arm mm.
Shoulder mm.
Thoracic mm.
Abdominal mm.
Lower limb mm. { Thigh / Leg / Foot }
Toe mm.
Mm. of urinary bladder and rectum

Putamen
Int. capsule, genu
Globus pallidus, med. and lat. parts
Int. capsule, post. limb
Claustrum
Extreme capsule
Insula, cortex
Ext. capsule

Thalamus: ant., med., intralaminar, ventral lat., post. nuclei

Corticospinal, corticorubral, corticoreticular, corticothalamic, thalamoparietal fibers, central thalamic radiation

Optic radiation
Auditory radiation } Sublenticular part

Fig. 541. Topography of the motor pathways in the internal capsule (cf. Fig. 514).

Organization of the Internal Capsule

Anterior limb of the internal capsule
Anterior thalamic radiation
Frontopontine fibers
Genu of the internal capsule
Corticobulbar fibers
Posterior limb of the internal capsule
Thalamolenticular part
Corticospinal fibers
Corticorubral fibers

Corticoreticular fibers
Corticothalamic fibers
Thalamoparietal fibers
Central thalamic radiation
Sublenticular part
Optic radiation
Auditory radiation
Corticotectal fibers
Temporopontine fibers
Retrolenticular part
Posterior thalamic radiation
Parieto-occipitopontine fibers

Note: Most of the long pathways between the cerebral cortex and the lower centers of the brain stem and spinal cord (descending or efferent and ascending or afferent pathways) course between the lenticular nucleus laterally and the thalamic and caudate nuclei medially. Together these fibers form the internal capsule in which can be distinguished an anterior limb, a genu, and a posterior limb. These designations are based upon the appearance of the internal capsule in horizontal section. In a three-dimensional view, one can also discern a sublenticular part (with optic and auditory radiations, corticotectal and temporopontine fibers) and a retrolenticular part (with a posterior thalamic radiation and parieto-occipitopontine fibers). The anterior limb contains corticothalamic and frontopontine fibers and thalamocortical fibers in the anterior thalamic radiation; the genu contains fascicles of corticobulbar fibers to the motor nuclei of the cranial nerves; and in the posterior limb, arranged from rostral to caudal, are the corticospinal, corticorubral, corticoreticular, corticothalamic, thalamoparietal fibers and the central thalamic radiation.

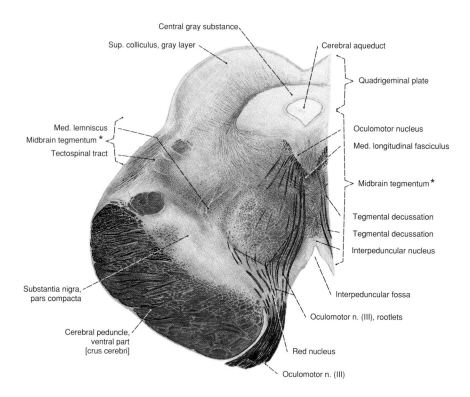

Fig. 542. Cross section through the midbrain [mesencephalon] at the level of the superior colliculi. WEIGERT's myelin stain (cf. Figs. 543, 546, 547, 550-555): white substance is stained black, gray substance lightly stained.

* Midbrain tegmentum = dorsal part of the cerebral peduncles. The tegmentum extends from the substantia nigra to the cerebral aqueduct (compare with Fig. 545).

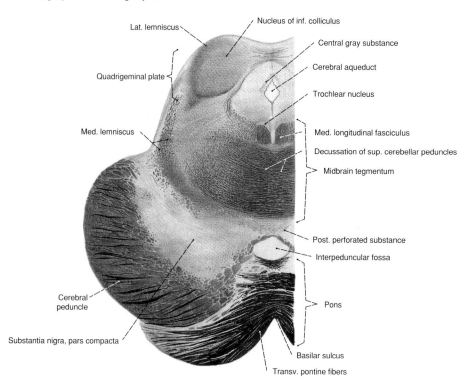

Fig. 543. Cross section through the midbrain [mesencephalon] at the level of the inferior colliculi.

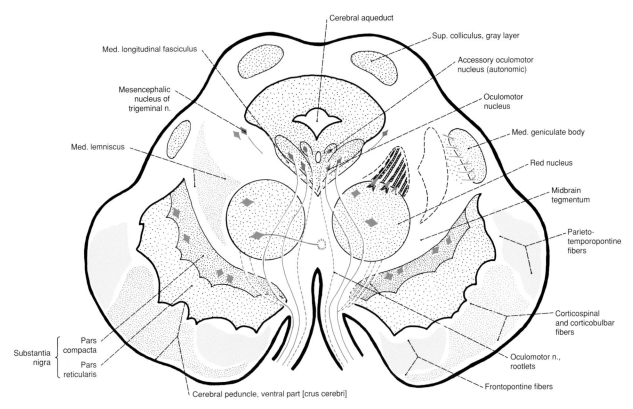

Fig. 544. Organization of the nuclei and fiber tracts of the midbrain [mesencephalon] at the level of the superior colliculi. Schematic cross section showing the descending pathways in red and yellow and the ascending in blue.

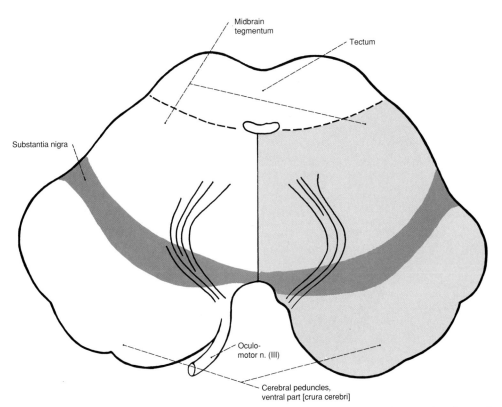

Fig. 545. Subdivisions of the midbrain [mesencephalon]: tectum, tegmentum and cerebral peduncles. The lightly shaded area indicates the two parts of the cerebral peduncle: a ventral part, the crus cerebri, and a dorsal part, the midbrain tegmentum.

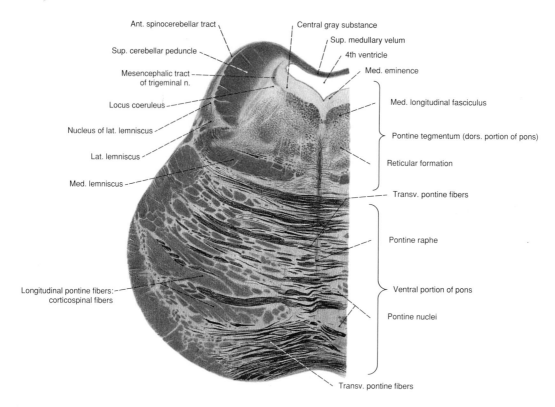

Fig. 546. Cross section through the middle region of the pons [metencephalon].

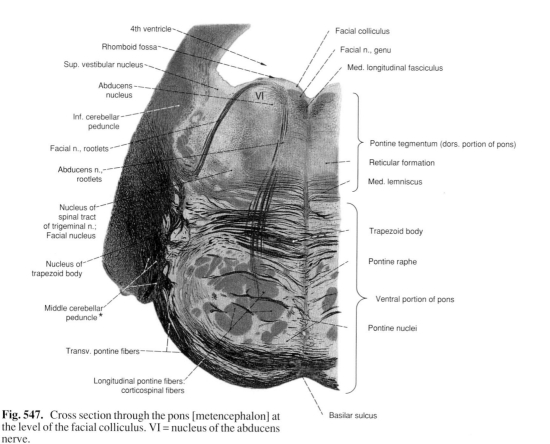

Fig. 547. Cross section through the pons [metencephalon] at the level of the facial colliculus. VI = nucleus of the abducens nerve.

* Brachium pontis

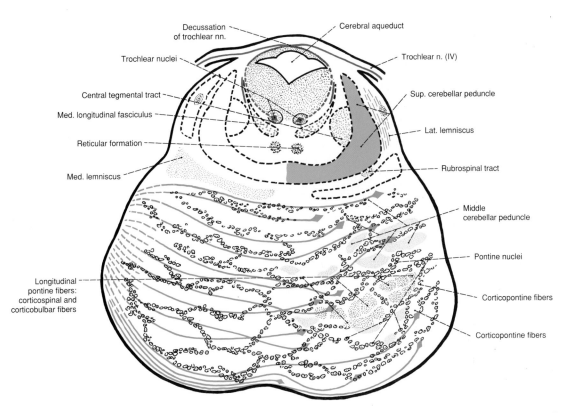

Fig. 548. Organization of the pontine nuclei and fibers at the level of the trochlear nuclei. Schematic cross section showing the descending pathways in red and yellow and the ascending in blue.

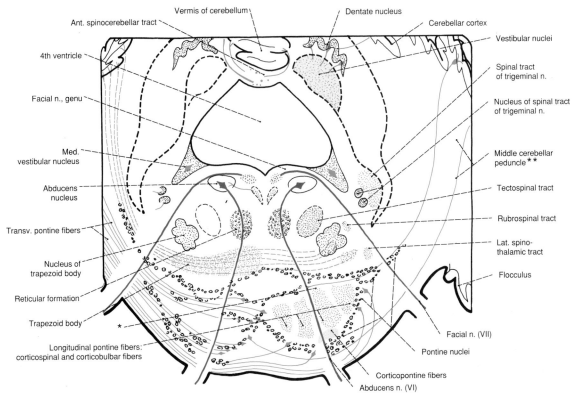

Fig. 549. Organization of the pontine nuclei and fibers at the level of the caudal portion of the pons, at the level of the abducens nuclei. Schematic cross section showing the descending pathways in red and yellow and the ascending in blue.

* Fibers from the cerebellar cortex to the pontine nuclei
** Brachium pontis

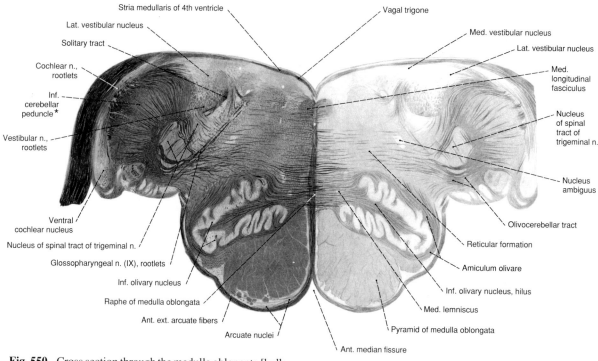

Fig. 550. Cross section through the medulla oblongata [bulb, myelencephalon] at the level of the middle of the rhomboid fossa.

* Restiform body

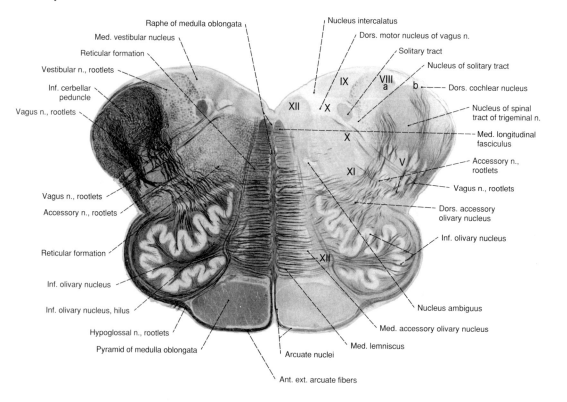

Fig. 551. Cross section through the medulla oblongata [bulb, myelencephalon] at the level of the caudal part of the rhomboid fossa and the hilus of the inferior olivary nucleus. The Roman numerals indicate the nuclei of the corresponding cranial nerves. VIIIa = vestibular nucleus, b = cochlear nucleus.

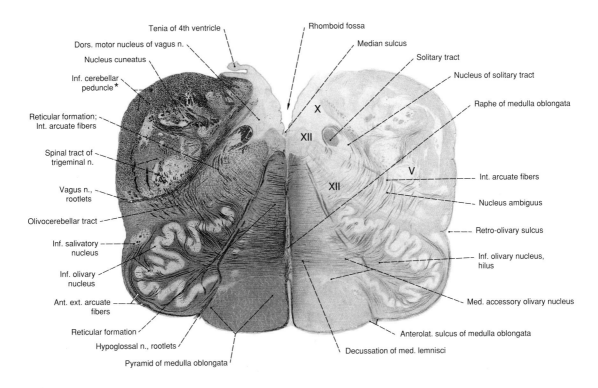

Tenia of 4th ventricle

Dors. motor nucleus of vagus n.

Nucleus cuneatus

Inf. cerebellar peduncle *

Reticular formation; Int. arcuate fibers

Spinal tract of trigeminal n.

Vagus n., rootlets

Olivocerebellar tract

Inf. salivatory nucleus

Inf. olivary nucleus

Ant. ext. arcuate fibers

Reticular formation

Hypoglossal n., rootlets

Pyramid of medulla oblongata

Rhomboid fossa

Median sulcus

Solitary tract

Nucleus of solitary tract

Raphe of medulla oblongata

X

XII

XII

V

Int. arcuate fibers

Nucleus ambiguus

Retro-olivary sulcus

Inf. olivary nucleus, hilus

Med. accessory olivary nucleus

Anterolat. sulcus of medulla oblongata

Decussation of med. lemnisci

Fig. 552. Cross section through the medulla oblongata [bulb, myelencephalon] at the level of the inferior salivatory nucleus.

* Restiform body

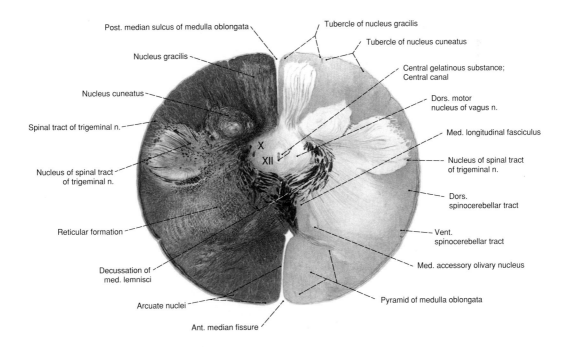

Post. median sulcus of medulla oblongata

Nucleus gracilis

Nucleus cuneatus

Spinal tract of trigeminal n.

Nucleus of spinal tract of trigeminal n.

Reticular formation

Decussation of med. lemnisci

Arcuate nuclei

Ant. median fissure

Tubercle of nucleus gracilis

Tubercle of nucleus cuneatus

Central gelatinous substance; Central canal

Dors. motor nucleus of vagus n.

Med. longitudinal fasciculus

Nucleus of spinal tract of trigeminal n.

Dors. spinocerebellar tract

Vent. spinocerebellar tract

Med. accessory olivary nucleus

Pyramid of medulla oblongata

X

XII

Fig. 553. Cross section through the medulla oblongata [bulb, myelencephalon] at the level of the decussation of the medial lemnisci (sensory decussation).

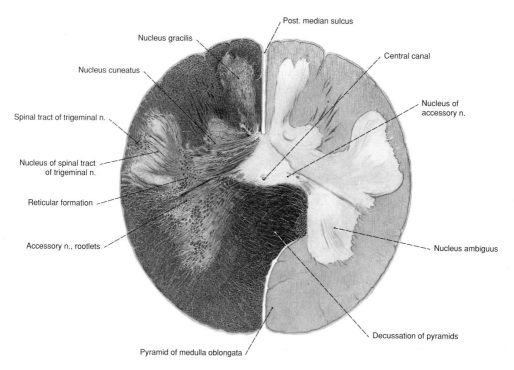

Post. median sulcus

Nucleus gracilis

Central canal

Nucleus cuneatus

Nucleus of
accessory n.

Spinal tract of trigeminal n.

Nucleus of spinal tract
of trigeminal n.

Reticular formation

Accessory n., rootlets

Nucleus ambiguus

Decussation of pyramids

Pyramid of medulla oblongata

Fig. 554. Cross section through the medulla oblongata [bulb, myelencephalon] at the level of the upper portion of the decussation of the pyramids (motor decussation). Roman numerals indicate the positions of the corresponding cranial nerve nuclei.

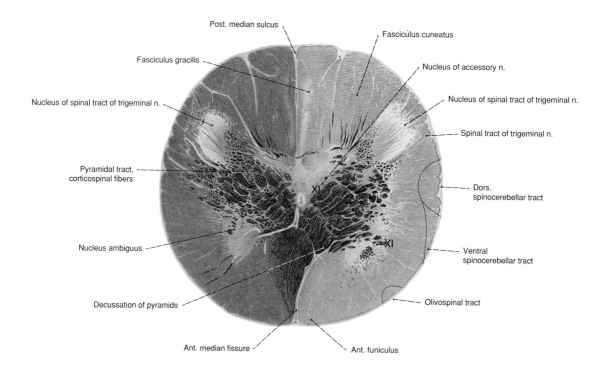

Post. median sulcus

Fasciculus cuneatus

Fasciculus gracilis

Nucleus of accessory n.

Nucleus of spinal tract of trigeminal n.

Nucleus of spinal tract of trigeminal n.

Spinal tract of trigeminal n.

Pyramidal tract,
corticospinal fibers

XI

Dors.
spinocerebellar tract

Nucleus ambiguus

XI

Ventral
spinocerebellar tract

Decussation of pyramids

Olivospinal tract

Ant. median fissure

Ant. funiculus

Fig. 555. Cross section through the medulla oblongata [bulb, myelencephalon] at the level of the lower portion of the decussation of the pyramids (motor decussation).

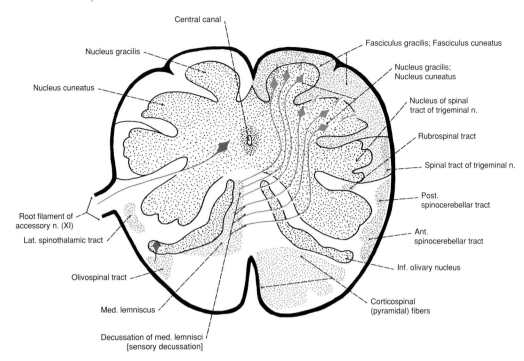

Central canal

Nucleus gracilis

Nucleus cuneatus

Fasciculus gracilis; Fasciculus cuneatus

Nucleus gracilis; Nucleus cuneatus

Nucleus of spinal tract of trigeminal n.

Rubrospinal tract

Spinal tract of trigeminal n.

Post. spinocerebellar tract

Ant. spinocerebellar tract

Root filament of accessory n. (XI)

Lat. spinothalamic tract

Olivospinal tract

Med. lemniscus

Inf. olivary nucleus

Corticospinal (pyramidal) fibers

Decussation of med. lemnisci [sensory decussation]

Fig. 556. Organization of the nuclei and fibers of the medulla oblongata [bulb, myelencephalon] at the level of the decussation of the medial lemnisci (sensory decussation). Schematic cross section.

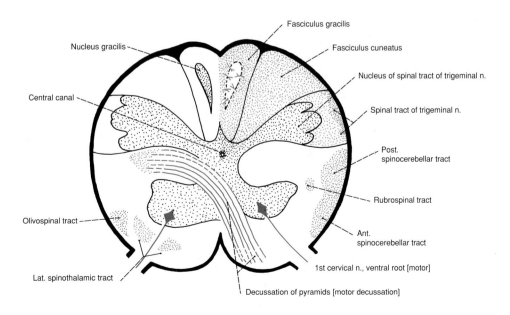

Fasciculus gracilis

Nucleus gracilis

Fasciculus cuneatus

Nucleus of spinal tract of trigeminal n.

Central canal

Spinal tract of trigeminal n.

Post. spinocerebellar tract

Rubrospinal tract

Olivospinal tract

Ant. spinocerebellar tract

1st cervical n., ventral root [motor]

Lat. spinothalamic tract

Decussation of pyramids [motor decussation]

Fig. 557. Organization of the nuclei and fibers of the medulla oblongata [bulb, myelencephalon] at the level of the decussation of the pyramids (motor decussation). Schematic cross section.

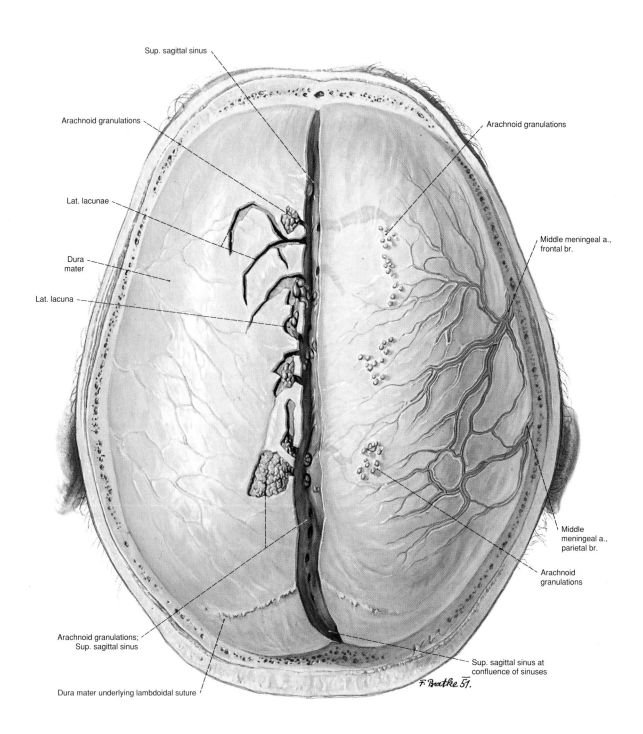

Sup. sagittal sinus

Arachnoid granulations

Lat. lacunae

Dura mater

Lat. lacuna

Arachnoid granulations

Middle meningeal a., frontal br.

Middle meningeal a., parietal br.

Arachnoid granulations

Sup. sagittal sinus at confluence of sinuses

Arachnoid granulations; Sup. sagittal sinus

Dura mater underlying lambdoidal suture

F. Brathe 51.

Fig. 558. Outer surface of the cranial dura mater after removal of the calvaria and opening of the superior sagittal sinus and some of its lateral lacunae. (From PERNKOPF: Atlas of Topographic and Applied Human Anatomy, Vol. 1, 3rd edition [W. PLATZER, Ed.], Urban & Schwarzenberg, Baltimore-Munich, 1989.)

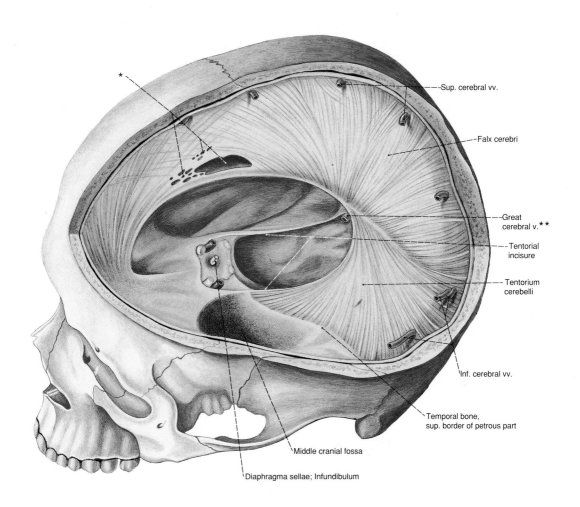

*

—Sup. cerebral vv.

—Falx cerebri

—Great
cerebral v.**

—Tentorial
incisure

—Tentorium
cerebelli

Inf. cerebral vv.

Temporal bone,
sup. border of petrous part

Middle cranial fossa

Diaphragma sellae; Infundibulum

Fig. 559. Skull with falx cerebri and tentorium cerebelli, viewed from the left and above. The cranial cavity is divided into three chambers: the supratentorial space for the right and left cerebral hemispheres and the infratentorial space for the hindbrain. The tentorial incisure surrounds the midbrain. (From FERNER/KAUTZKY: Handbook of Neurosurgery, Vol. 1, Springer, Heidelberg, 1959.)

* Gaps in the falx cerebri
** Vein of GALEN

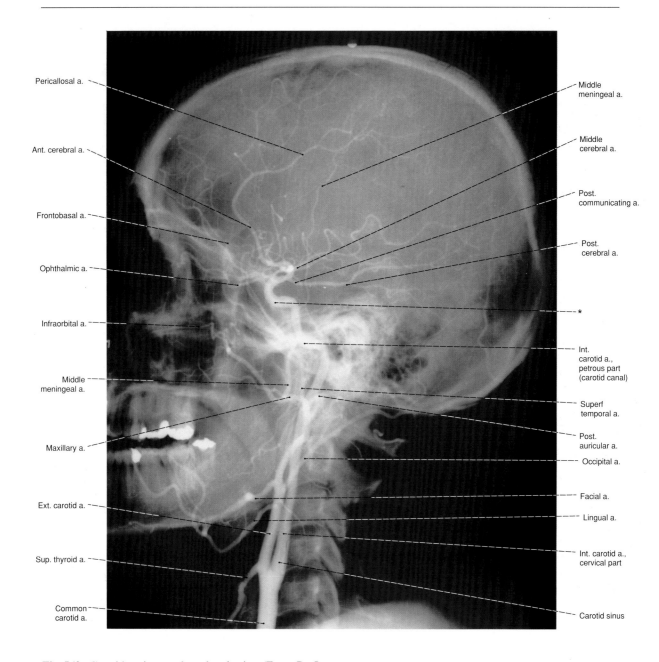

Pericallosal a.

Ant. cerebral a.

Frontobasal a.

Ophthalmic a.

Infraorbital a.

Middle meningeal a.

Maxillary a.

Ext. carotid a.

Sup. thyroid a.

Common carotid a.

Middle meningeal a.

Middle cerebral a.

Post. communicating a.

Post. cerebral a.

*

Int. carotid a., petrous part (carotid canal)

Superf temporal a.

Post. auricular a.

Occipital a.

Facial a.

Lingual a.

Int. carotid a., cervical part

Carotid sinus

Fig. 560. Carotid angiogram, lateral projection. (From: Dr. J. C. DEMBSKI, Marienhof Hospital, Koblenz).

* Clinically: Carotid siphon

The **internal carotid artery**, in its petrous part, penetrates the base of the skull through the carotid canal in the petrous portion of the temporal bone. After leaving the bony canal near the apex of the petrous portion, it is separated from the trigeminal ganglion by a thin bony or connective tissue septum. It ascends in a sulcus at the lateral surface of the body of the sphenoid bone, where it is situated close to the frontal pole of the trigeminal ganglion (**"ganglionic segment"**). The artery then bends rostrally and ascends in the sagittal direction forward toward the base of the anterior clinoid process. This is the **"cavernous sinus segment"**, which extends from the apex of the petrous portion of the temporal bone to the base of the

anterior clinoid process. At this point the artery lies in a shallow sulcus in the lateral wall of the body of the sphenoid bone. Below the base of the anterior clinoid process, the artery takes a sharp bend, with its convexity directed anteriorly (**"carotid genu"**), pierces the dura mater and the arachnoid, and running posteriorly, as the cerebral portion, comes to lie below the optic nerve (II) as it enters the optic canal (Figs. 184-186). From here the artery courses within the subarachnoid space (**"cisternal segment"**), and finally divides into its terminal branches (**"dividing segment"**). In radiology, these tortuous segments are designated as the **"carotid siphon"** (Fig. 560).

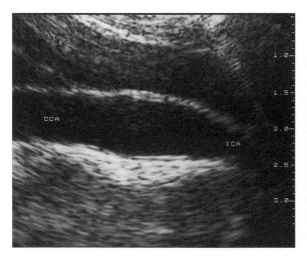

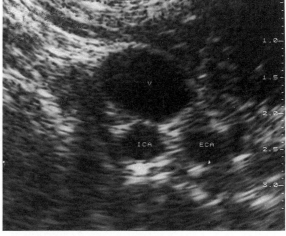

a) Transition of the common carotid artery (CCA) into the internal carotid artery (ICA) (2-fold enlargement). Caudal = left, cranial = right. Reflections from the vessel walls appear light, those from the vessel lumen appear dark. Note the enlargement at the origin of the internal carotid artery = carotid sinus ("bulb"). The plane of the section is such that the external carotid artery is not present.

b) Cross section of the internal carotid artery (ICA), external carotid artery (ECA) and internal jugular vein (V) (2-fold enlargement), about 2 cm cranial to the carotid bifurcation ("carotid fork").

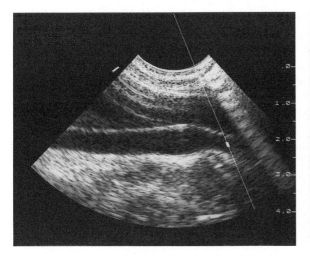

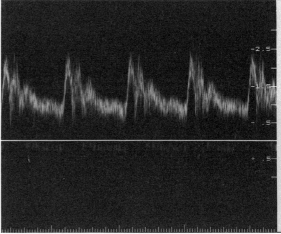

c) Doppler ultrasonography of the common carotid artery and the internal carotid artery (actual size). The direction of the Doppler beam is superimposed.

d) The spectrum of Doppler frequencies measured in Fig. 561c (ordinate: Doppler frequencies in kHz; abscissa: time). The systolic acceleration and the continuous diastolic flow are distinguishable. The profile of the pulse is typical for a cerebral artery. (From: Dr. G.-M. v. REUTERN, Neurology Clinic, University of Freiburg.)

Figs. 561 a-d.   Sonograms of the carotid artery. Combined ultrasonography and Doppler ultrasonography.

High resolution ultrasonographic equipment is used for diagnosis of vascular diseases. It produces a sectional image of the arterial wall. By comparison radiologic angiography records the shadow cast by the blood vessel. Since ultrasonography by itself can only present the anatomical relations of the vessel wall, it is combined with Doppler ultrasonography, which can measure the velocity of the blood flow. In Doppler ultrasonography the vessels are irradiated transcutaneously and the frequency received by reflection is compared with the originating frequency. Because the blood corpuscles are moving relative to the axis of the beam, a shift in frequency occurs (Doppler shift) that is proportional to the blood-flow velocity.

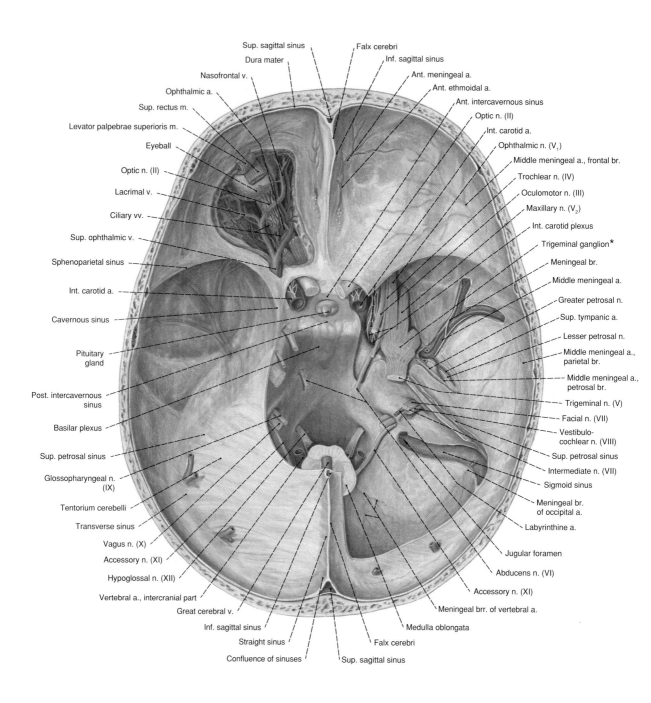

Sup. sagittal sinus
Dura mater
Nasofrontal v.
Ophthalmic a.
Sup. rectus m.
Levator palpebrae superioris m.
Eyeball
Optic n. (II)
Lacrimal v.
Ciliary vv.
Sup. ophthalmic v.
Sphenoparietal sinus
Int. carotid a.
Cavernous sinus
Pituitary gland
Post. intercavernous sinus
Basilar plexus
Sup. petrosal sinus
Glossopharyngeal n. (IX)
Tentorium cerebelli
Transverse sinus
Vagus n. (X)
Accessory n. (XI)
Hypoglossal n. (XII)
Vertebral a., intercranial part
Great cerebral v.
Inf. sagittal sinus
Straight sinus
Confluence of sinuses

Falx cerebri
Inf. sagittal sinus
Ant. meningeal a.
Ant. ethmoidal a.
Ant. intercavernous sinus
Optic n. (II)
Int. carotid a.
Ophthalmic n. (V$_1$)
Middle meningeal a., frontal br.
Trochlear n. (IV)
Oculomotor n. (III)
Maxillary n. (V$_2$)
Int. carotid plexus
Trigeminal ganglion*
Meningeal br.
Middle meningeal a.
Greater petrosal n.
Sup. tympanic a.
Lesser petrosal n.
Middle meningeal a., parietal br.
Middle meningeal a., petrosal br.
Trigeminal n. (V)
Facial n. (VII)
Vestibulo-cochlear n. (VIII)
Sup. petrosal sinus
Intermediate n. (VII)
Sigmoid sinus
Meningeal br. of occipital a.
Labyrinthine a.
Jugular foramen
Abducens n. (VI)
Accessory n. (XI)
Meningeal brr. of vertebral a.
Medulla oblongata
Falx cerebri
Sup. sagittal sinus

Fig. 562.   Internal aspect of the base of the skull, viewed from above, showing the cranial dura mater, intracranial venous sinuses, meningeal arteries, and cranial nerves. The roof of the left orbit has been removed. The tentorium cerebelli has been mostly removed on the right. The right sigmoid sinus and cavernous sinus have been opened. The trigeminal ganglion and the middle meningeal artery are exposed.

* Semilunar or GASSERIAN ganglion

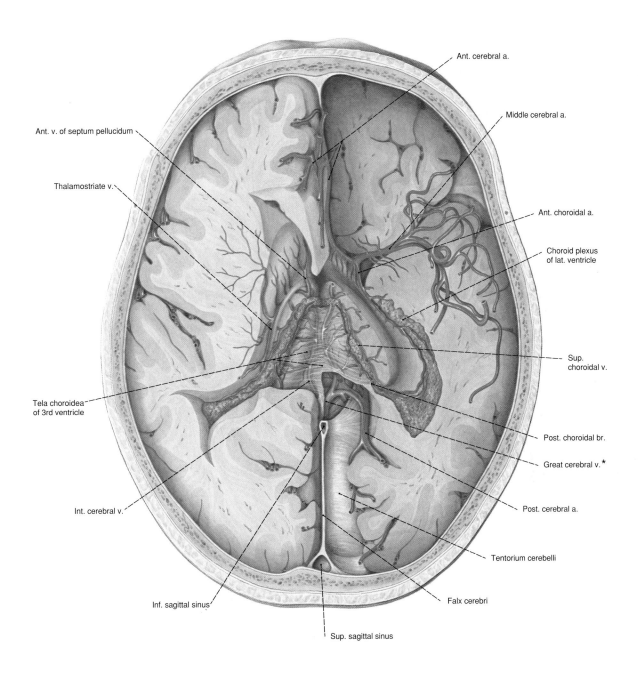

Ant. cerebral a.

Middle cerebral a.

Ant. v. of septum pellucidum

Ant. choroidal a.

Thalamostriate v.

Choroid plexus of lat. ventricle

Sup. choroidal v.

Tela choroidea of 3rd ventricle

Post. choroidal br.

Great cerebral v. *

Int. cerebral v.

Post. cerebral a.

Tentorium cerebelli

Inf. sagittal sinus

Falx cerebri

Sup. sagittal sinus

Fig. 563. Horizontal section through the brain showing the branches of the anterior, middle and posterior cerebral arteries and the position and course of the internal cerebral veins. (From PERNKOPF: Atlas of Topographic and Applied Human Anatomy, Vol. 1, 3rd edition [W. PLATZER, Ed.], Urban & Schwarzenberg, Baltimore-Munich, 1989.)

* Vein of GALEN

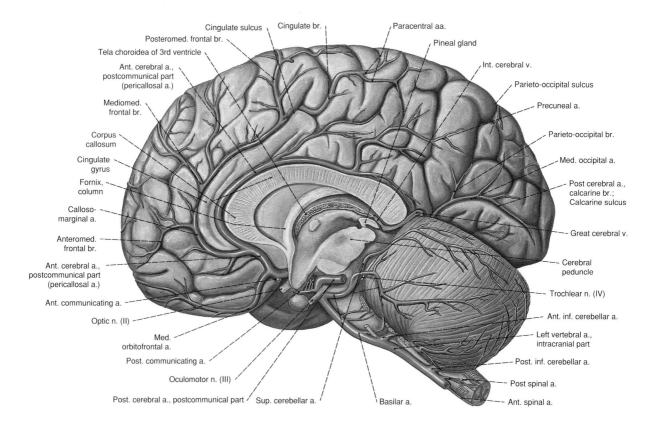

Cingulate sulcus
Cingulate br.
Paracentral aa.
Pineal gland
Posteromed. frontal br.
Tela choroidea of 3rd ventricle
Int. cerebral v.
Ant. cerebral a., postcommunical part (pericallosal a.)
Parieto-occipital sulcus
Mediomed. frontal br.
Precuneal a.
Corpus callosum
Parieto-occipital br.
Cingulate gyrus
Med. occipital a.
Fornix, column
Post cerebral a., calcarine br.; Calcarine sulcus
Calloso-marginal a.
Great cerebral v.
Anteromed. frontal br.
Cerebral peduncle
Ant. cerebral a., postcommunical part (pericallosal a.)
Trochlear n. (IV)
Ant. communicating a.
Ant. inf. cerebellar a.
Optic n. (II)
Left vertebral a., intracranial part
Med. orbitofrontal a.
Post. inf. cerebellar a.
Post. communicating a.
Oculomotor n. (III)
Post spinal a.
Post. cerebral a., postcommunical part
Sup. cerebellar a.
Basilar a.
Ant. spinal a.

Fig. 564. Arteries of the medial surface of the cerebral hemisphere and of the surface of the cerebellum in a medial view. The left hemisphere has been removed.

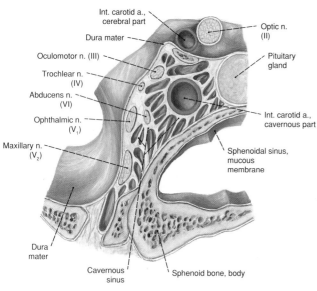

Int. carotid a., cerebral part
Optic n. (II)
Dura mater
Pituitary gland
Oculomotor n. (III)
Trochlear n. (IV)
Abducens n. (VI)
Ophthalmic n. (V₁)
Int. carotid a., cavernous part
Maxillary n. (V₂)
Sphenoidal sinus, mucous membrane
Dura mater
Cavernous sinus
Sphenoid bone, body

Fig. 565. Frontal section through the left cavernous sinus at the level of the pituitary gland [hypophysis cerebri]. The cavernous portion of the internal carotid artery is surrounded by the venous chambers of the sinus. After forming the "carotid genu" from which the ophthalamic artery arises (cf. Figs. 184-187), the carotid artery perforates the dura mater and appears below the optic nerve (II) as the latter enters the optic canal. This segment of the artery ("cisternal segment") is surrounded by cerebrospinal fluid. The two cross sections of the internal carotid artery have different diameters as a result of the exit of the ophthalamic artery.

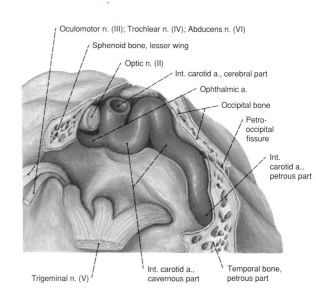

Oculomotor n. (III); Trochlear n. (IV); Abducens n. (VI)
Sphenoid bone, lesser wing
Optic n. (II)
Int. carotid a., cerebral part
Ophthalmic a.
Occipital bone
Petro-occipital fissure
Int. carotid a., petrous part
Trigeminal n. (V)
Int. carotid a., cavernous part
Temporal bone, petrous part

Fig. 566. Intracranial course of the left internal carotid artery. The artery lies in a sulcus in the lateral wall of the sphenoid bone. Note the origin of the ophthalamic artery at the "carotid genu" and the S-shaped curvature of the carotid artery ("carotid siphon") in the carotid sulcus next to the sella turcica. The trigeminal (V), oculomotor (III), trochlear (IV) and abducens (VI) nerves have been deflected laterally.

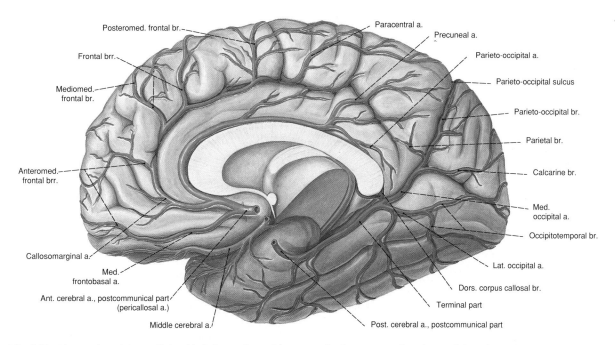

Posteromed. frontal br.

Frontal brr.

Mediomed. frontal br.

Anteromed. frontal brr.

Callosomarginal a.

Med. frontobasal a.

Ant. cerebral a., postcommunical part (pericallosal a.)

Middle cerebral a.

Paracentral a.

Precuneal a.

Parieto-occipital a.

Parieto-occipital sulcus

Parieto-occipital br.

Parietal br.

Calcarine br.

Med. occipital a.

Occipitotemporal br.

Lat. occipital a.

Dors. corpus callosal br.

Terminal part

Post. cerebral a., postcommunical part

Fig. 567. The arteries of the medial and inferior surface of the right cerebral hemisphere in a medial view. Note: Branches of the anterior and posterior cerebral arteries extend over the superior medial border of the cerebral cortex on their way to the superior lateral surface of the hemisphere. The anterior cerebral artery supplies the medial surface up to the parieto-occipital sulcus, the posterior cerebral artery supplies the inferior surface of the hemisphere and the cuneus with the exception of the upper surface of the base of the frontal lobe.

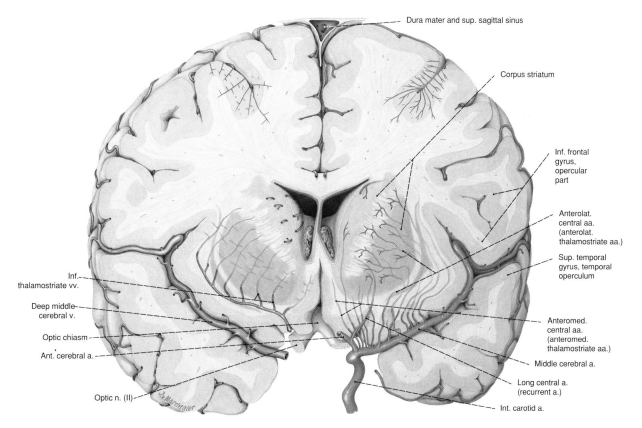

Dura mater and sup. sagittal sinus

Corpus striatum

Inf. frontal gyrus, opercular part

Anterolat. central aa. (anterolat. thalamostriate aa.)

Sup. temporal gyrus, temporal operculum

Anteromed. central aa. (anteromed. thalamostriate aa.)

Middle cerebral a.

Long central a. (recurrent a.)

Int. carotid a.

Inf. thalamostriate vv.

Deep middle cerebral v.

Optic chiasm

Ant. cerebral a.

Optic n. (II)

Fig. 568. Vascularization of the telencephalon in a semidiagrammatic frontal section anterior to the optic chiasm. Note the central branches from the proximal parts of the middle and anterior cerebral arteries for the basal ganglia and the internal capsule. The precommunical part of the anterior cerebral artery gives origin to a number of central arteries, including the anteromedial central arteries [anteromedial thalamostriate arteries], which usually divide into short central arteries and the long central artery [recurrent artery (of HEUBNER)].

346   Brain

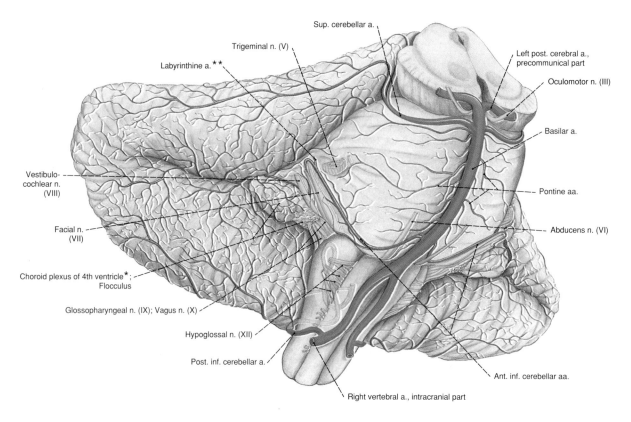

Sup. cerebellar a.
Trigeminal n. (V)
Labyrinthine a.**
Left post. cerebral a., precommunical part
Oculomotor n. (III)
Basilar a.
Vestibulo-cochlear n. (VIII)
Pontine aa.
Facial n. (VII)
Abducens n. (VI)
Choroid plexus of 4th ventricle*; Flocculus
Glossopharyngeal n. (IX); Vagus n. (X)
Hypoglossal n. (XII)
Post. inf. cerebellar a.
Ant. inf. cerebellar aa.
Right vertebral a., intracranial part

Fig. 569. Arteries of the hindbrain [rhombencephalon]. (From: Dr. TSCHABITSCHER, Vienna.) On the left side of the figure, the anterior inferior cerebellar artery courses over the abducens nerve (= 79%); on the right, however, it courses under the nerve (=16%).

* BOCHDALEK's "flower basket"
** In 15% of cases this artery exits from the basilar artery

Cerebellar arteries

The cerebellum is supplied by three paired arteries:
1. posterior inferior cerebellar artery from the vertebral artery, intracranial part,
2. anterior inferior cerebellar artery,
3. superior cerebellar artery, the latter two from the basilar artery. They are interconnected by anastomoses.
Characteristically the cerebellar arteries course in an arch around the brain stem, from ventral to dorsal. The branches course along the folia of the cerebellum to penetrate into the fissures. The arteries generally underlie the larger veins.

1. Posterior inferior cerebellar artery. It originates 1-3 cm caudal to the union of the two vertebral arteries and lies between the root fibers of the vagal group, bends around the medulla oblongata and reaches the tela choroidea of the fourth ventricle (branches supply the choroid plexus of the fourth ventricle), from where it passes to the inferior surface of the cerebellum and loops around the tonsil. It supplies the lateral parts of the medulla oblongata (occlusion of the artery leads to WALLENBERG's syndrome) and the inferior surface of the cerebellum.

2. Anterior inferior cerebellar artery. Originating from the basilar artery, it crosses the abducens nerve (VI), and forms a loop, whose vertex may reach to the internal acoustic meatus. The distal limb of the loop parallels the facial (VII) and vestibulocochlear (VIII) nerves. In most cases, it gives rise to the labyrinthine artery. Its end branches reach the flocculus and the choroid plexus in the lateral recess of the fourth ventricle.

3. Superior cerebellar artery. It arises from the anterior part of the basilar artery just caudal to its division, courses within the cisterna ambiens along the anterior border of the pons around the cerebral peduncle dorsally and, in its initial portion, is separated from the posterior cerebral artery by the oculomotor nerve (III). The superior cerebellar artery divides into two main branches which both parallel the trochlear nerve (IV) and reach the quadrigeminal (tectal) plate. The artery supplies the superior surface of the cerebellum and the cerebellar nuclei.

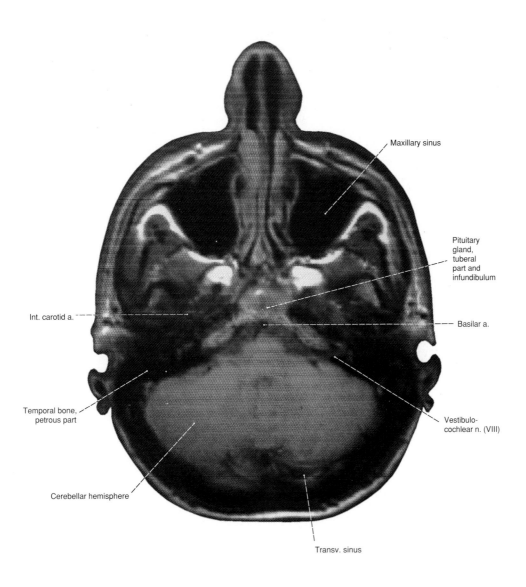

Maxillary sinus

Pituitary gland, tuberal part and infundibulum

Int. carotid a.

Basilar a.

Temporal bone, petrous part

Vestibulo-cochlear n. (VIII)

Cerebellar hemisphere

Transv. sinus

Fig. 570. Magnetic resonance image (MRI). Horizontal section through the skull at the level of the cerebellum, pituitary gland [hypophysis cerebri], and the maxillary sinus. Note the vestibulocochlear nerve (VIII) in the internal acoustic meatus. (From: Dr. H. FRIEDBURG, Radiology Clinic, Division of Diagnostic Radiology, University Hospital, Freiburg i. Br.)

Refer to the block of text in Vol. II, page 146.

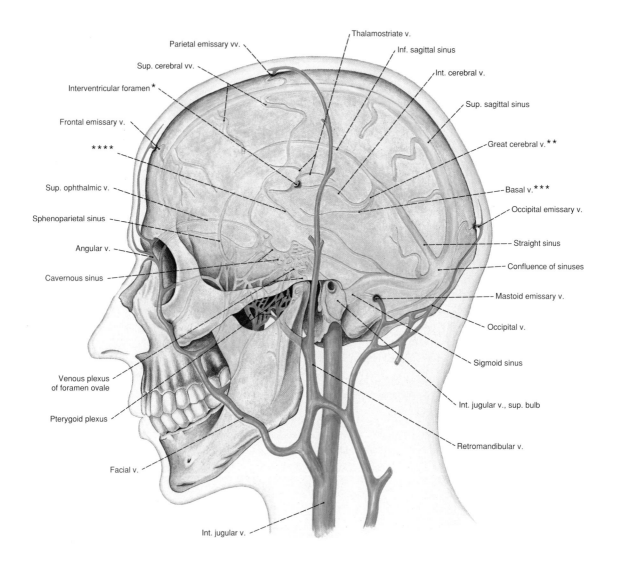

Fig. 571. The large veins of the head, the sinuses of the dura mater and their interconnections. The dural sinuses and veins covered by bone are drawn transparently and colored light blue. (From H. FERNER/R. KAUTZKY: Handbook of Neurosurgery, Vol. 1, Springer, Heidelberg, 1959.)

    *   Foramen of MONRO
    ** Vein of GALEN
  *** ROSENTHAL'S vein
**** Anastomosis of LABBÉ

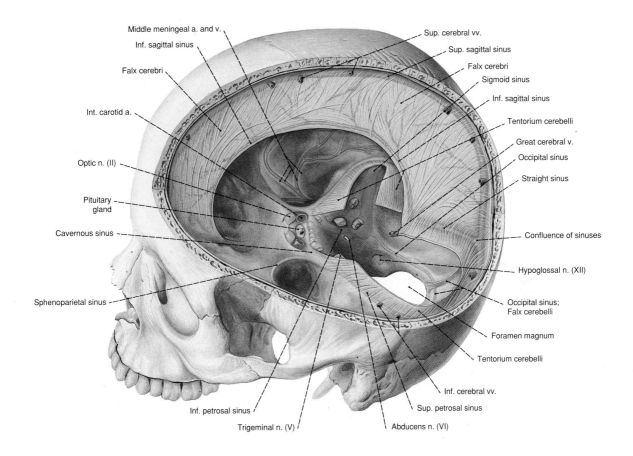

Middle meningeal a. and v.

Inf. sagittal sinus

Falx cerebri

Int. carotid a.

Optic n. (II)

Pituitary gland

Cavernous sinus

Sphenoparietal sinus

Inf. petrosal sinus

Trigeminal n. (V)

Sup. cerebral vv.

Sup. sagittal sinus

Falx cerebri

Sigmoid sinus

Inf. sagittal sinus

Tentorium cerebelli

Great cerebral v.

Occipital sinus

Straight sinus

Confluence of sinuses

Hypoglossal n. (XII)

Occipital sinus;
Falx cerebelli

Foramen magnum

Tentorium cerebelli

Inf. cerebral vv.

Sup. petrosal sinus

Abducens n. (VI)

Fig. 572. The intracranial dura mater and the dural sinuses viewed from above and left. A large portion of the tentorium cerebelli has been removed on the left side and a narrow strip from the right side.

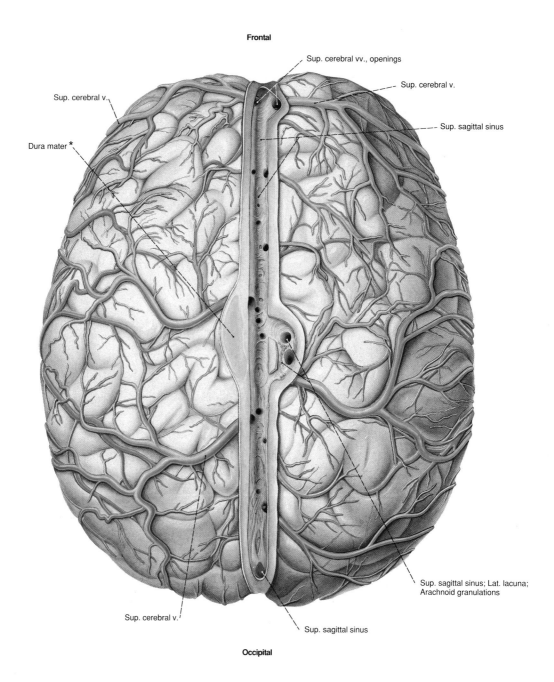

Frontal

Sup. cerebral vv., openings

Sup. cerebral v.

Sup. cerebral v.

Sup. sagittal sinus

Dura mater *

Sup. sagittal sinus; Lat. lacuna;
Arachnoid granulations

Sup. cerebral v.

Sup. sagittal sinus

Occipital

Fig. 573. Veins and arteries of the brain, the superior sagittal sinus and its lateral lacunae viewed from above. A strip of dura mater has been retained along the superior sagittal sinus, which has been opened longitudinally. The branches of the middle cerebral artery ascend along the convexity of the hemisphere but do not reach its crest. Here, the terminal branches of the anterior cerebral artery extend about a finger's width over the cortical crest onto the convexity. The superior cerebral veins empty into the superior sagittal sinus and its lateral lacunae.

* A subdural lateral lacuna of the superior sagittal sinus

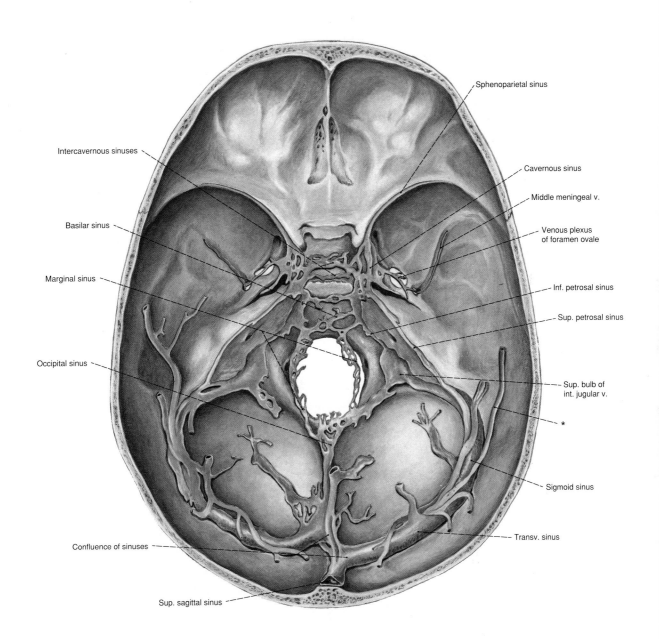

Sphenoparietal sinus

Intercavernous sinuses

Cavernous sinus

Middle meningeal v.

Basilar sinus

Venous plexus of foramen ovale

Marginal sinus

Inf. petrosal sinus

Sup. petrosal sinus

Occipital sinus

Sup. bulb of int. jugular v.

*

Sigmoid sinus

Confluence of sinuses

Transv. sinus

Sup. sagittal sinus

Fig. 574. Sinuses of the dura mater at the base of the skull. (From: Dr. H. FERNER, Vienna.)

* Anastomosis of LABBÉ

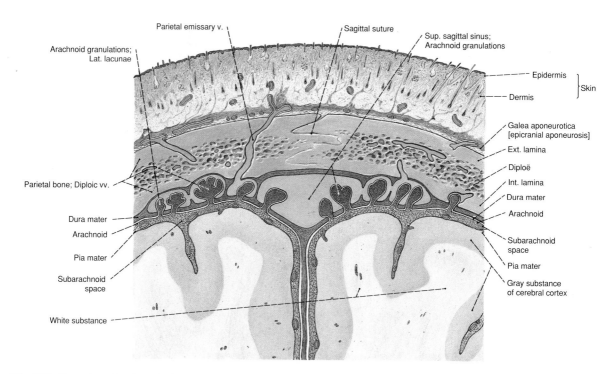

Fig. 575. Frontal section through the neurocranium showing the scalp (skin + galea aponeurotica), bones of the skull, meninges and cerebral cortex.

Parietal emissary v.

Sagittal suture

Sup. sagittal sinus;
Arachnoid granulations

Arachnoid granulations;
Lat. lacunae

Epidermis

Dermis

Skin

Galea aponeurotica
[epicranial aponeurosis]

Ext. lamina

Diploë

Int. lamina

Parietal bone; Diploic vv.

Dura mater

Arachnoid

Dura mater

Arachnoid

Pia mater

Subarachnoid
space

Subarachnoid
space

Pia mater

Gray substance
of cerebral cortex

White substance

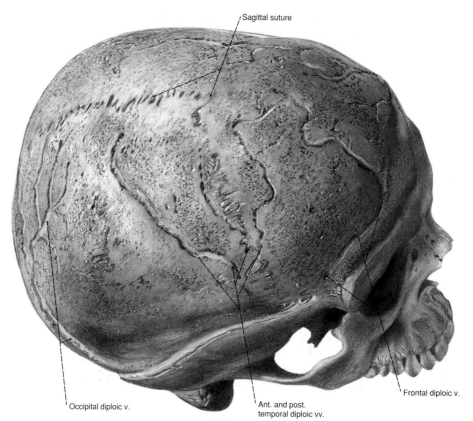

Fig. 576. Diploic canals and veins of the neurocranium exposed by removing the outer table (external lamina) of the bones of the skull.

Sagittal suture

Occipital diploic v.

Ant. and post.
temporal diploic vv.

Frontal diploic v.

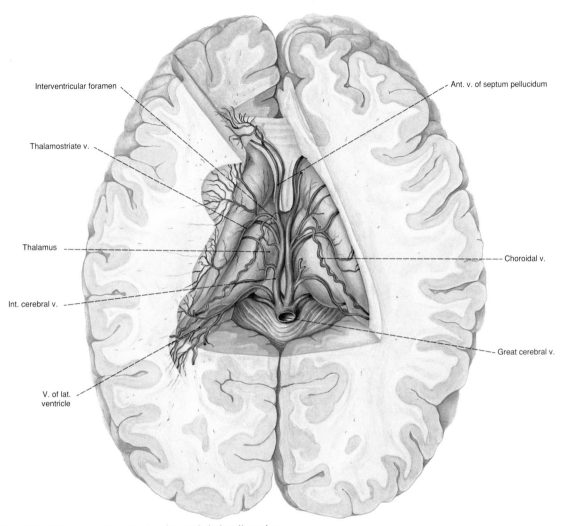

Interventricular foramen

Thalamostriate v.

Thalamus

Int. cerebral v.

V. of lat. ventricle

Ant. v. of septum pellucidum

Choroidal v.

Great cerebral v.

Fig. 577.   The internal cerebral veins and their tributaries draining the deep cerebral structures. (From: H. FERNER, Z. Anat. 120, 1958.)

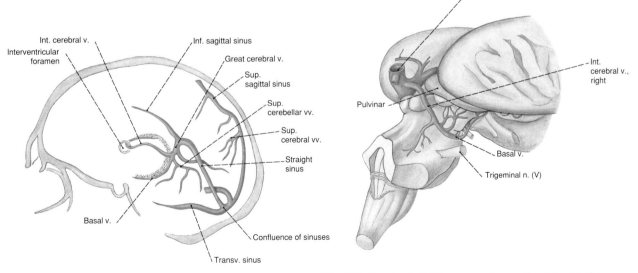

Int. cerebral v.

Interventricular foramen

Inf. sagittal sinus

Great cerebral v.

Sup. sagittal sinus

Sup. cerebellar vv.

Sup. cerebral vv.

Straight sinus

Basal v.

Confluence of sinuses

Transv. sinus

Great cerebral v.

Int. cerebral v., right

Pulvinar

Basal v.

Trigeminal n. (V)

Fig. 578.   Schematic of a phlebosinogram, lateral projection.

Fig. 579.   Basal vein (of ROSENTHAL) ascending lateral to the midbrain and the great cerebral vein (of GALEN). (From TOLDT/HOCHSTETTER: Anatomic Atlas, Vol. 2, 27th edition, Urban & Schwarzenberg, Munich-Vienna-Baltimore, 1979.)

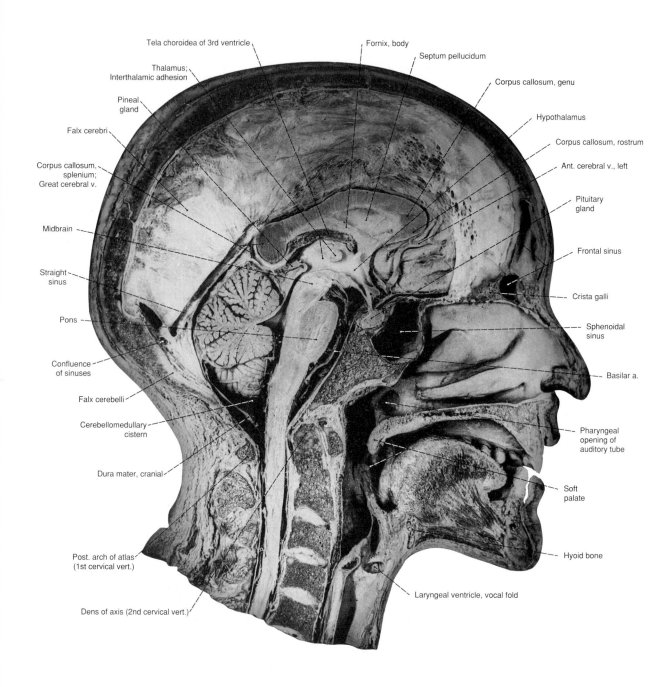

Tela choroidea of 3rd ventricle

Thalamus;
Interthalamic adhesion

Pineal
gland

Falx cerebri

Corpus callosum,
splenium;
Great cerebral v.

Midbrain

Straight
sinus

Pons

Confluence
of sinuses

Falx cerebelli

Cerebellomedullary
cistern

Dura mater, cranial

Post. arch of atlas
(1st cervical vert.)

Dens of axis (2nd cervical vert.)

Fornix, body

Septum pellucidum

Corpus callosum, genu

Hypothalamus

Corpus callosum, rostrum

Ant. cerebral v., left

Pituitary
gland

Frontal sinus

Crista galli

Sphenoidal
sinus

Basilar a.

Pharyngeal
opening of
auditory tube

Soft
palate

Hyoid bone

Laryngeal ventricle, vocal fold

Fig. 580.   Median sagittal section through the head.

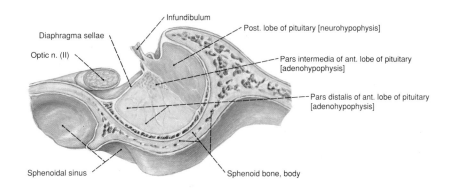

Fig. 581. Median section through the pituitary gland [hypophysis cerebri] and its position in the hypophyseal fossa of the sella turcica.

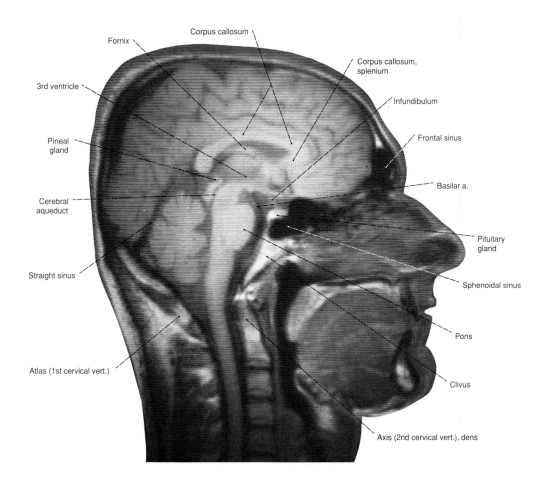

Fig. 582. Magnetic resonance (MR) image showing a paramedian sagittal section through the head and the upper cervical vertebral column. Note in particular the clarity of the infundibulum (pituitary stalk), the fornix and the cerebral aqueduct. (From: Dr. H. FRIEDBURG, Radiology Clinic, University of Freiburg i. Br.)
Refer to the block of text in Vol. II, page 146.

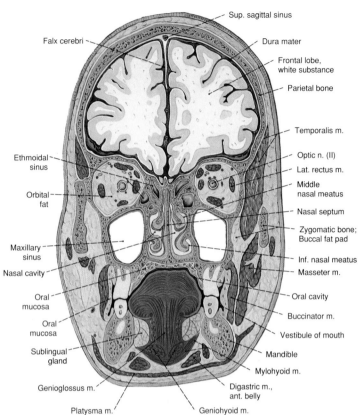

Sup. sagittal sinus

Falx cerebri

Dura mater

Frontal lobe, white substance

Parietal bone

Temporalis m.

Ethmoidal sinus

Optic n. (II)

Lat. rectus m.

Middle nasal meatus

Orbital fat

Nasal septum

Zygomatic bone; Buccal fat pad

Maxillary sinus

Inf. nasal meatus

Masseter m.

Nasal cavity

Oral mucosa

Oral cavity

Buccinator m.

Oral mucosa

Vestibule of mouth

Sublingual gland

Mandible

Mylohyoid m.

Genioglossus m.

Digastric m., ant. belly

Platysma m.

Geniohyoid m.

Fig. 583a. Frontal section through the head directly behind the orbital cavities. Note the gross spatial organization of the head. Upper third: cranial cavity with cerebrum; middle section: nasal cavity, with ethmoidal air cells adjacent to the orbits, beneath which are the maxillary sinuses; lower third, separated by the palate from the nasal cavity, and consisting of the oral cavity, tongue and sublingual region with the sublingual salivary gland. The cervical end is formed by the platelike mylohyoid muscle (diaphragm of the mouth).

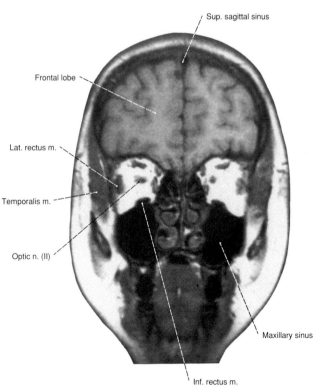

Sup. sagittal sinus

Frontal lobe

Lat. rectus m.

Temporalis m.

Optic n. (II)

Maxillary sinus

Inf. rectus m.

Fig. 583b. Magnetic resonance (MR) image (Magnetic Resonance System, General Electric). Frontal section through the skull at about the level of Fig. 583a. (From: Dr. M. T. McNamara, Princess Grace Hospital, Monte Carlo, Monaco.)

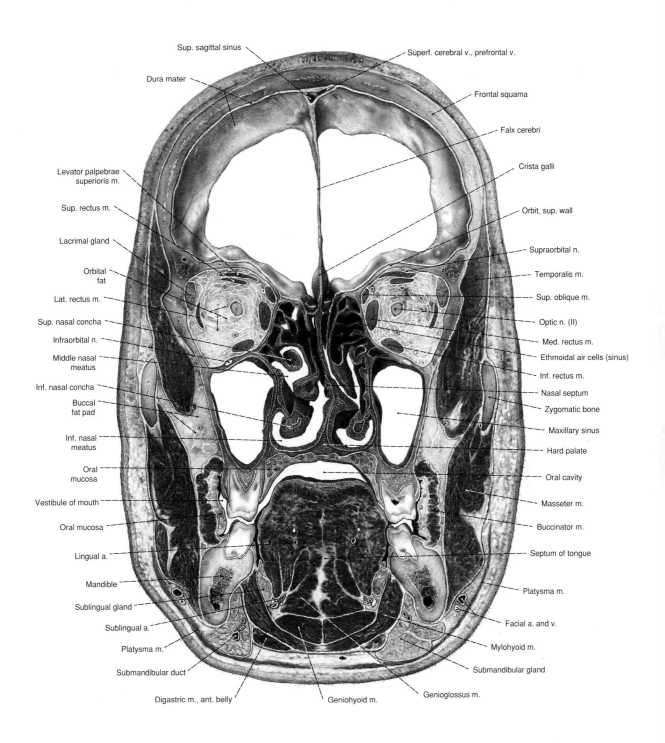

Sup. sagittal sinus

Dura mater

Levator palpebrae superioris m.

Sup. rectus m.

Lacrimal gland

Orbital fat

Lat. rectus m.

Sup. nasal concha

Infraorbital n.

Middle nasal meatus

Inf. nasal concha

Buccal fat pad

Inf. nasal meatus

Oral mucosa

Vestibule of mouth

Oral mucosa

Lingual a.

Mandible

Sublingual gland

Sublingual a.

Platysma m.

Submandibular duct

Digastric m., ant. belly

Superf. cerebral v., prefrontal v.

Frontal squama

Falx cerebri

Crista galli

Orbit, sup. wall

Supraorbital n.

Temporalis m.

Sup. oblique m.

Optic n. (II)

Med. rectus m.

Ethmoidal air cells (sinus)

Inf. rectus m.

Nasal septum

Zygomatic bone

Maxillary sinus

Hard palate

Oral cavity

Masseter m.

Buccinator m.

Septum of tongue

Platysma m.

Facial a. and v.

Mylohyoid m.

Submandibular gland

Geniohyoid m.

Genioglossus m.

Fig. 584. Frontal section through the head behind the eyeballs. The frontal lobes of the cerebral hemispheres have been removed (cf. Fig. 583).

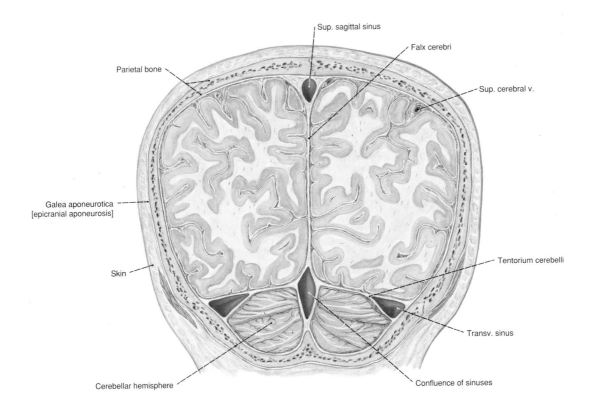

Sup. sagittal sinus

Falx cerebri

Parietal bone

Sup. cerebral v.

Galea aponeurotica
[epicranial aponeurosis]

Tentorium cerebelli

Skin

Transv. sinus

Cerebellar hemisphere

Confluence of sinuses

Fig. 585. Frontal section through the neurocranium at the intersection of the superior sagittal and straight sinuses at the confluence of the sinuses.

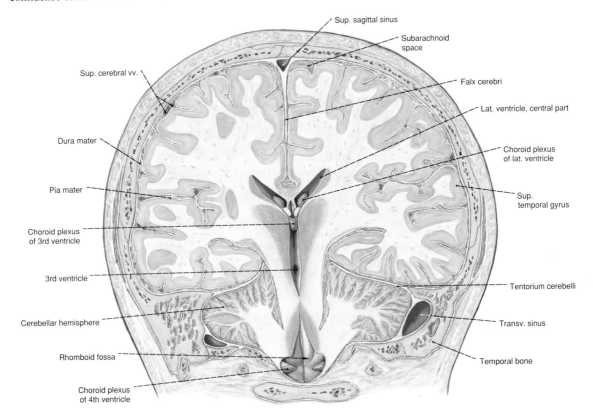

Sup. sagittal sinus

Subarachnoid space

Sup. cerebral vv.

Falx cerebri

Lat. ventricle, central part

Dura mater

Choroid plexus of lat. ventricle

Pia mater

Sup. temporal gyrus

Choroid plexus of 3rd ventricle

3rd ventricle

Tentorium cerebelli

Cerebellar hemisphere

Transv. sinus

Rhomboid fossa

Temporal bone

Choroid plexus of 4th ventricle

Fig. 586. Frontal section through the neurocranium at the level of the third ventricle.

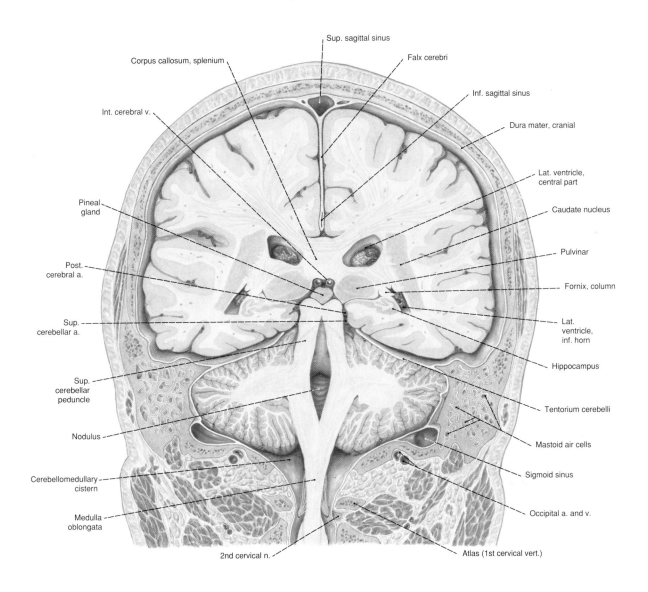

Sup. sagittal sinus

Corpus callosum, splenium

Falx cerebri

Inf. sagittal sinus

Int. cerebral v.

Dura mater, cranial

Pineal gland

Lat. ventricle, central part

Caudate nucleus

Post. cerebral a.

Pulvinar

Sup. cerebellar a.

Fornix, column

Sup. cerebellar peduncle

Lat. ventricle, inf. horn

Nodulus

Hippocampus

Cerebellomedullary cistern

Tentorium cerebelli

Medulla oblongata

Mastoid air cells

Sigmoid sinus

Occipital a. and v.

2nd cervical n.

Atlas (1st cervical vert.)

Fig. 587. Frontal section through the neurocranium at the level of the central part of the lateral ventricle, showing the dural septa, the cerebrum in the supratentorial space above the tentorium cerebelli, and the hindbrain (cerebellum, pons and medulla oblongata) in the infratentorial space. The dural septa (falx cerebri and tentorium cerebelli) prevent gross movements of the soft substance of the brain in transverse and vertical directions. Pressure differences may cause constriction of parts of the cerebrum in the tentorial incisure, such that the parahippocampal gyrus, the splenium of the corpus callosum, etc., may be affected and the cerebellar tonsils forced into the foramen magnum.

Note the triangular contour of the superior sagittal sinus. The dural sinuses are incompressible and noncontractile (there are no muscles in their walls). The sigmoid sinus is protected from compression by its position deep within the bony substance of the petrous portion of the temporal bone. (From PERN-KOPF: Atlas of Topographic and Applied Human Anatomy, Vol. 1, 3rd edition [W. PLATZER, Ed.], Urban & Schwarzenberg, Baltimore-Munich, 1989.)

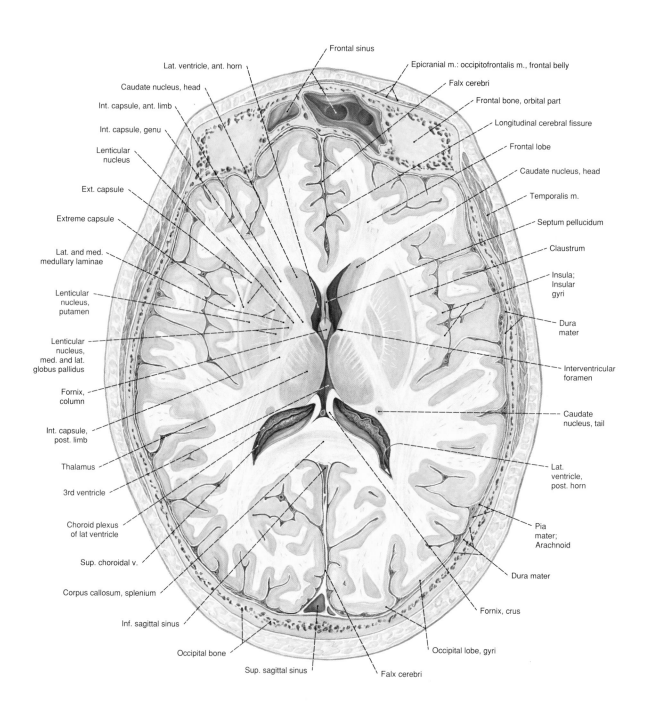

Frontal sinus

Lat. ventricle, ant. horn

Caudate nucleus, head

Int. capsule, ant. limb

Int. capsule, genu

Lenticular nucleus

Ext. capsule

Extreme capsule

Lat. and med. medullary laminae

Lenticular nucleus, putamen

Lenticular nucleus, med. and lat. globus pallidus

Fornix, column

Int. capsule, post. limb

Thalamus

3rd ventricle

Choroid plexus of lat ventricle

Sup. choroidal v.

Corpus callosum, splenium

Inf. sagittal sinus

Occipital bone

Sup. sagittal sinus

Epicranial m.: occipitofrontalis m., frontal belly

Falx cerebri

Frontal bone, orbital part

Longitudinal cerebral fissure

Frontal lobe

Caudate nucleus, head

Temporalis m.

Septum pellucidum

Claustrum

Insula; Insular gyri

Dura mater

Interventricular foramen

Caudate nucleus, tail

Lat. ventricle, post. horn

Pia mater; Arachnoid

Dura mater

Fornix, crus

Occipital lobe, gyri

Falx cerebri

Fig. 588. Horizontal section through the neurocranium at the level of the basal ganglia and the internal capsule.

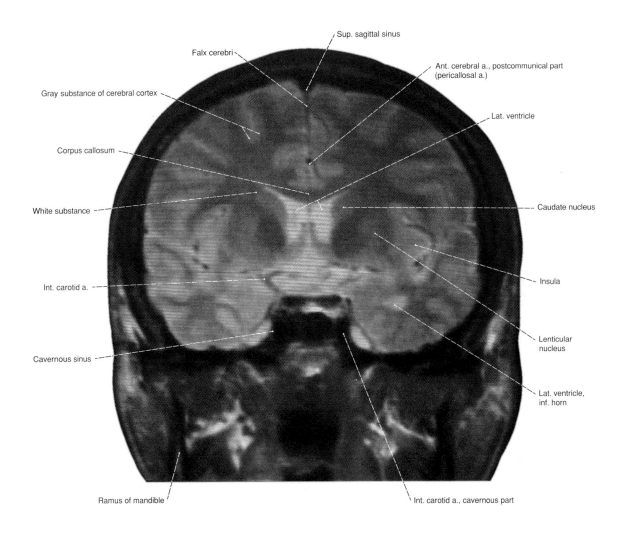

Sup. sagittal sinus

Falx cerebri

Ant. cerebral a., postcommunical part
(pericallosal a.)

Gray substance of cerebral cortex

Lat. ventricle

Corpus callosum

White substance

Caudate nucleus

Int. carotid a.

Insula

Cavernous sinus

Lenticular
nucleus

Lat. ventricle,
inf. horn

Ramus of mandible

Int. carotid a., cavernous part

Fig. 589. Magnetic resonance (MR) image. Frontal section
through the brain at the level of the caudate nucleus, the insula,
the lateral ventricle and the cavernous sinus. Note the distinct
difference between the gray and white substance of the brain.
(From: Dr. H. FRIEDBURG, Radiology Clinic of the University
of Freiburg i. Br.)

Refer to the block of text in Vol. II, page 146.

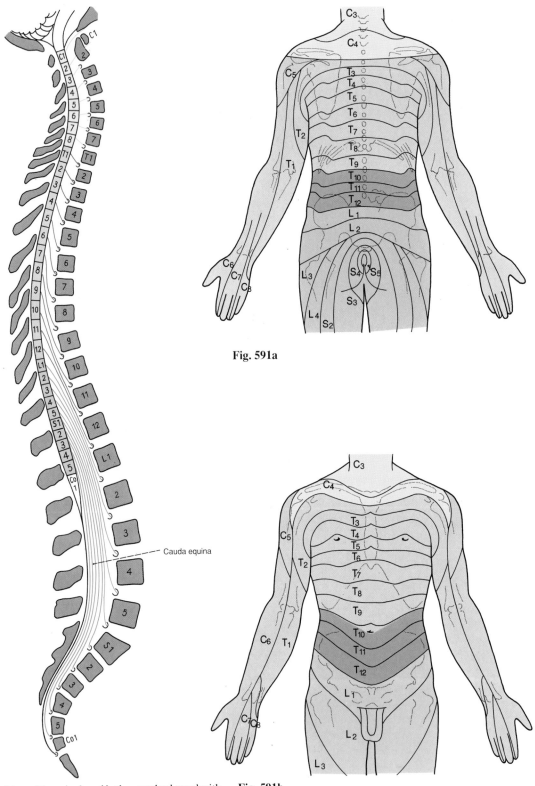

Fig. 591a

Fig. 591b

Fig. 590. Position of the spinal cord in the vertebral canal with the spinal cord segments and the segmental spinal nerves characterized by different colors. Note the relations of the spinal cord segments to the vertebral segments, the increasingly descending course of the segmental spinal nerve roots to their respective intervertebral foramina and the formation of the cauda equina. Cervical segments C 1-8 are yellow, thoracic segments T 1-12 are pink, lumbar segments L 1-5 and sacral segments S 1-5 are blue and the coccygeal segment is white.

Figs. 591a, b. Segmental distribution of the dermatomes on the ventral and dorsal aspect of the body. The letters and numerals refer to the corresponding segmental spinal nerves (altered schema of HANSEN and SCHLIACK).

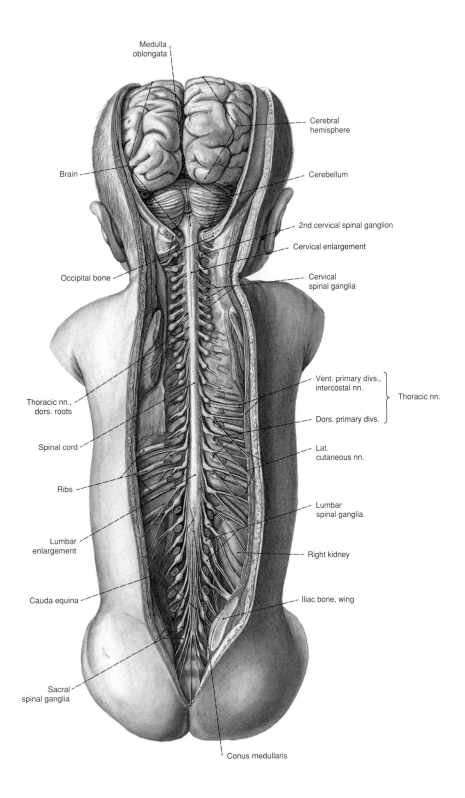

Medulla
oblongata

Cerebral
hemisphere

Brain

Cerebellum

2nd cervical spinal ganglion

Cervical enlargement

Occipital bone

Cervical
spinal ganglia

Vent. primary divs.,
intercostal nn.

Thoracic nn.

Dors. primary divs.

Thoracic nn.,
dors. roots

Spinal cord

Lat.
cutaneous nn.

Ribs

Lumbar
spinal ganglia

Lumbar
enlargement

Right kidney

Cauda equina

Iliac bone, wing

Sacral
spinal ganglia

Conus medullaris

Fig. 592. Central nervous system (brain and spinal cord) of
a newborn, dorsal exposure. The spinal dura mater is completely removed. Dissection of spinal (dorsal root) ganglia and
spinal nerves.

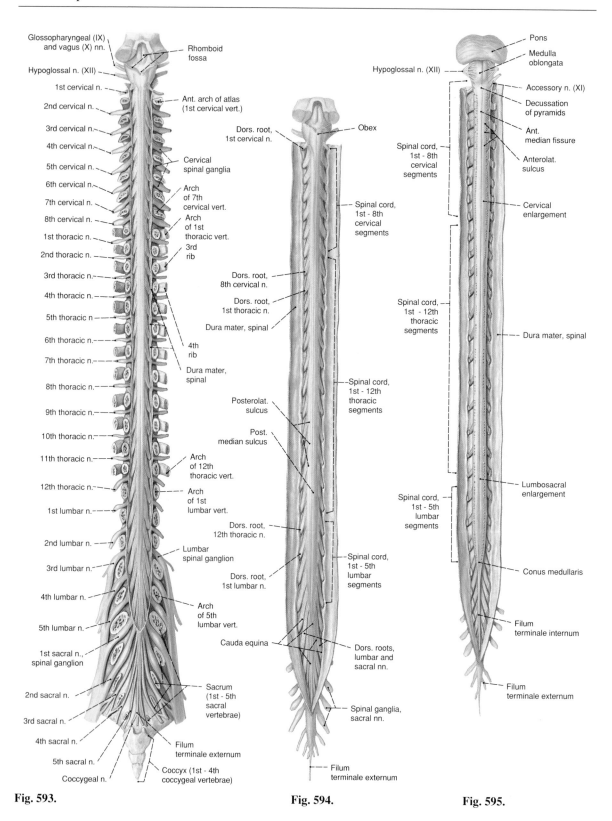

Fig. 593.

Fig. 594.

Fig. 595.

Fig. 593. Spinal cord within the vertebral canal, dorsal view. Vertebral arches and dura mater have been extensively removed.

Fig. 594. Spinal cord and rootlets (fila radicularia) of spinal nerves in a dorsal view. The dura mater has been split longitudinally and reflected.

Fig. 595. Spinal cord in a ventral view. The dura mater has been split longitudinally. The ventral roots of the spinal nerves have been severed from the spinal cord at their origins; the dorsal roots are visible.

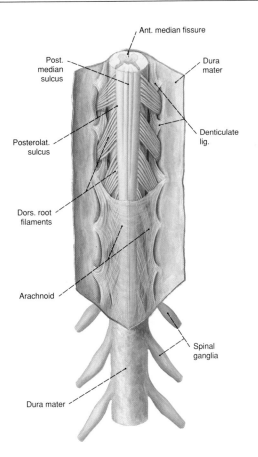

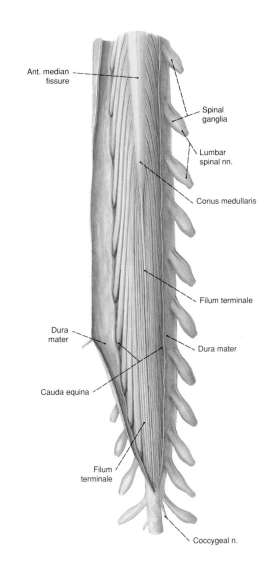

Fig. 596. Spinal cord and meninges, dorsal view. In the lower third the dura mater remains unopened; in the upper third the arachnoid has been removed.

Fig. 597. Caudal portion of the spinal cord, ventral view. The ▶ dura mater has been opened longitudinally. Cauda equina = ventral and dorsal roots of the lumbar and sacral spinal nerves.

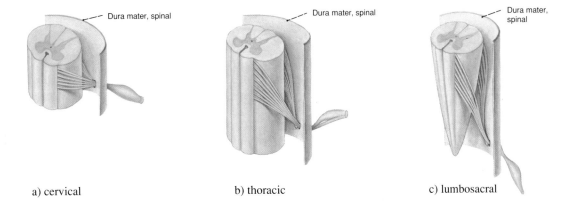

a) cervical   b) thoracic   c) lumbosacral

Figs. 598a-c. The relationship of the ventral (anterior, motor) and dorsal (posterior, sensory) spinal nerve roots to the dura mater at various levels.

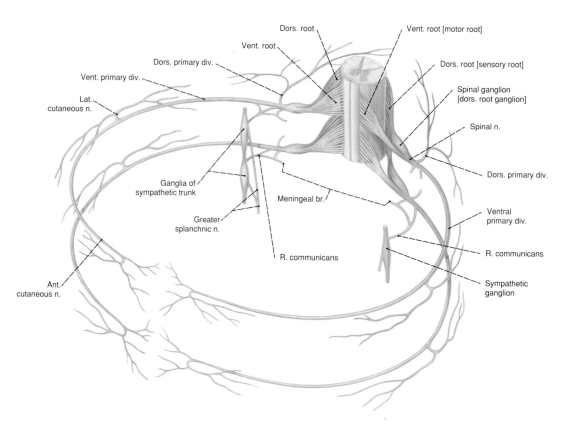

Fig. 599.   Representation of two thoracic spinal cord segments with corresponding segmental spinal nerves and their branches.

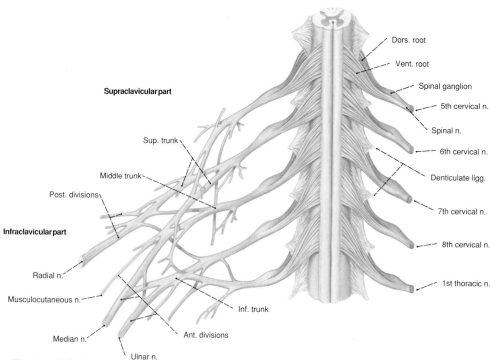

Fig. 600.   The brachial plexus and corresponding segments of the cervical and thoracic spinal cord in a ventral view (cf. Fig. 283).

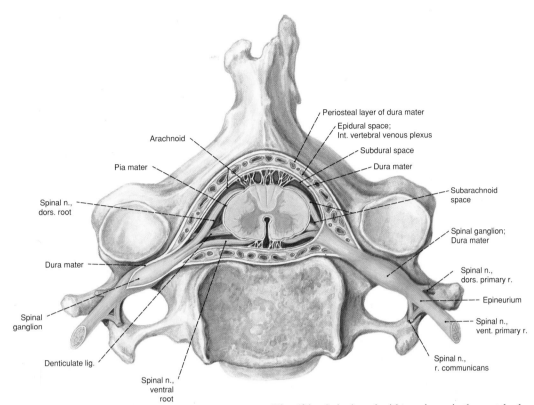

Periosteal layer of dura mater

Epidural space;
Int. vertebral venous plexus

Subdural space

Dura mater

Arachnoid

Pia mater

Subarachnoid
space

Spinal n.,
dors. root

Spinal ganglion;
Dura mater

Spinal n.,
dors. primary r.

Dura mater

Epineurium

Spinal n.,
vent. primary r.

Spinal
ganglion

Spinal n.,
r. communicans

Denticulate lig.

Spinal n.,
ventral
root

Fig. 601. Spinal cord with meninges in the vertebral canal, cervical level, in transverse section. On the left side of the figure, the dura mater has been opened to expose the spinal (dorsal root) ganglion. The dura mater has been tinted yellow and the periosteum of the vertebral canal white.

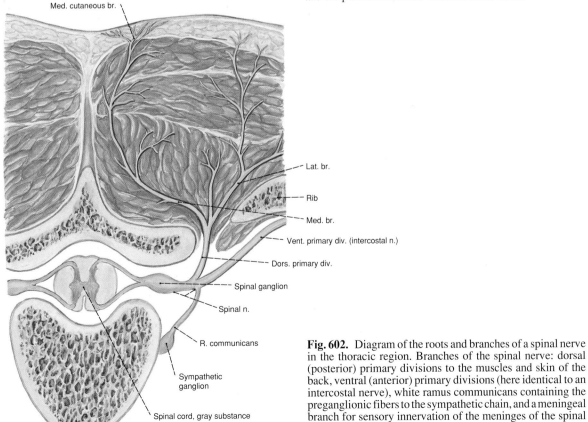

Med. cutaneous br.

Lat. br.

Rib

Med. br.

Vent. primary div. (intercostal n.)

Dors. primary div.

Spinal ganglion

Spinal n.

R. communicans

Sympathetic
ganglion

Spinal cord, gray substance

Fig. 602. Diagram of the roots and branches of a spinal nerve in the thoracic region. Branches of the spinal nerve: dorsal (posterior) primary divisions to the muscles and skin of the back, ventral (anterior) primary divisions (here identical to an intercostal nerve), white ramus communicans containing the preganglionic fibers to the sympathetic chain, and a meningeal branch for sensory innervation of the meninges of the spinal cord (cf. Fig. 599).

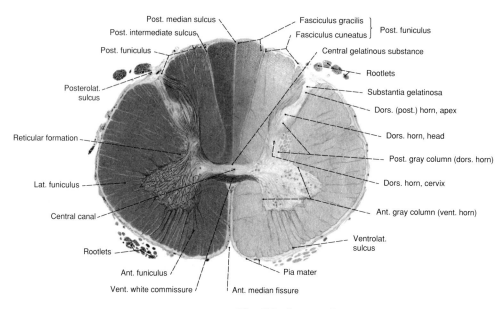

Post. median sulcus

Post. intermediate sulcus

Post. funiculus

Posterolat. sulcus

Reticular formation

Lat. funiculus

Central canal

Rootlets

Ant. funiculus

Vent. white commissure

Fasciculus gracilis

Fasciculus cuneatus } Post. funiculus

Central gelatinous substance

Rootlets

Substantia gelatinosa

Dors. (post.) horn, apex

Dors. horn, head

Post. gray column (dors. horn)

Dors. horn, cervix

Ant. gray column (vent. horn)

Ventrolat. sulcus

Pia mater

Ant. median fissure

Fig. 603. Cross section through the spinal cord at the cervical level (silver technique).

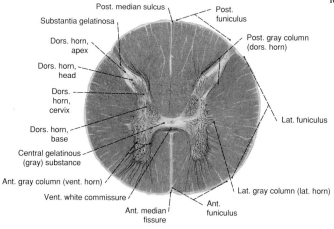

Post. median sulcus

Substantia gelatinosa

Dors. horn, apex

Dors. horn, head

Dors. horn, cervix

Dors. horn, base

Central gelatinous (gray) substance

Ant. gray column (vent. horn)

Vent. white commissure

Ant. median fissure

Post. funiculus

Post. gray column (dors. horn)

Lat. funiculus

Lat. gray column (lat. horn)

Ant. funiculus

Fig. 604. Cross section through the spinal cord at the thoracic level.

Post. median sulcus

Dors. root of spinal n.

Post. gray column (dors. horn)

Central gelatinous substance

Ant. median fissure

Fig. 605. Cross section through the spinal cord at the level of the lumbar enlargement.

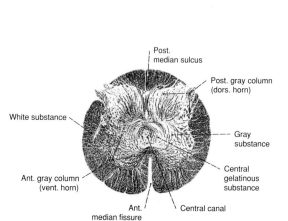

Post. median sulcus

Post. gray column (dors. horn)

White substance

Gray substance

Central gelatinous substance

Ant. gray column (vent. horn)

Ant. median fissure

Central canal

◀ **Fig. 606.** Cross section through the spinal cord at the level of the conus medullaris.

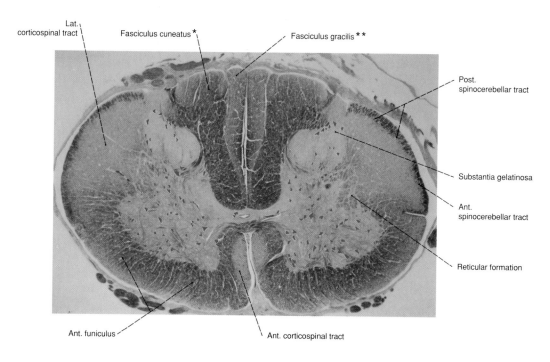

Lat. corticospinal tract

Fasciculus cuneatus *

Fasciculus gracilis * *

Post. spinocerebellar tract

Substantia gelatinosa

Ant. spinocerebellar tract

Reticular formation

Ant. funiculus

Ant. corticospinal tract

Fig. 607. Cervical spinal cord of a newborn, WEIGERT'S myelin stain (nuclear areas yellow, tracts brown-black). The lateral and anterior corticospinal (pyramidal) tracts are not yet myelinated and therefore appear as lighter areas. Example of myelinogenesis. (From: Dr. H. FERNER, Vienna.)

* BURDACH's column
** GOLL's column

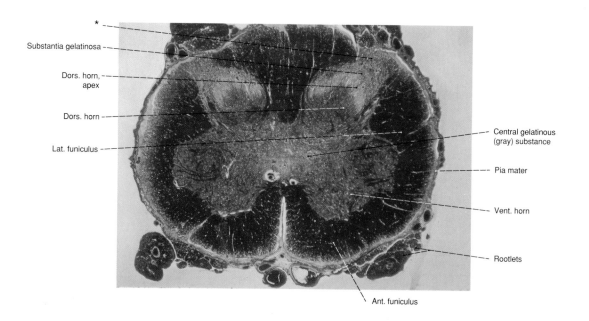

*

Substantia gelatinosa

Dors. horn, apex

Dors. horn

Lat. funiculus

Central gelatinous (gray) substance

Pia mater

Vent. horn

Rootlets

Ant. funiculus

Fig. 608. Cross section of an adult spinal cord at the level of the lumbar enlargement. WEIGERT's myelin stain. (From: Dr. H. FERNER, Vienna.)

* Zona spongiosa

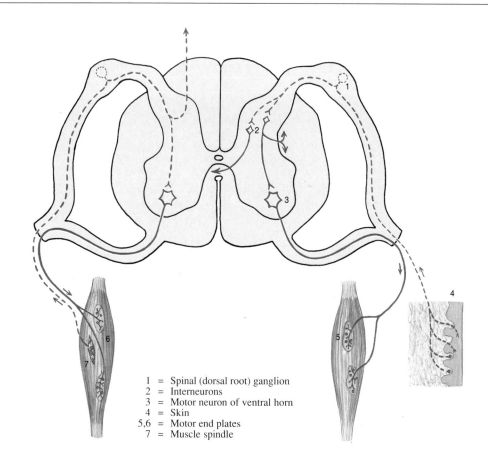

1 = Spinal (dorsal root) ganglion
2 = Interneurons
3 = Motor neuron of ventral horn
4 = Skin
5,6 = Motor end plates
7 = Muscle spindle

Fig. 609

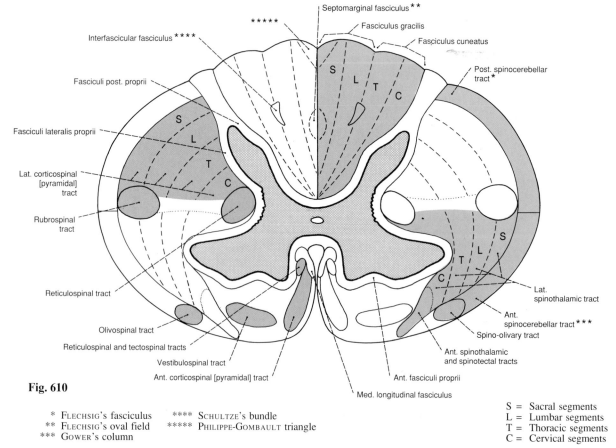

Septomarginal fasciculus **
Fasciculus gracilis
Fasciculus cuneatus
Interfascicular fasciculus ****
*****
Post. spinocerebellar tract *
Fasciculi post. proprii
Fasciculi lateralis proprii
Lat. corticospinal [pyramidal] tract
Rubrospinal tract
Reticulospinal tract
Olivospinal tract
Reticulospinal and tectospinal tracts
Vestibulospinal tract
Ant. corticospinal [pyramidal] tract
Med. longitudinal fasciculus
Ant. fasciculi proprii
Ant. spinothalamic and spinotectal tracts
Spino-olivary tract
Ant. spinocerebellar tract ***
Lat. spinothalamic tract

Fig. 610

   * Flechsig's fasciculus   **** Schultze's bundle
  ** Flechsig's oval field   ***** Philippe-Gombault triangle
 *** Gower's column

S = Sacral segments
L = Lumbar segments
T = Thoracic segments
C = Cervical segments

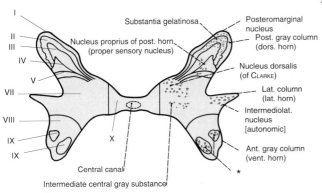

I
II
III
IV
V
VII
VIII
IX
IX

Substantia gelatinosa
Nucleus proprius of post. horn
(proper sensory nucleus)

Posteromarginal
nucleus
Post. gray column
(dors. horn)

Nucleus dorsalis
(of CLARKE)

Lat. column
(lat. horn)

Intermediolat.
nucleus
[autonomic]

Ant. gray column
(vent. horn)

X

*

Central canal
Intermediate central gray substance

Fig. 611. Cell groups of the gray substance of the spinal cord and REXED's laminae. The section corresponds to the 10th thoracic segment. According to REXED (1952), the gray substance of the spinal cord can be divided into 10 regions (laminae I-X). The first 8 form sequential layers extending from dorsal to ventral, roughly paralleling the dorsal and ventral surfaces of the spinal cord. They are not organized segmentally, but extend continuously throughout the length of the spinal cord such that their size varies according to level. Lamina IX is not a separate layer, but instead surrounds the region of the motor neurons in the ventral (anterior) horn. Lamina X is the gray substance surrounding the central canal. Lamina VI is not indicated since it is only well-formed in several cervical and lumbar segments. (From W. ZENKER: The Internal Structure of the Spinal Cord. In BENNINGHOFF: Anatomy, Vol. 3, 13th/14th edition [W. ZENKER, Ed.], Urban & Schwarzenberg, Munich-Vienna-Baltimore, 1985.)

* Motor neurons of the ventral horn
I-X REXED's laminae

Fig. 609. Diagram of a monosynaptic reflex arc (left) and a polysynaptic reflex arc (right) of the spinal cord. The broken line indicates the afferent limb, and the solid line the efferent limb. Included among the basic mechanisms of the spinal cord are: 1. the direct bineuronal monosynaptic proprioceptive reflex (e.g., knee jerk reflex); 2. the indirect multineuronal polysynaptic reflex in which several neurons of the ipsi- or contralateral side participate (e.g., plantar, cremasteric, abdominal, and mucosal reflexes).

Fig. 610. Position of the individual tracts within the white substance of the spinal cord. In this diagram of a "typical" cross section, the locations of the bulk of the fibers in any individual tract are indicated. On the left side of the figure, the motor efferent tracts (red) and on the right, the sensory afferents (blue) are illustrated. The broken lines indicate the segmental arrangement of the various fiber systems. In the posterior funiculus, note that the positions of the interfascicular fasciculus (= SCHULTZE's bundle), the septomarginal fasciculus (= FLECHSIG's oval field) and the PHILIPPE-GOMBAULT triangle are indicated, despite their occurrence only in specific segments. (From BENNINGHOFF: Anatomy, Vol. 3, 13th/14th edition [W. ZENKER, Ed.], Urban & Schwarzenberg, Munich-Vienna-Baltimore, 1985.)

Fig. 612. The thoracic intercostal nerves (T_1 - T_{12}) descending ▶ ventrally to innervate the body wall (skin and muscles of the thoracic and abdominal walls). The tenth and eleventh intercostal nerves reach the navel (umbilicus). On the left half of the body, the segmental sensory supply of the viscera is indicated. When the viscera are diseased, hyperesthesia and hyperalgesia frequently occur in the corresponding segmental cutaneous areas. This occurs because the dorsal afferent roots of the spinal nerves contain not only fibers from the dermatomes but also visceral afferent fibers from the corresponding visceral regions. The hypersensitive and hyperalgesic cutaneous regions are known as HEAD's zones (HEAD, English neurologist, 1861-1940). These are important for diagnosis of diseases of the viscera.

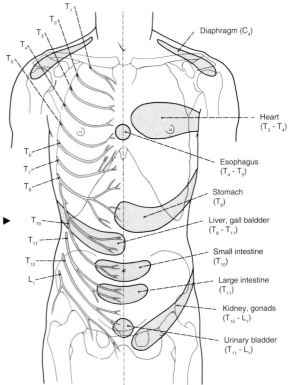

Diaphragm (C₄)

Heart
(T_3 - T_4)

Esophagus
(T_4 - T_5)

Stomach
(T_8)

Liver, gall bladder
(T_8 - T_{11})

Small intestine
(T_{10})

Large intestine
(T_{11})

Kidney, gonads
(T_{10} - L_1)

Urinary bladder
(T_{11} - L_1)

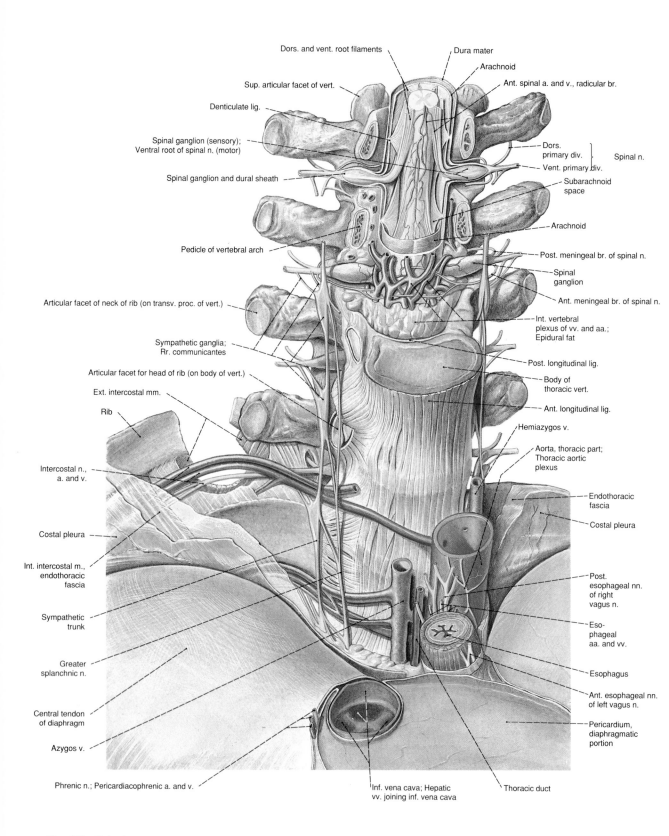

Dors. and vent. root filaments

Dura mater

Arachnoid

Sup. articular facet of vert.

Ant. spinal a. and v., radicular br.

Denticulate lig.

Spinal ganglion (sensory);
Ventral root of spinal n. (motor)

Dors. primary div.

Spinal n.

Vent. primary div.

Spinal ganglion and dural sheath

Subarachnoid space

Arachnoid

Pedicle of vertebral arch

Post. meningeal br. of spinal n.

Spinal ganglion

Ant. meningeal br. of spinal n.

Articular facet of neck of rib (on transv. proc. of vert.)

Int. vertebral plexus of vv. and aa.; Epidural fat

Sympathetic ganglia; Rr. communicantes

Post. longitudinal lig.

Articular facet for head of rib (on body of vert.)

Body of thoracic vert.

Ext. intercostal mm.

Ant. longitudinal lig.

Rib

Hemiazygos v.

Aorta, thoracic part; Thoracic aortic plexus

Intercostal n., a. and v.

Endothoracic fascia

Costal pleura

Costal pleura

Int. intercostal m., endothoracic fascia

Post. esophageal nn. of right vagus n.

Sympathetic trunk

Eso- phageal aa. and vv.

Greater splanchnic n.

Esophagus

Ant. esophageal nn. of left vagus n.

Central tendon of diaphragm

Pericardium, diaphragmatic portion

Azygos v.

Phrenic n.; Pericardiacophrenic a. and v.

Inf. vena cava; Hepatic vv. joining inf. vena cava

Thoracic duct

Fig. 613. Spinal cord, spinal meninges and the other components of the vertebral canal in the lower thoracic region. View from ventral and above.

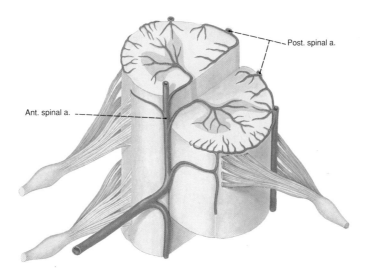

Fig. 614. Arterial supply of the spinal cord. (From K. Piscol: Blood Supply of the Spinal Cord and Its Clinical Relevance. Neurology Series, Vol. 8, Springer, Munich-Berlin-New York, 1972.)

A **spinal segment** (or **neuromere**) is that part of the cord which gives rise to each pair of spinal nerves. Each segment corresponds to the rootlets of one spinal nerve. There are 8 cervical, 12 thoracic, 5 lumbar, 5 sacral and 1 coccygeal pairs of spinal nerves, which are generally referred to numerically (C_1 - C_8, T_1 - T_{12}, L_1 - L_5, S_1 - S_5 and Co_1). The coccygeal spinal segment is situated at the level of the second lumbar vertebra.

In contrast, the **vertebral segments** correspond to the vertebral bodies. Because the adult spinal cord reaches only to the second lumbar vertebra, the spinal segments do not correspond topically to the vertebral segments. This positional discrepancy increases as one progresses caudally. Thus, the eighth thoracic vertebra corresponds to the tenth thoracic spinal segment and the fifth sacral spinal segment to the position of the first lumbar vertebra. The nerve roots of a spinal segment must, therefore, increasingly lengthen as they descend caudally to their segmentally appropriate intervertebral foramina.

The **cauda equina**. The dural and arachnoidal sac does not terminate with the spinal cord, but instead extends as the filum terminale externum (dural) and as the filum terminale internum (pial) to the second sacral vertebra. The descending ventral (auterior) and dorsal (posterior) roots of lumbar and sacral segments lie within this sac and are bathed in cerebrospinal fluid. This collection spinal nerve roots extending beyond the termination of the neural substance of the cord is called the cauda equina.

The peripheral segments are designated according to the body region that is supplied by a particular spinal nerve, as follows:

1. The area of skin supplied by one spinal nerve = **dermatome**. Dermatomes are band-shaped areas of the skin, most easily identifiable in the trunk region. There the course of the ventral primary rami (divisions) of the thoracic nerves (intercostal nerves) corresponds to the dermatomes because these ventral roots do not form a plexus.

2. A group of skeletal muscles innervated by one spinal nerve = **myotome**.

3. The viscera innervated by one spinal nerve = **enterotome**. Most skeletal muscles are innervated by several spinal nerves; only an occasional few situated in the center of the myotome are supplied by a single spinal nerve; e.g., the diaphragm by the third and fourth cervical nerves, the deltoid muscle by the fifth cervical nerve, the extensor hallucis longus muscle by the fifth lumbar nerve.

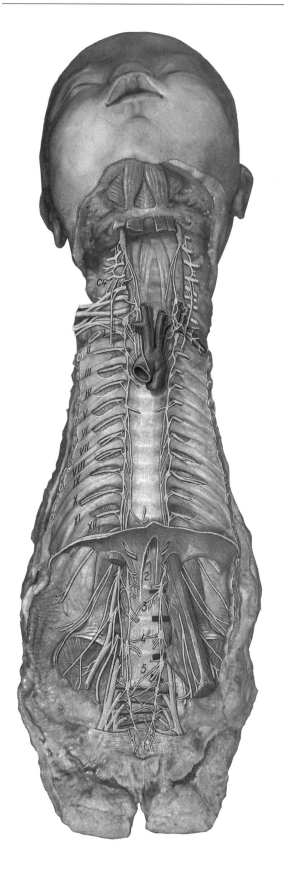

The **sympathetic trunk** is a paired neural chain extending on both sides of the vertebral column from the base of the skull to the coccyx. It is characterized by nodular enlargements, the central or chain ganglia. At the base of the skull it continues cephalad as a sympathetic fiber network, the internal carotid plexus, around the internal carotid artery and its branches and establishes connections with the cranial nerves. Sympathetic nerve fibers reach the orbit and the eye ball via the internal carotid and ophthalmic arteries.

The sympathetic trunk is located in the neck behind the prevertebral fascia on the longus colli muscle. In the thoracic region it is covered by the parietal (costal) pleura and in the abdomen and pelvis by the parietal peritoneum.

The **cervical portion** of the sympathetic trunk has three ganglia, the superior, middle and inferior cervical ganglia. The superior ganglion is about 2 cm long and lies at the level of the atlas and axis (1st and 2nd cervical vertebrae, respectively). From its inferior pole originates the superior cardiac nerve which courses along the common carotid artery to the heart. The middle cervical ganglion lies in the area of the transverse portion of the inferior thyroid artery and is variable in size, form, position and occurrence. It gives rise to the largest of the three cervical cardiac nerves, the middle cardiac nerve. The inferior cervical ganglion is situated at the level of the superior thoracic aperture behind the prevertebral part of the vertebral artery and sends fibers to the heart via the inferior cardiac nerve. (The branches of the vagus nerve [X] to the heart are named cardiac **branches**). Between the middle and inferior cervical ganglia, the sympathetic trunk forms a loop around the subclavian artery, the ansa subclavia (of VIEUSSENS). In most cases, the inferior cervical ganglion is fused with the first thoracic ganglion to form a large common ganglion, the stellate ganglion, on the anterior surface of the head of the first rib. The **12 thoracic sympathetic ganglia** lie near the head of the ribs and are interconnected by interganglionic branches. They are connected to the segmental spinal nerves by one or more rami communicantes. Additionally, the **greater splanchnic nerve** is formed by medial branches from the 5th to the 10th thoracic ganglia, and the **lesser splanchnic nerve** from the 10th to the 12th ganglia. The splanchnic nerves pierce the crus of the diaphragm and end in the abdominal cavity in the prevertebral **celiac ganglion**. Direct branches reach the aorta and the bronchial tree.

In the abdominal and pelvic areas, the sympathetic trunk lies in the groove between the vertebral bodies and the psoas major muscle. The lumbar and sacral ganglia are not in a precise segmental arrangement. The lumbar and sacral splanchnic branches reach the prevertebral ganglia of the aorta and its branches. At the level of the coccyx the two sympathetic trunks unite and end with a common ganglion impar (Figs. 616, 617).

Fig. 615a. The sympathetic trunk in a newborn. On the right side of the dissection the brachial and lumbosacral plexuses are exposed; on the left side, the sympathetic trunk and its ganglia at lumbar and sacral levels are displayed upon black markers.

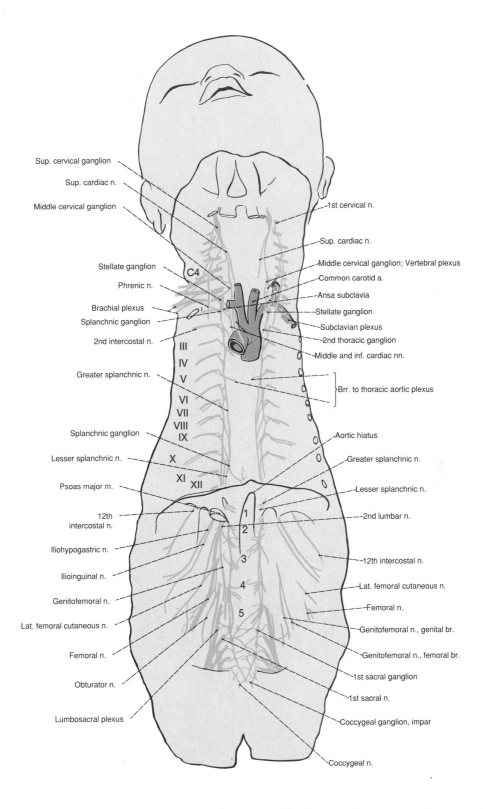

Sup. cervical ganglion

Sup. cardiac n.

Middle cervical ganglion

1st cervical n.

Sup. cardiac n.

Middle cervical ganglion; Vertebral plexus

Stellate ganglion

Common carotid a.

Phrenic n.

Ansa subclavia

Brachial plexus

Stellate ganglion

Splanchnic ganglion

Subclavian plexus

2nd intercostal n.

2nd thoracic ganglion

Middle and inf. cardiac nn.

Greater splanchnic n.

Brr. to thoracic aortic plexus

Splanchnic ganglion

Aortic hiatus

Lesser splanchnic n.

Greater splanchnic n.

Psoas major m.

Lesser splanchnic n.

12th intercostal n.

2nd lumbar n.

Iliohypogastric n.

12th intercostal n.

Ilioinguinal n.

Lat. femoral cutaneous n.

Genitofemoral n.

Femoral n.

Lat. femoral cutaneous n.

Genitofemoral n., genital br.

Femoral n.

Genitofemoral n., femoral br.

Obturator n.

1st sacral ganglion

1st sacral n.

Lumbosacral plexus

Coccygeal ganglion, impar

Coccygeal n.

C4

III
IV
V
VI
VII
VIII
IX
X
XI
XII

1
2
3
4
5

Fig. 615b. The sympathetic trunk and its connections with the spinal nerves of a newborn. Explanatory diagram for Fig. 615a.

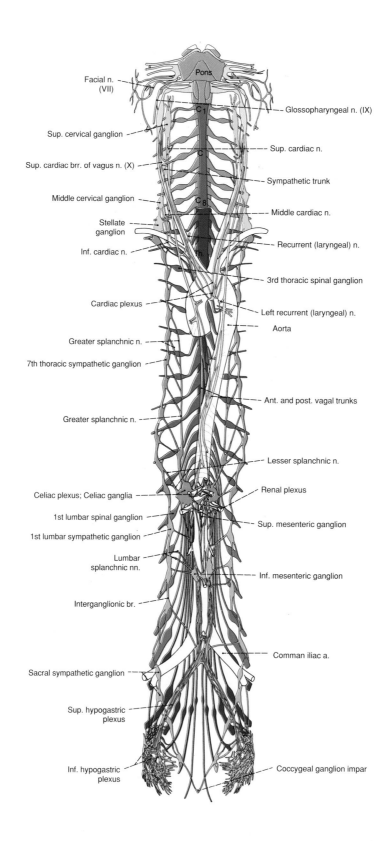

Fig. 616. The sympathetic trunk and its relationship to the cranial and spinal nerves.

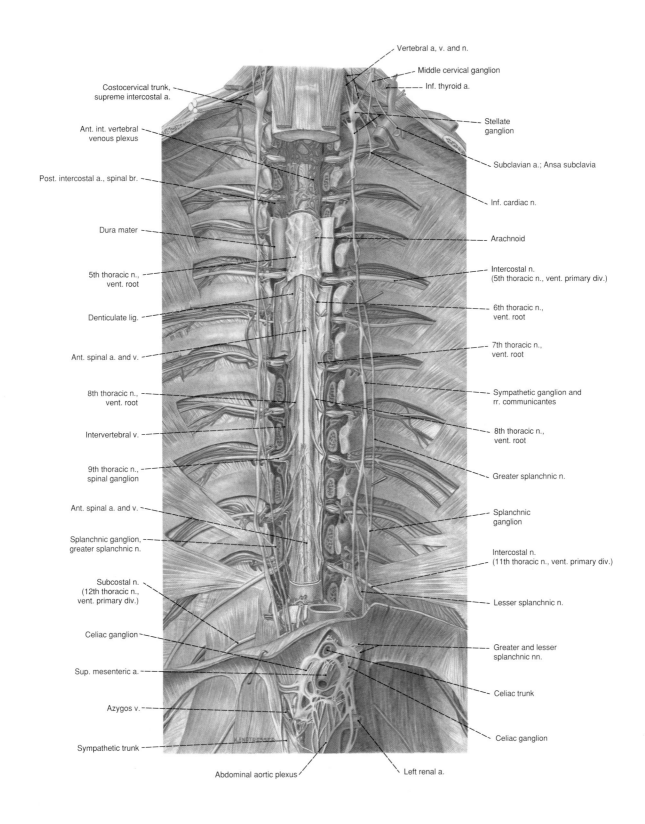

Fig. 617. The thoracic vertebral canal opened anteriorly. The dura mater has been cut open to expose the spinal nerves, their roots, primary divisions (rami) and spinal (dorsal root) ganglia, and their connections (rami communicantes) with the sympathetic trunk. (From PERNKOPF: Atlas of Topographic and Applied Human Anatomy, Vol. 2, 3rd edition [H. FERNER, Ed.], Urban & Schwarzenberg, Baltimore-Munich, 1989.)

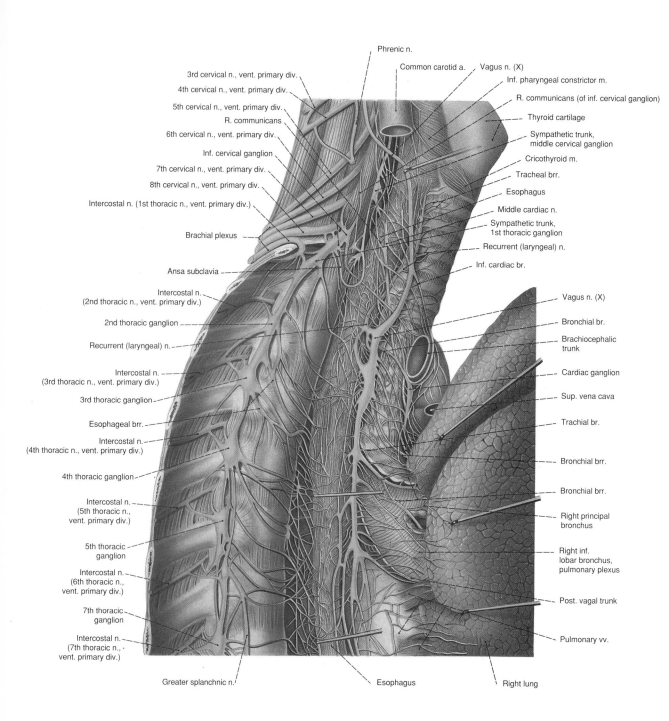

Fig. 618. Lower cervical and upper thoracic parts of the autonomic nervous system (from Braeucker). The superficial cervical muscles and blood vessels have been removed, the vagal trunk retracted somewhat anteriorly, and the thoracic wall removed to the posterior parts of the ribs and vertebral column. The right lung has been extirpated to expose the entire length of the esophagus. The great vessels up to the origin of the brachiocephalic trunk and a portion of the superior vena cava have been resected.

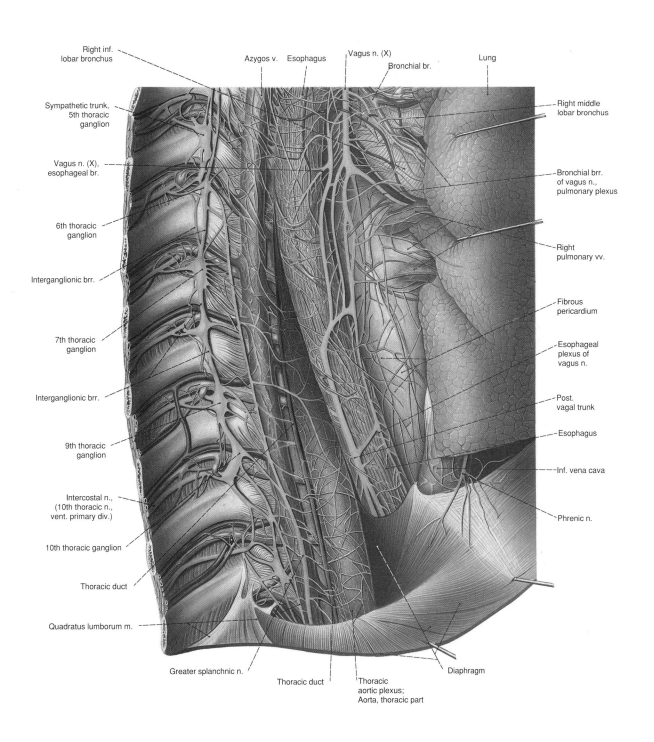

Right inf. lobar bronchus

Sympathetic trunk, 5th thoracic ganglion

Vagus n. (X), esophageal br.

6th thoracic ganglion

Interganglionic brr.

7th thoracic ganglion

Interganglionic brr.

9th thoracic ganglion

Intercostal n., (10th thoracic n., vent. primary div.)

10th thoracic ganglion

Thoracic duct

Quadratus lumborum m.

Azygos v.   Esophagus   Vagus n. (X)   Bronchial br.   Lung

Right middle lobar bronchus

Bronchial brr. of vagus n., pulmonary plexus

Right pulmonary vv.

Fibrous pericardium

Esophageal plexus of vagus n.

Post. vagal trunk

Esophagus

Inf. vena cava

Phrenic n.

Greater splanchnic n.   Thoracic duct   Thoracic aortic plexus; Aorta, thoracic part   Diaphragm

Fig. 619. Lower thoracic portions of the autonomic nervous system (from BRAEUCKER). This figure is a direct caudal continuation of Fig. 618. The only difference between this dissection and that of Fig. 618 is that the azygos vein, the thoracic aorta and the thoracic duct have been preserved.

Fig. 620. Arch, no triradius.

Fig. 621. Loop, one triradius.

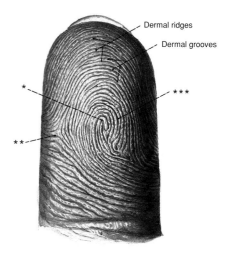

Fig. 622. Whorl, two triradii.

Figs. 620-622. Basic forms of papillary ridge patterns.

Fig. 625. The palmar surface of the finger tip. The grooves and ridges of the papillary layer of the dermis form patterns on the fingers and toes that are determined by the number of triradii and the number of papillary ridges between the center of the pattern and the nearest triradius.

   * Central point of pattern
  ** Triradius
*** Sweat glands = sudoriferous glands

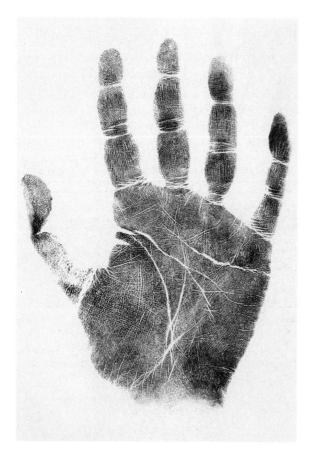

Fig. 623. Imprint of a right hand.

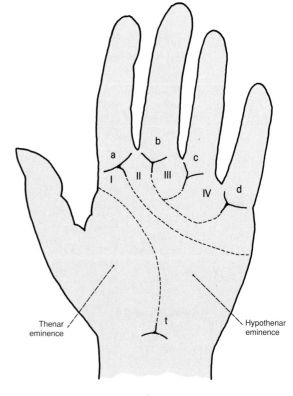

Fig. 624. Diagram of the skin crease patterns of the hand, of the compartments of the palm (thenar, hypothenar, interdigital spaces I-IV), of the palmar digital triradii (a-d) and axial triradius (t).

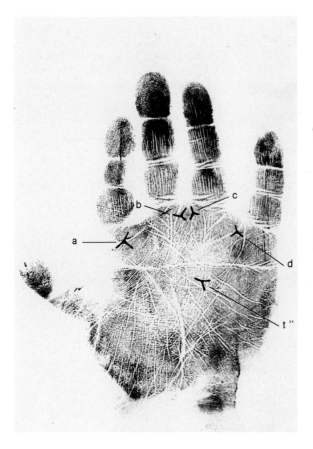

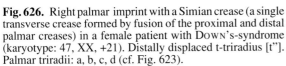

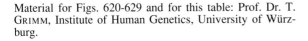

Fig. 627. Normal palmar creases.

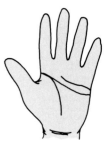

Fig. 628. SYDNEY crease.

Fig. 629. Simian crease.

Fig. 626. Right palmar imprint with a Simian crease (a single transverse crease formed by fusion of the proximal and distal palmar creases) in a female patient with DOWN's-syndrome (karyotype: 47, XX, +21). Distally displaced t-triradius [t"]. Palmar triradii: a, b, c, d (cf. Fig. 623).

Material for Figs. 620-629 and for this table: Prof. Dr. T. GRIMM, Institute of Human Genetics, University of Würzburg.

Frequency of Simian and SYDNEY creases

	DOWN's Syndrome	Normal
Simian crease	58.8%	5.5%
SYDNEY crease	21.9%	8.1%

The skin of the palmar surface of the hand and fingers and of the plantar surface of the foot and toes differs from the rest of the skin primarily in that it possesses no hairs or sebaceous glands, but is crenated with papillary grooves and ridges. On the papillary ridges there are numerous openings of sweat (sudoriferous) glands (Fig. 625). The ridges originate from tactile elevations during the third month of gestation. The papillary pattern (Figs. 620-622) and the number of ridges remain constant for life. Because of the very large number of possible pattern combinations it is very unlikely that two humans will ever have the same pattern. Therefore, fingerprinting can establish individual identities. On the tips of the fingers and toes, three main types of patterns may be distinguished: arches, loops and whorls (Figs. 620-622), wherein the number of ridges between the central point of a pattern and the nearest triradius determines the quantitative value of the pattern (Fig. 625). In chromosomal abnormalities (e.g., DOWN's syndrome: 47, XY, +21 or 47, XX, +21 or TURNER's syndrome: 45, XO) there are typical alterations in the distribution frequency of patterns (see the above table). The total finger ridge count (TFRC) of an individual correlates with the number of X- and Y-chromosomes.

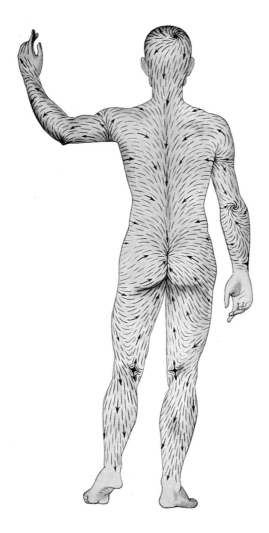

Fig. 630a. Dorsal view of the hair patterns (streams, whorls, crosses).

The extent of body hair varies considerably among the different races of mankind. There is a process of hair regression which, by examination of the hairiness of the middle phalanx, can be followed from the "old races" such as the Ainu, the Australian aborigines and Caucasians to the "younger races" such as the Mongoloid and the Negroid peoples.

The anthropogenetic development of the furry covering is an important step for homeothermy. The hairs of the epidermis trap a standing layer of air which reduces heat loss through convection. The basis for the evolutionary decrease in hirsuteness may be found in the confinement of human activity to the day-time hours (hunting in the heat of the day) and the use of fire and of skin covering. Likewise, hypotrichosis can be a natural selection advantage in the case of an ectoparasitic condition. The regressive change in hair covering is not comparable to individual balding

which is caused by degeneration of the hair follicles. Those hirsute individuals (e.g., congenital hypertrichosis lanuginosa congenita) who show excessive hairiness over the entire body provide a glimpse into our genetic history. An autosomal dominant disorder of the responsible gene complex is the basis for their lack of hair regression.

The **axillary hair region** in Caucasians is rich in .apocrine glands. Their initially odorless secretion devel-ops its olfactory stimulatory effect after oxidation. Dense axillary hairiness supports odor intensification. Negroids have slight axillary hairiness, Mongoloids none at all. The secretion itself demonstrates ethnic differences. In Negroids a musk-like fragrance is detectable which is due to the content of acetylcholinesterase. The secretion of Pygmies seems to repel insects. Related to the lack of hairiness, the axillary region of Mongoloids has no apocrine glands, resulting in minimal body odor.

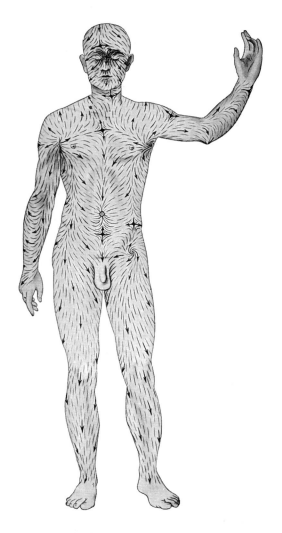

Fig. 630b. Ventral view of the hair patterns (streams, whorls, crosses).

Pubic hair, which, like axillary hair, signals puberty, begins to appear at 6 to 10 years of age. Both types of hair are under complex hormonal control. For the rest of the body, masculine terminal hair follows the pattern of hairlines and whorls established embryonically/fetally. Females can retain the fine down-like hair (langugo) of infancy; when this is replaced by dark terminal hair hirsutism occurs. Body hairs vary in length between 3 and 60 millimeters. Among the heavily haired populations are the Ainu, followed by the Azerbaijanis, North Indians, Europeans, Australian aborigines and the Papuans. Reduction of the human hairy covering is most advanced on the back. The thick-ness of human thoracic hair increases with sexual maturity and is maximal between ages 40 and 60. Only at about 50 years of age do gray thoracic hairs appear. In those older than 90 years, thoracic hairs contain no pigment whatsoever, even if the scalp hairs are not yet completely gray.

Regression of human hair covering is especially evident on the middle phalanx of the fingers and toes. While the proximal phalanges are still predominantly covered with hair, the evolutionary process is most advanced in the distal phalanges. The middle phalanges demonstrate the greatest variability, genetically as well as environmentally conditioned. It is presumed that this evolutionary regressive trend will not continue. The pressure of natural selection to reduce body hairiness has decreased worldwide and is closely related to technological development (clothing, heating, cooling). (Cf. L.K. Ekkes: Body Hairiness: Atavistic Relic? Der Hautarzt 38: 125-130, 1987.)

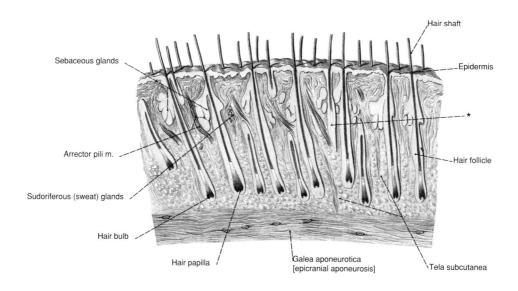

Fig. 631. Vertical section through the scalp (microscopic enlargement).

* Club hair

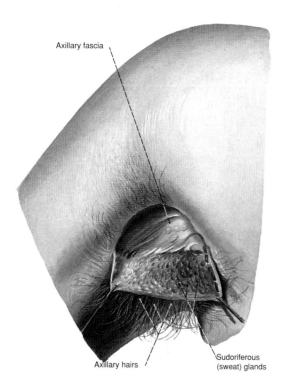

Fig. 632. Merocrine sweat glands of the axilla.

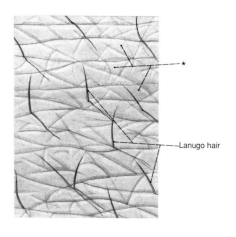

Fig. 633. Lanugo (fetal) hairs of the human skin.

* Epidermal scales

The Figures

For didactic purposes, contrasts have been intensified in the multicolored figures to enhance the recognition of areas that are difficult to differentiate. Thus, the colors used for various tissues (e.g., tendons, cartilage, bones, muscles), vessels (arteries, veins, lymphatics), and nerves differ from those found in living or dead bodies or in a preserved cadaver (e.g., arteries are colored red, veins are blue, nerves are yellow, lymphatic vessels and lymph nodes are colored green).

Together with Prof. SOBOTTA and the subsequent editors, Profs. BECHER and FERNER, the artists who created the basic collection of illustrations were: K. HAJEK, Prof. E. LEPIER, H. v. EICKSTEDT, J. KOSANKE. Others contributed to subsequent editions, the artists of the present edition are: Ms. Ulrike BRUGGER, Munich; Mr. Jonathan DIMES, Baltimore and Munich; Mr. Nikolaus LECHENBAUER, Reutte, Austria, and Munich; Ms. Katharina SCHUMACHER, Munich; Ms. Kirsten SIEDEL, Bernried near Munich.

The following figure numbers indicate newly developed illustrations as well as those significant revised from previous editions:

U. BRUGGER: 64; 73; 75; 78; 79; 90; 93; 95; 96; 97; 98; 99; 107; 111; 112; 113; 115; 116; 117; 118; 119; 120; 121; 122; 125; 126; 127; 128; 137; 138; 139; 140; 148; 152; 159; 168; 210; 213; 214; 224; 226; 234; 235; 236; 238; 240; 249; 285; 294; 301; 326; 327; 328; 332b; 342; 343; 344; 345; 347; 355; 356; 358; 359; 360; 366; 367; 368; 369; 376; 392; 405a, b, c; 420; 421; 424; 425; 429; 430; 431; 432; 433; 434; 439; 440; 444; 447; 448; 449; 452; 453; 455; 457; 458; 465; 466; 473; 474; 475; 479; 507; 544; 545; 548; 549; 565; 566; 567; 577; 578; 579; 581; 585; 586; 588; 593; 594; 595; 596; 597; 598; 599; 600; 601; 612; 614.

J. DIMES: 77.

N. LECHENBAUER: 258; 259; 260; 261; 262; 263; 264; 265; 266; 267; 269; 292; 318a, b; 333; 613; 631; 633.

L. SCHNELLBÄCHER: 288; 329; 330; 618; 619.

K. SCHUMACHER: 135; 166; 188; 189; 206; 298; 302; 303; 304; 305; 306; 307; 308; 309; 310; 313; 357; 363; 495; 500; 508; 509; 534; 535; 556; 557; 569; 602; 609.

Index

Numbers following entries refer to figures